OXFORD GEOGRAPHICAL AND
ENVIRONMENTAL STUDIES

General *Editors: Gordon Clark, Andrew Goudie, and Ceri Peach*

# INDIGENOUS LAND MANAGEMENT IN WEST AFRICA

Editorial Advisory Board

Professor Kay Anderson (United Kingdom)
Professor Felix Driver (United Kingdom)
Professor Rita Gardner (United Kingdom)
Professor Avijit Gupta (Singapore)
Professor Christian Kesteloot (Belgium)
Professor David Thomas (United Kingdom)
Professor Billie Lee Turner II (USA)
Professor Michael Watts (USA)
Professor James Wescoat (USA)

# Indigenous Land Management in West Africa

An Environmental Balancing Act

Kathleen M. Baker

OXFORD
UNIVERSITY PRESS

# OXFORD
UNIVERSITY PRESS

Great Clarendon Street, Oxford OX2 6DP

Oxford University Press is a department of the University of Oxford.
It furthers the University's objective of excellence in research, scholarship,
and education by publishing worldwide in

Oxford  New York

Athens  Auckland  Bangkok  Bogotá  Buenos Aires  Calcutta
Cape Town  Chennai  Dar es Salaam  Delhi  Florence  Hong Kong  Istanbul
Karachi  Kuala Lumpur  Madrid  Melbourne  Mexico City  Mumbai
Nairobi  Paris  São Paulo  Shanghai  Singapore  Taipei  Tokyo  Toronto  Warsaw
and associated companies in  Berlin  Ibadan

Oxford is a registered trade mark of Oxford University Press
in the UK and certain other countries

Published in the United States
by Oxford University Press Inc., New York

© Kathleen M. Baker 2000

The moral rights of the author have been asserted
Database right Oxford University Press (maker)

First published 2000

All rights reserved. No part of this publication may be reproduced,
stored in a retrieval system, or transmitted, in any form or by any means,
without the prior permission in writing of Oxford University Press,
or as expressly permitted by law, or under terms agreed with the appropriate
reprographics rights organizations. Enquiries concerning reproduction
outside the scope of the above should be sent to the Rights Department,
Oxford University Press, at the address above

You must not circulate this book in any other binding or cover
and you must impose the same condition on any acquirer

British Library Cataloguing in Publication Data
Data available

Library of Congress Cataloging in Publication Data
Data available

ISBN 0–19–823393–0

1  3  5  7  9  10  8  6  4  2

Typeset by Best-set Typesetter Ltd., Hong Kong
Printed in Great Britain
on acid-free paper by
Biddles Ltd
Guildford & Kings Lynn

# EDITORS' PREFACE

Geography and environmental studies are two closely related and burgeoning fields of academic inquiry. Both have grown rapidly over the past two decades. At once catholic in its approach and yet strongly committed to a comprehensive understanding of the world, geography has focused upon the interaction between global and local phenomena. Environmental studies, on the other hand, have shared with the discipline of geography, an engagement with different disciplines addressing wide-ranging environmental issues in the scientific community and the policy community of great significance. Ranging from the analysis of climate change and physical processes to the cultural dislocations of postmodernism and human geography these two fields of inquiry have been in the forefront of attempts to comprehend transformations taking place in the world, manifesting themselves in a variety of separate but interrelated spatial processes.

The new 'Oxford Geographical and Environmental Studies' series aims to reflect this diversity and engagement. It aims to publish the best and original research studies in the two related fields and in doing so, to demonstrate the significance of geographical and environmental perspectives for understanding the contemporary world. As a consequence, its scope will be international and will range widely in terms of its topics, approaches, and methodologies. Its authors will be welcomed from all corners of the globe. We hope the series will assist in redefining the frontiers of knowledge and build bridges within the fields of geography and environmental studies. We hope also that it will cement links with topics and approaches that have originated outside the strict confines of these disciplines. Resulting studies will contribute to frontiers of research and knowledge as well as representing individually the fruits of particular and diverse specialist expertise in the traditions of scholarly publication.

*Gordon Clark*
*Andrew Goudie*
*Ceri Peach*

# ACKNOWLEDGEMENTS

Many people have helped me greatly with the preparation of this book by sharing their ideas, providing information and reading and commenting on drafts. I would especially like to thank Philip Stott, Sarah Jewitt, Debby Potts, Bob Bradnock, Justine Dunn, Kate Schreckenberg, Peter Thackray, W. T. Dale, Richard D'Silva, Alan Lepper, and officials at the Met. Office for their generous help. I would also like to acknowledge the considerable contribution of Dominic Thackray who has taken such care in drawing the maps and diagrams.

Much of the material for this book has been obtained from libraries and the help I have received from librarians at SOAS, Wye College, ODI, DFID and NRI cannot go unrecognized. Fieldwork has also been essential to this volume and I would like to thank SOAS for funding the greater part of my field research in Africa.

As a working mother, writing a book has proved an additional burden on family life and I am grateful to many personal friends who have helped me a great deal by entertaining my children, when they were younger, in school holidays. In particular, I would like to acknowledge the kindness and help of Liz and Clive Marsh, Yvonne and Graham Salt, and not least of all my parents-in-law whose support has been invaluable.

The support and encouragement I have received from my parents deserves special attention. Sadly, my father died before this book was completed but not before he had shared with me much of his knowledge and experience of growing coffee and rubber in South Asia. Support from my mother too has been invaluable. She has read and commented on countless drafts of the text, compiled the index, discussed a variety of issues over and over again and acted as child minder whenever her help was needed. Finally, I would like to thank my husband Robert and my children, Anna and Peter for their extreme tolerance, for their love, and constant support, and also for their frequent and gentle reminders that there is life beyond academic work, and 'The Book'.

K. M. B.

# CONTENTS

| | |
|---|---|
| *List of Figures* | xi |
| *List of Tables* | xiii |
| 1. Environmental Equilibrium and Non-equilibrium | 1 |
| 2. Problems and Prognosis: Perspectives on Agriculture at a Regional Scale | 39 |
| 3. Smallholder adaptation: the humid domain | 79 |
| 4. Non-equilibrium and the cocoa sector | 117 |
| 5. Unpredictable savanna environments | 149 |
| 6. Farming in the semi-arid domain: adaptation to an uncertain environment | 183 |
| 7. Rangeland livestock management | 217 |
| 8. Conclusions | 245 |
| *Index* | 253 |

# LIST OF FIGURES

| | | |
|---|---|---|
| 1.1 | Representation of eight equilibrium conditions | 5 |
| 1.2 | Possible influences on environmental variability | 9 |
| 1.3 | Spectrum of states from non-equilibrium to equilibrium | 10 |
| 1.4 | Departures from the 1961–90 mean annual rainfall in the sahel for the region 10–20°N, 15–30°E: June to September | 23 |
| 1.5 | Distribution of mean monthly rainfall at nine stations in West Africa, 1978–97 | 24 |
| 1.6 | Mean monthly rainfall and standard deviation for each of eight stations in West Africa, 1978–97 | 25 |
| 1.7 | Range of rainfall and mean in each month at nine stations in West Africa 1978–97 in millimetres | 28 |
| 1.8 | Variations in rainfall during 4 months at four stations in West Africa | 29 |
| 2.1 | Production, area, and yield of millet and sorghum in West Africa, 1965–96 | 43 |
| 2.2 | Production, area, and yield of maize and paddy in West Africa, 1965–96 | 49 |
| 2.3 | Production, area, and yield of groundnuts and cassava in West Africa, 1965–96 | 52 |
| 2.4 | Fertilizer consumption in West Africa, 1983/4–1994/5 | 57 |
| 3.1 | Factors influencing characteristics of indigenous farming | 81 |
| 3.2 | Location of sites under discussion in the humid zone | 82 |
| 3.3 | Bush-fallow cultivation phase one—the progressive decline in energy during the cultivation phase | 105 |
| 3.4 | Bush-fallow cultivation phase two—the accumulation of energy during the fallow period | 106 |
| 3.5 | The flow of energy through a compound farm/garden | 107 |
| 3.6 | Hierarchy of influences on decision-making at field level | 108 |
| 4.1 | Cocoa—annual average, daily price, and end of season stocks | 131 |
| 4.2 | World market price of cocoa and producer price in Nigeria, 1970–86 | 133 |
| 5.1 | Vegetation zones of West Africa | 150 |
| 5.2 | Idealized scheme of vegetation types and woody plant density along a gradient of rainfall | 152 |
| 5.3(a) | The disposition of East African savanna types within the PAM/PAN plane | 160 |
| (b) | The suggested phenological and physiognomic disposition of South American savanna types within the PAM/PAN plane | 160 |
| 5.4 | Diagrammatic cross section of the intertropical convergence zone over West Africa | 161 |
| 5.5 | Framework of a tree community dynamic model for a humid savanna of the Côte d'Ivoire | 175 |
| 6.1 | Paddy fields along the Lower Gambia (1981) | 202 |

6.2 Former paddy fields after inundation with saline water (1991) 202
7.1 Cattle stocks ('000 head) in West Africa, 1961–96 225
7.2 Reduced access for Fulani to grazing lands due to presence of irrigated areas and National Park near Kainji Lake, Nigeria 227

# LIST OF TABLES

| | | |
|---|---|---:|
| 1.1 | Coefficients of variability of rainfall for stations in West Africa | 27 |
| 2.1 | Millet: changes in production, area and yield 1965–7 to 1994–6 | 42 |
| 2.2 | Sorghum: changes in production, area, and yield 1965–7 to 1994–6 | 46 |
| 2.3 | Maize: changes in production, area, and yield 1965–7 to 1994–6 | 48 |
| 2.4 | Paddy: changes in production, area, and yield 1965–7 to 1994–6 | 50 |
| 2.5 | Groundnuts: changes in production, area and yield 1965–7 to 1994–6 | 53 |
| 2.6 | Cassava: changes in production, area and yield 1965–7 to 1994–6 | 55 |
| 2.7 | Changes in the relative importance of yam and cassava in West Africa | 56 |
| 4.1 (*a*) | Changes in world cocoa production | 118 |
| (*b*) | Cocoa production in West Africa | 118 |

# 1

# Environmental Equilibrium and Non-equilibrium

## Introduction

Nature's amazing balancing act! Is this too frivolous a start to a book with serious intentions? Not if it helps to focus the reader's mind on two things: first, a fluctuating physical environment in tropical Africa causing ecological non-equilibrium punctuated, perhaps, by periods of equilibrium; and secondly, the practical apprehension of this by the indigenous smallholders of Africa. An autecological[1] approach to the production of crops and livestock, based on a deep, almost instinctive understanding of a non-equilibrial environment, is fundamental to their survival strategies. It is the argument of this book that smallholder methods of land management should be scrutinized much more closely by development experts with a view to applying elements of indigenous knowledge to development strategies.

Over the past forty years West Africa has experienced numerous approaches to agricultural development. These range from the largest scale, capital intensive schemes planned and implemented by foreign aid agencies to the smallest schemes run by NGOs (de Wilde, 1967; Wallace, 1979; Bates, 1981; Baker, 1982, 1985; Adams, W. M., 1985; Iyegha, 1988; Bräutigam, 1998). Levels of success among these have varied, but generally, they have been low. Development schemes have contributed far less than anticipated to the region's agricultural production, the bulk of which still emanates from smallholders. In spite of their contribution, Mortimore (1998: xv) notes that smallholders have been dismissed consistently as 'anachronistic survivors, certain to disappear in the rush to modernisation, or as quite malignant in their treatment of the African environment'. Academic research mostly over the past twenty to thirty years has increasingly highlighted the merits of smallholder production in terms of its efficiency and its skilful use of the environment (Jones, 1960; Pélissier, 1966; Upton, 1967; Netting, 1968, 1993; Benneh, 1972; Norman, 1972, 1974; Helleiner, 1975; Ruthenberg, 1980, 1985; Richards, 1985, 1986; Smith, 1992; Toulmin, 1992; Fairhead and Leach, 1996; Reij *et al.*, 1996; Sullivan, 1998). However, it is questionable whether such views are fully

---

[1] *autecological*: the ecology of an individual organism or species.

accepted by those concerned with the practical implementation of development strategies. This is not to say that smallholder agriculture is without its problems as there is clearly much room for improvement. Nevertheless, smallholder farming has continued to survive in spite of some of the severest environmental conditions (Mortimore, 1989, 1998), and in spite of being hampered by decades of discriminatory policies and actions by governments throughout the region (World Bank, 1981; Bates, 1981; Onimode, 1982; Andrae and Beckman, 1985; Berg, 1986; Hinderink and Sterkenburg, 1987; Iyegha, 1988).

Smallholder success in the production of crop plants and animals, often in the virtual absence of external inputs, is dependent on an autecological approach in which great care and attention is taken to meet the specific needs of plants and animals. African smallholders are not unique in this because farming everywhere is essentially autecological. However, what is different about West African farmers is that their methods reflect close adaptation to a fluctuating, non-equilibrial physical environment. This marks a major difference from many development schemes which are masterminded, or very greatly influenced by the West where assumptions have long been based on equilibrial ecology (see below). In such schemes, land, and its capacity to produce is very often the basic unit of development planning (Moss, 1992). In view of the variation and variability in abiotic components of tropical ecosystems, particularly the rainfall, the ability of the plant or animal to cope with local environmental conditions should be the starting point for development initiatives—not just the capacity of land to produce under average conditions. A location-specific autecological approach is an option seldom taken up by development planners, not least because of the cost and time involved in acquiring the necessary environmental knowledge. Rarely have development schemes made significant use of the wealth of environmental knowledge possessed by local people. Instead, development schemes have been based on externally constructed knowledges of ecological conditions—both human and physical. It is this misunderstanding of the African environment that could be avoided by the greater involvement of indigenous people in development strategies.

If, then, the first step in agricultural development planning focused on the ecology of the basic unit of production, be it plant or animal, the second step would need to examine the available ways of managing this basic unit and incorporating it into the wider agricultural system of which it was, or was to become, one of the cornerstones. At this point practical problems begin to emerge and interdisciplinary cooperation at different levels becomes necessary to ensure that appropriate institutions are established in order that development in a wider sense may occur (Moss, 1992; Bräutigam, 1998). There are many stages and elements in the development process. This book focuses predominantly on factors relating to primary production.

In addition to ecology being a key concept in development, it is also asserted in this book that indigenous people should be much more involved in development as it is they who have the necessary ecological knowledge which could be combined with the expertise and funds of development agencies. Although there is increasing recognition of the potential role that indigenous knowledge could play in development planning, there is, as yet, very little evidence that such acknowledgement is being put into practice.

This chapter focuses on concepts of ecological equilibrium and non-equilibrium. After a discussion of definitions, the evolution of ecological thought is reviewed to demonstrate why the equilibrium paradigm, very much a product of the temperate zone, has so dominated ecological thinking. With evidence to suggest that non-equilibrium is more likely to prevail in the environment than equilibrium, the concept of non-equilibrium is considered in relation to the tropics.

## Defining non-equilibrium, equilibrium, and disequilibrium

The literature is far from precise in its definition in an ecological context, of the terms equilibrium, non-equilbrium, and disequilibrium. For the purposes of this text the term 'non-equilibrium' is taken to suggest that no equilibrium exists while disequilibrium suggests the movement away from equilibrial conditions. Where a system appears to be unstable, we cannot know whether this is an aberration from a stable state or whether the system has ever been in equilibrium. Where natural systems such as the human body have measurable periods of relative stability, compared with periods of marked instability, the term 'disequilibrium' may seem more appropriate. However, much depends on the scale of investigation for even when the human body appears to be relatively stable at one scale, at a smaller scale such stability may not be replicated, suggesting that equilibrium is perceived rather than real. As we can never be sure whether environmental instability reflects non-equilibrium or disequilibrium, in this text, the term 'non-equilibrium' is preferred.

### Concepts of equilibrium

Definitions of equilibrium are no more straightforward although the literature on the subject is abundant. Concepts of equilibrium have been woven into Western thinking since antiquity (De Angelis and Waterhouse, 1987; Ellis *et al.*, 1993). They can be identified in the ideas of Malthus at the end of the eighteenth century and they underpin Darwin's theories of evolution where natural selection results ultimately in optimally structured communities (Wiens, 1984). Concepts of equilibrium were fundamental to certain scientific principles developed in the eighteenth and nineteenth centuries and were later incorporated into ecology (Bramwell, 1989). The equilibrium paradigm not only survived into the mid-twentieth century but was alive and well in von Bertalanffy's work on General Systems Theory (von Bertalanffy, 1956, 1962). Systems theory has been widely adopted and adapted to many disciplines and subjects and has been drawn into common parlance where definition of terms such as 'systems', 'equilibrium', and 'stability' have become unclear. (It is noteworthy that the term 'ecology' has suffered much the same fate.)

Chorley and Kennedy (1971) observe that equilibrium 'is a highly ambiguous state which presents many different aspects and is the subject of a wide variety of definitions' (Chorley and Kennedy, 1971: 201). This view is echoed by De Angelis and Waterhouse (1987: 2), who observe that 'Equilibrium and stability are not sharply defined concepts when applied to real systems.' According to Wiens (1984), equilibrium has been defined in a variety of ways which divide essentially into two groups. The first involves considerations of stability or 'steadiness' of community components, and the second, the capacity of systems to return to their previous state in the wake of perturbation. Chorley and Kennedy (1971) identify eight different equilibrium states, the most relevant to this text being dynamic equilibrium and dynamic metastable equilibrium (Fig. 1.1). Dynamic equlibrium is defined as

balanced fluctuations about a constantly changing system condition which has a trajectory of unrepeated 'average' states through time. The rate of change of fluctuations is so much greater than that of the average system that, when observed instantaneously or on a short time scale, the former masks the latter to give the appearance of a steady state equilibrium. . . . Where thresholds allow occasionally great fluctuations to initiate a new regime of dynamic equilibrium a more complex state of 'dynamic metastable equilibrium' exists (Chorley and Kennedy, 1971: 203).

What is critically important is that equilibrium is rarely perceived as being static, a view echoed by Wiens (1984). It is constantly changing and may be moving along a trajectory but fluctuating around a mean. Implicit in the discussion of equilibrium is the existence of some feedback mechanism which is constantly trying to maintain balance in the system.

The ambiguity of the language involved creates problems in defining equilibrium in practical terms. For example, the term 'balance' is frequently used synonymously with equilibrium, but balance relates to the condition of different variables in the system. It is rare for all variables in a system to be in equilibrium. Any equilibrial state must, therefore, be defined in terms of the nature and scale of the variables involved (Wiens, 1984). In consideration of the practical manifestation of equilibrium, much depends on whether the focus is on open or on closed physical systems. It could be argued that equilibrium only has any practical meaning in terms of isolated systems and that success in terms of explanation of real world systems can only be attained by reducing complex closed and open systems to numerous sub-systems. Each reduction brings the system closer to a virtually isolated state which can then be evaluated

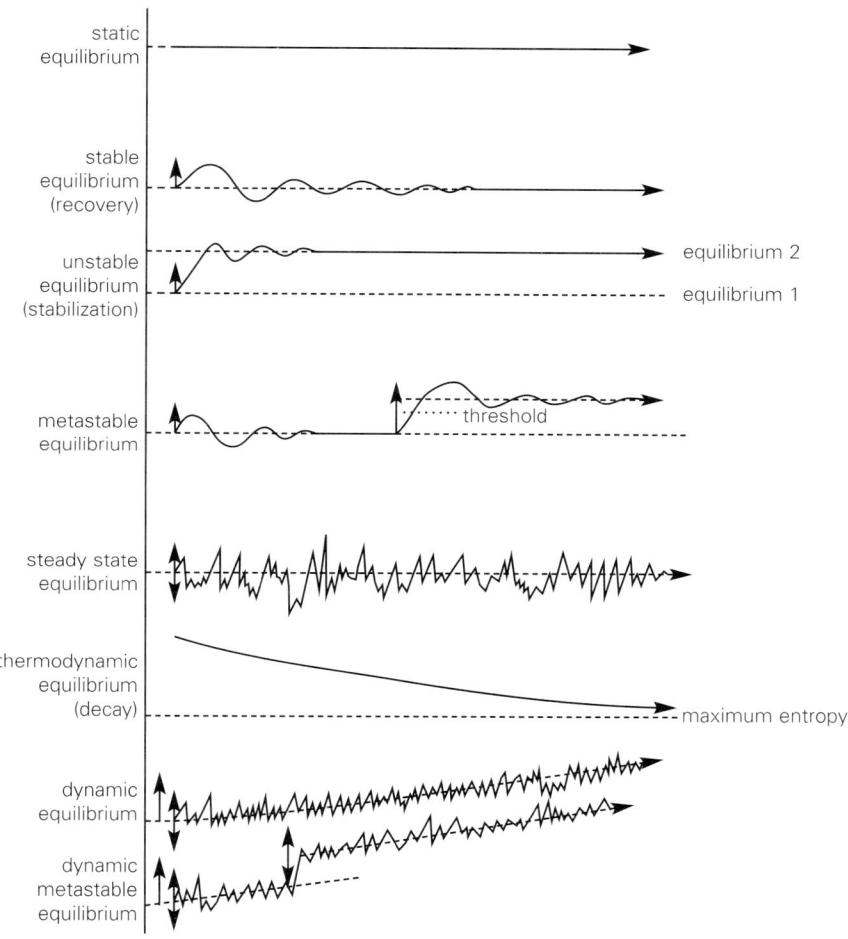

**Fig. 1.1.** Representation of eight equilibrium conditions
*Source*: Chorley and Kennedy (1971), 202.

in terms of its equilibrium position. Measurement of scientific systems under laboratory conditions will allow an assessment of equilibrium and an evaluation of feedback factors but in systems identified in physical geography, for example, which are much more open-ended than many a laboratory experiment, identification of equilibrium may be possible, though more difficult and elusive than in a laboratory. Where environmental systems involve human impact, then the situation becomes increasingly complex and equilibrium appears to be more of a theoretical concept than a practical reality. Furthermore, the evaluation of feedback processes becomes even more complicated as there is no clear-cut 'norm' or equilibrium from which deviation can be measured. The 'fleeting' nature of equilibrial states was captured very early on in empirical work by Pound and Clements (1898, cited in McIntosh, 1985). They recognized that plant communities are in stable equilibrium but rarely, and when they are, it is for a brief time only, after which it would seem that non-equilibrial conditions would prevail. This contrasts sharply with Clements's later work of 1916 and 1936 which was firmly based on the concept of equilibrium and which for half a century influenced ecological thought.

## The possibility of more than one stable state

It has been hypothesized from empirical observation and supported by mathematical models that systems may stabilize at more than one level. May (1976) raised the question of whether the dynamic behaviour of a multidimensional system results in its stabilization at a single level, a 'single valley' or 'global attractor' as he describes it, or are there many valleys or levels at which a system may stabilize?

### Stable limit cycles

A system may, for example, cycle between two stable limits such as that described by Coppock (1993) in the semi-arid range lands of the Borana Plateau in southern Ethiopia. Here settlements are maintained around a well until the area is overgrazed and the productive capacity of the cattle is in jeopardy. Woody vegetation encroaches on to the overgrazed area limiting herbaceous growth. As woody material is not suitable forage, such sites are abandoned and grass recovers in the absence of grazing. The ensuing development of the herbaceous layer reflects further recovery of the system but accumulating biomass which is potential fuel in this semi-arid environment increases the likelihood of dry season fires. Following such events, open grassland is re-established and pastoralists may re-colonize the site in a cycle taking 60–100 years to reach completion. The system thus cycles between relatively stable states.

### Multiple stable equilibrium systems

Both empirical research and mathematical models support the concept of multiple stable state systems (Noy-Meir, 1975; May, 1976; Walker and Noy-Meir, 1982). Pressures such as grazing, drought, fire, disease or a combination of these among others could have the effect of stabilizing or destabilizing the system. Disturbance may cause oscillation around the nearest equilibrium position but if fluctuations are too great, for example, through persistent drought, the system might conceivably collapse into a new stable state (Noy-Meir, 1975; Walker and Noy-Meir, 1982). Once a new equilibrium position has been established the system is unlikely to return to its former state even if the effect of the original force diminishes quite significantly. The system will remain in its new equilibrium position until perturbations are sufficient to propel it to a new equilibrium position, probably quite different from the former system. Feedback is ever important but far less predictable than in single equilibrium systems. The magnitude of the disturbance and the capacity of the system to absorb change will also influence the capacity of the system to recover. The complexity of such a multiple stable equilibrium system makes management very difficult, particularly as management methods appear to aim to minimize deviation from the equilibrium position. In such cases identifying the equilibrium position is a challenge in itself, let alone devising an appropriate method of management, and confirms yet again that hypotheses concerning equilibrium and non-equilibrium are more easily defined in theory than in practice.

Scale complicates the issue even further in multiple stable equilibrium systems. May (1976) discusses studies which, since the 1970s, have revealed that the nonlinear properties of such models, particularly at small scales, have displayed a wide range of behaviour. These have been typical of both multiple stable systems, driven by seemingly random fluctuations, as well as single, stable equilibrium systems. While the readjustment of these systems may seem to represent movements towards equilibria, they may also reflect the reaction of populations to random environmental effects. It is at this point, where complexity is at its greatest, and in environments such as the arid and semi-arid lands of West Africa where the abiotic components of ecosystems are unstable, that it becomes insupportable to accept a paradigm which argues for the establishment of equilibria in a system which assumes stability in the abiotic phase where clearly instability is more the norm. Opinion is thus divided as to whether systems are governed by predictable natural laws on the one hand, such as exemplified by the Clementsian view, or whether the major determinants of the characteristics of ecological communities are random changes, often of considerable significance in the abiotic components of the system (Ellis *et al.*, 1993; May, 1976; Noy-Meir, 1975; Scoones, 1993). The truth is probably somewhere in between these two extremes (Hubbell and Foster, 1987), though probably tending towards the latter.

## Concepts of non-equilibrium

Since the 1970s there has been growing acceptance that non-equilibrium, rather than equilibrium, is the norm in ecological systems (Connell, 1987; Holling, 1973; Hubbell and Foster, 1987; Sousa, 1984; Wiens, 1984). The non-equilibrium model refutes underlying assumptions of 'traditional' ecology, namely that the characteristics of environmental systems are determined predominantly by the biota. This may be the case where variability is low in the abiotic components of the system (Wiens, 1984) but in the tropics it is the abiotic components which are subject to stochastic effects and which are increasingly believed to be the regulators of the system, rather than the biota (Fig. 1.2). The fluctuations in rainfall patterns in both semi-arid and humid West Africa, discussed below, lend weight to this assumption. Where abiotic factors are prone to fluctuation, this can cause significant changes in the biota and where ecosystems are perceived to be rarely, or ever, in equilibrium, non-equilibrium is likely to prevail. Such a non-equilibrial situation can also be initiated by external factors such as plagues of insect pests or the destructive effects of war. In both the above examples the term 'disequilibrium' could be used in place of non-equilibrium but as there is little evidence to suggest that ecosystems return to an equilibrium after disturbance, the term 'non-equilibrium' is preferred. It is also logical to assume that non-equilibrium may be initiated by changes in biotic factors, which in turn may bring about changes in abiotic variables, which then may have a destabilizing effect on the local ecology. For example, forest clearance by humans can lead to changes in abiotic factors such as the substrate, which can fundamentally alter relations between living organisms and the environment in the vicinity of the deforested area, and possibly beyond. Although driven by biotic factors, it can be argued that loss of stability occurs as a result of changes wrought in abiotic factors. Fig. 1.2 summarizes these possible causes of non-equilibrium and demonstrates that instability, be it non-equilibrium or disequilibrium, is much more likely to prevail in environmental systems than stability or equilibrium.

Although theoretical arguments increasingly support the dominance of non-equilibrial conditions, a major problem remains the identification of either of these conditions in the landscape. If environmental conditions are perceived as stressful, it would be difficult to know whether the conditions represented an equilibrium system in dynamic equilibrium, which is equilibrium theory's method of coping with instability, or whether non-equilibrium prevailed. This raises the question of whether there may be any real distance between proponents of dynamic equilibrium and proponents of disequilibrium. Theoretically, the differences are clear but in practice they may be more difficult to discern. Mortimore (1989) aptly summarizes the practical problem of distinguishing between 'uncertainty-as-aberration' and 'uncertainty-as-norm' (Mortimore, 1989: 214). So much would depend on the time available for observation and on the scale of the analysis (see below). An alternative perception of the relationship between equilibrium and non-equilibrium comes from Wiens (1984) whose perception places the two states at different ends of a continuum (Fig. 1.3). One end of the continuum is characterized by systems of low abiotic variability where the system is largely regulated by the interaction of biotic elements (equilibrium systems). At the other end of the continuum, systems are characterized by much greater variability in the

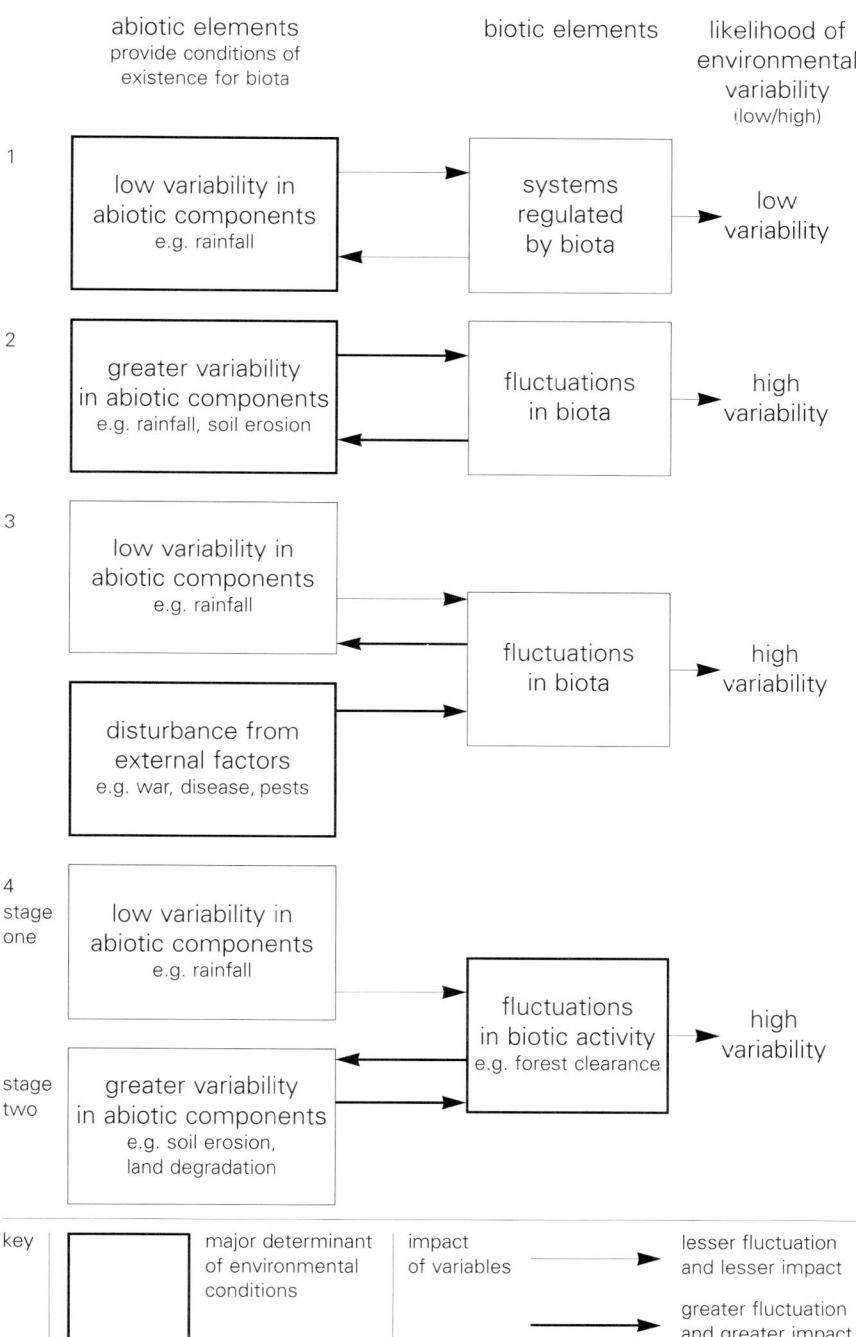

**Fig. 1.2.** Possible influences on environmental variability

**Natural communities may be arrayed along a spectrum of states from equilibrium to non-equilibrium. At either extreme, several attributes of community structuring or dynamics can be anticipated, as shown**

| Non-equilibrium | Equilibrium |
|---|---|
| biotic decoupling | biotic coupling |
| species independence | competition |
| unsaturated | saturated |
| abiotic limitation | resource limitation |
| density independence | density dependence |
| opportunism | optimality |
| large stochastic effects | few stochastic effects |
| loose patterns | tight patterns |

Fig. 1.3. Spectrum of states from non-equilibrium to equilibrium
*Source*: After Wiens (1984), 451.

abiotic component, particularly the climate, with the result that this, rather than the interaction of the biotic elements of the system, is the main determinant of the nature of the ecosystem (non-equilibrium systems). Conditions of equilibrium and non-equilibrium are not necessarily mutually exclusive as natural communities may exist along a spectrum of states ranging from equilibrium to non-equilibrium. As Fig. 1.3 suggests, they may assume many intermediate states between equilibrium and non-equilibrium. It thus seems that non-equilibrium in ecological systems is more common than equilibrium and yet, traditional ecology has been based on the assumptions of equilibrium. The next section examines the evolution of concepts of equilibrium and non-equilibrium and shows how the former may have come to dominate ecological thought.

## The evolution of concepts of non-equilibrium and equilibrium in ecology

This section reveals that concepts of non-equilibrium were supported by empirical evidence before the beginning of the twentieth century and yet they were submerged by the overwhelming acceptance of equilibrium theory, which was based more on assumption than on empirical evidence. Although the tropics barely receive a mention here, nevertheless, the inclusion of this section in some detail is justified because of the consequences that the development of ecological thought has had on perceptions and understanding of tropical ecology.

Ecology was given coherence by Ernst Haeckel (1866), the German biologist/zoologist who coined the term '*oekologie*'. His was the first anti-mechanistic, holistic approach to biology (Bramwell, 1989; Odum, 1971; McIntosh, 1985). He saw the universe as a unified and balanced organism where space and organic beings were made of the same atoms, where mind and matter were one. It is conceivable that Haeckel's balanced universe was the product of his deep involvement in the sciences where concepts of balance and equilibrium were evident. Continuing the work of Haeckel, Schimper (1898, 1903) developed the organismic concept of the unit of vegetation (Stott, 1999) and introduced the basis of the classificatory system later used by Clements and Tansley (see below).

While Haeckel's observations were frequently focused at a global scale, it was plant ecologists working at field level, at the scale of the plant community, who were among the first to examine and challenge concepts of stability and equilibrium. One of the earliest challenges was by Eugenius Warming in 1895 whose observations suggested that plant communities were neither static, nor were they in equilibrium. Any equilibrium which might be attained, Warming argued, would not be sustained and could be disturbed by physical factors, by animals, by fungi and by the relentless struggle within and between plant communities. Warming's recognition of competition among the biota reflected the influence of Darwin and stands in marked contrast to Haeckel's notions of a balanced universe (McIntosh, 1985). Perhaps it is unjust to compare the findings of Haeckel and Warming, not least because their scientific backgrounds, their research objectives, and even the scale at which they were generalizing were so very different. Nevertheless, it emphasizes that before the advent of the twentieth century the equilibrium/non-equilibrium debate was under way among ecologists.

Early in the twentieth century, the non-equilibrium element of the debate was eclipsed by the work in North America of F. E. Clements (1916) who changed the nature of plant ecology. Clements placed emphasis on dynamism in plant ecology. He focused on changes in plant communities over time, changes in the constituent plant populations, and changes in the sites they occupied. Using the analogy of the organism, Clements argued that plant communities grew and developed in an ordered manner, a process which he termed 'plant succession'. Clements identified primary succession which was the development of vegetation from bare ground to the climax state, and secondary succession which was the successional pattern that followed disturbance of the vegetation, either when it was in a climax state, or before it had reached it. The effects of fire, for example, could initiate the process of secondary succession. Through plant succession, the tendency of vegetation in an area of consistent climate would be to converge to a 'climax' stage. This climatic climax vegetation would remain stable, or in a state of dynamic equilibrium, and was capable of reproducing itself in the absence of external disturbance. In the event of disturbance, feedback mechanisms would come into play to restore an equilibrium.

By 1936 the generalizations embodied in Clements's work on dynamic plant ecology had been taken to greater extremes. It was claimed that the vegetation of vast areas of the globe was in a climax state, in the absence of human disturbance (Weaver and Clements, 1938, cited in McIntosh, 1985). According to Miles (1987), Clements's theory was 'a mix of scholarship and sound observation curiously blended with a purely deductive and hypothetical classification of succession based on the concept of "climax" vegetation, an attempt to create a unifying theoretical framework of succession' (Miles, 1987: 4).

Clements's theory of vegetation development may not have been entirely substantiated by field data, but none the less, it had widespread appeal for a variety of reasons. First, Clements's holistic approach to ecology and his analogy of the organism were 'tidy-minded' and have been pedagogically useful to a range of people, not just ecologists. Secondly, the concept of the organism derived from the nineteenth century was widely accepted in natural history and was developed by Clements. To some extent, the concept of the organism persists in ecology today. Thirdly, Clements's views on the persistence of climax vegetation and on the convergence of plant successions to the regional climax fitted in with views prevalent in related subjects. Geomorphology, for example, was greatly influenced by the work of W. M. Davis (1909) who argued that landscapes cycled through phases of youth, maturity, and old age, in many ways a parallel to the organism and equilibrium theory of vegetation development. Finally, theories of vegetation development were accepted in the absence of information on the dynamism of climate over time (Miles, 1987).

Clements's fame extended beyond North America. His work received some recognition in Britain in the early part of the twentieth century though in a modified form. Aspects of Clements's monoclimax theory were incorporated into a polyclimax theory of vegetation which was developed by British ecologists led by Sir Arthur Tansley (Tansley, 1916, 1920). While Clements's ideas received a cool reception from some European ecologists, his monoclimax concept greatly influenced the major Zürich–Montpellier school (Whittaker, 1953, cited in Golley, 1977). In spite of the criticisms it received, knowledge of Clements's theory is widespread largely because it was incorporated into leading texts on both plant and animal ecology which introduced it to a wide range of students throughout the world (McIntosh, 1985). The temperate zone bias in Clements's theory is inevitable because it was based largely on his experience. Nevertheless, it still risked generalizing, often inaccurately, at a global level. In spite of its shortcomings, Clements's theory of vegetation has probably reached, if not influenced most ecologists and plant geographers in the world.

Clements's theory of vegetation development has also had its critics amongst ecologists and is now largely discredited. In his vegetation theory of

1916 Clements even contradicted his own, earlier research which stated that 'In nature both formations and subordinate groups are in stable equilibrium only rarely and usually for a comparatively short time' (Pound and Clements, 1898, 1900: 313). Such findings were based on field data and the need for abundant field data was assigned considerable importance by Clements. In spite of his commitment to field evidence, Clements's theory of vegetation published in 1916, was allegedly based heavily on inductive reasoning and not on empirical evidence (Miles, 1987).

At about the same time that Clements began working on plant succession, Cowles, a physiographic geographer from Chicago, identified the role of climate, topography, and other physical and biotic factors in the development of vegetation. Cowles argued that variables influencing climate all moved independently of each other at different rates, at different times and in different directions. Following extensive field observation, Cowles concluded that succession was not necessarily a straight-line process and that a state of equilibrium was never reached (Cowles, 1901, cited in McIntosh 1985: 83). Similar observations on succession were made by Gleason in 1927 when he too criticized Clements's model, arguing that succession could take plant communities in a variety of directions including a reverse direction (Gleason, 1917, 1927).

W. S. Cooper, a student of Cowles, conducted extensive field studies on the dynamics of the forest of Isle Royale, Michigan (Cooper, 1913, cited in McIntosh, 1985). Here he identified an equilibrium not based on Clements's linear succession model, but on a much more complex analogy to a braided stream. In sample stands in the forest Cooper mapped the trees and aged them by counting annual rings. In contrast to the prevailing image of a plant community as essentially homogenous, Cooper demonstrated that the forest was a mosaic of trees of different ages where each patch of the mosaic was in a state of flux. The forest as a whole remained at equilibrium, the changes in the various parts balancing each other. Similarly, Watt's work (1919, 1947) on small-scale disturbance in plant communities and the importance of the gap-phase in community reproduction added weight to the findings of Cooper, and indeed confirmed the observations of Eugenius Warming in 1895, that plant communities are in a continuous state of flux. It is noteworthy that different researchers measuring different variables have reached similar conclusions on the existence of non-equilibrium in plant communities.

Early in the twentieth century ecologists had recognized that vegetation patterns in most areas were in a state of flux and by the mid-1920s it was increasingly accepted that the concept of climax vegetation was subjective (Cooper, 1926) and equilibrium, though frequently a convenient concept, had little meaning in reality. According to Wiens (1984), equilibrium has been verified by little more than unquestioning faith in the concept. There was reliable empirical evidence which argued against the equilibrium hypothesis but in spite of this, the concept of equilibrium and of climax vegetation has persisted and in mathematical ecology equations have been developed around the concept that an equilibrium state exists (De Angelis and Waterhouse, 1987). With the application of ecological theory to development, the problems associated with the unrealistic and possibly incorrect assumptions of the equilibrium model are becoming apparent, and the predominantly non-equilibrial nature of tropical ecology is emerging powerfully.

## The relevance of the equilibrium/non-equilibrium debate to the tropics

The virtual dismissal of concepts of non-equilibrium by European ecologists has had disastrous implications for development strategies in the tropics. Essentially, it has meant that development projects into which expatriates have had a significant input have been underpinned by ecological assumptions inappropriate to tropical environments. Although ecology has a higher profile now than ever before in the International Community and although Barber Conable (1987), then President of the World Bank, stated that 'sound ecology is good economics',[2] a statement which seems fundamentally correct, there is little evidence of research being done to ensure that appropriate ecological assumptions are being used in development planning. What emerged from the section above was that equilibrium theory might possibly be applicable where environmental fluctuations were relatively limited, but where they were considerable, non-equilibrial conditions were more likely to prevail. This has been recognized by academic writers for some 30 years but there is little evidence of it being accepted by development practitioners. However, the work of Scoones on livestock management in Southern Africa (see below) suggests that such ideas are now starting to be accepted by those involved in the practical aspects of development, but changing minds is a slow process.

## Non-equilibrium in the semi-arid tropics

In the semi-arid tropics the availability of moisture is a major determinant of the quality of life. Rainfall is highly variable both in time and space. Normally, there are distinct wet and dry seasons in a year so variability occurs between seasons and also from one year to the next. In an environment driven by the uncertainty of rainfall and other physical factors such as fire and dust-laden winds, development initiatives based on concepts of stability and equilibrium have had little hope of success. Indeed, some of the worst failures to increase primary productivity are to be found in the world's drylands. It would seem

---

[2] Address by Barber B. Conable, then President of the World Bank and the International Finance Corporation, to the World Resources Institute, 5 May 1987, Washington DC.

that development experts have barely noticed that the biota, plant and animal—including humans—in these semi-arid environments cope with fluctuations in the physical environment. Although the equilibrium/ non-equilibrium debate has been in existence for at least a century, albeit with regard to plant communities in the main, it is only since the 1970s that the body of evidence has increased to suggest that ecological communities in the drylands of the tropics are more influenced by stochastic events than by predictable changes resulting from fluctuations in the biota (Ellis *et al.*, 1993).

In 1975 Noy-Meir challenged the equilibrium concept in dryland ecosystems, observing that the abiotic component was highly variable and that ecological conditions in tropical rangelands were driven more by climatic events than by the interaction of the biotic elements, although the latter could be responsible for the movement of the ecosystem from one equilibrium position to another (Noy-Meir, 1975; Walker and Noy-Meir, 1982). Strong support for the non-equilibrium model has been provided by Sullivan (1996; unpubl. Ph.D. thesis, 1998), from her work on the degradation of open woodland by the alleged misuse of resources by local herders in north-west Namibia. She demonstrates the importance of abiotic factors in the resilience of the environment and argues that there have been errors of judgement in the analysis of environmental conditions because of unjustified adherence to the equilibrium paradigm. Sullivan argues that concepts of community dynamics which place the density-dependent effects of people and livestock well above the effects of abiotic variables as determinants of primary productivity have compounded the problems of analysing and understanding environmental conditions (Sullivan, 1996).

Scoones's work (1993), which is also concerned with separating equilibrium and non-equilibrium factors in shaping grazing systems in Zimbabwe, focuses on livestock rather than vegetation. Analysing livestock population data covering 60 years, Scoones (1993) argues that in the long run non-equilibrial factors, the result of environmental fluctuation, tend to be the major influence on the size of cattle populations, resulting in populations of below 'equilibrium' density. However, in the years between episodic events equilibrial processes have been significant in influencing cattle numbers. In other words, interaction of the biotic elements may be the dominant regulator of the ecological community when physical disturbance is low. This accords with Fig. 1.2. Scoones concludes that over the long term in semi-arid, southern Zimbabwe, ecological communities are regulated by both non-equilibrium and equilibrium conditions at different times (Scoones, 1993).

Based on field evidence from arid and semi-arid areas of North America, Wiens (1984) has concluded similarly that climate, rather than the interaction of biotic components, is a major factor in the regulation of ecosystems. He has also extended the argument which combines equilibrium and non-equilibrium situations at different times in the same ecosystem by postulating that ecosystems exist along a continuum (Fig. 1.3). Even the strongest proponents of non-equilibrial ecology speak of the existence of equilibrial phases (Hubbell and Foster, 1987; Noy-Meir, 1975; Scoones, 1993). Thus it is possible that ecological communities in the tropics are regulated by a combination of equilibrial and non-equilibrial factors, though more empirical evidence is needed to support this. Also of importance is the scale of investigation.

## Scale, equilibrium, and non-equilibrium

Scale is of critical importance in the equilibrium/non-equilibrium debate. There is a world of difference between the functioning of the ecosystem at the quadrat scale and at the landscape scale (May, 1994). Sousa (1984) argues that few natural populations or communities persist at or near an equilibrium condition at a local scale. However, viewed at a larger scale, equilibrial conditions may appear to prevail. Using the example of rangelands in the savannas and sahel of West Africa, at a regional level there may appear to be instability on the northern fringes of the rangelands, particularly where dunes have encroached on grazing land. But apart from this fringe area, the continued existence of the rangelands at a regional scale may be seen to be an expression of environmental stability.

When viewed at field level, studies of rangelands have demonstrated that a mosaic of different systems can be identified. On account of the high variability of abiotic factors over space and time, rainfall in particular, at any one time some patches of the mosaic may reflect stable ecological conditions whereas others, where rainfall has been low, may suggest the existence of environmental stress and ecological non-equilibrium. If the scale of observation is expanded to incorporate many patches which have been disturbed at different times, Connell (1987) argues that even though disturbance at a local scale is causing change in species composition, the average species composition of the larger area may remain constant even though each patch within it is changing. Evidence from the range lands of southern Africa suggest that non-equilibrium in patch dynamics may be necessary for the character of the system to be maintained at a larger scale (Fuls and Bosch, 1991; Fuls, 1992). Similar conclusions have been reached by Sullivan (unpubl. Ph.D. thesis, 1998) working on perceived environmental degradation in open woodland in northwest Namibia. She argues that perturbations may occur at smaller scales both in space and time, than at larger scales, and that increasing the spatial or temporal scale of observation may reveal a significant degree of ecosystem persistence. Sullivan also notes that shortlived or transient 'patch dynamics' observed at smaller scales may be of critical importance to the persistence of systems at larger scales or at 'higher' levels in the hierarchy.

Attempting to model changes at multiple scales has led to theoretical work on hierarchies. O'Neill (1989) argues that if a model explicitly considers two time-scales, then the expected pattern would be that the slower time-scale establishes a slowly changing trajectory compared with which the faster dynamics are asymptotically stable. O'Neill (1989) hypothesizes that if the system moves towards a major area of instability, the rate of return of fluctuating variables to positions on the original trajectory increases as the point of instability approaches. He suggests that these and other models could form the basis of a system monitoring large-scale changes at global level. By setting up regular field level experiments, it may be possible to detect whether slow changes in selected environmental variables are leading the system to a point of instability by monitoring short-term recovery times. If recovery time increases, it may suggest that the entire system is moving towards instability and remedial action is necessary. O'Neill (1989) observes that the application of theoretical work on multiple scales could be of value in monitoring changes in African grasslands. Such hierarchical work echoes the views of May (1994) who argues that ecologists have, for too long, worked at the level of the quadrat and that conservation biology, for example, may be greatly aided by studies focused at a larger scale where, to some extent, stability does exist. The critical importance of not extrapolating conclusions from one scale to another is clearly evident from all the above works and should be noted by development practitioners.

If perceptions of equilibrium and non-equilibrium vary according to scale, two thoughts emerge: first, development experts' perceptions of the environment may have been focused at too large a scale and so have not been sufficiently sensitive to variations at a local level. This leads to the second point which is the very great need for interdisciplinary work in development rather than the comparatively narrow focus on economic costs and benefits. There is evidence that change is on the way, but for many projects funded by international agencies, economic aspects continue to prevail in spite of growing rhetoric about the need to involve local expertise.

## Stability, resistance, and resilience

Stability may be defined as the ability of an ecosystem to maintain or return to its original condition following a natural or human-induced disturbance. Not surprisingly, the definition is more easily applied in theory than in practice. Stability may be disturbed at any scale by the disruption, usually, of abiotic factors. What constitutes a disturbance to an ecosystem is difficult to define but causes are physical and biological and include, for example, fires, floods, droughts, high winds, landslides, waves, and desiccation. Causes of biological disturbance include everything from predation to the non-intentional destruction of biota by other organisms (Sousa, 1984). Changes in abiotic factors could conceivably be brought about by human impact. If the disturbance is minimal, organisms act as buffers and adjust to compensate. But where the disturbance in the physical environment is extreme, the limits of effective physiological or behavioural response will be exceeded. Initially, this may cause reductions in growth and in the case of agroecosystems, crop yield. With greater extremity of the disturbance, the biotic components of the system may die and instability may follow. This suggests that the system has changed from one of stability, but whether a steady state or equilibrium has ever existed in semi-arid West Africa is debatable when both ecological and social systems in this part of the world reflect uncertainty (Mortimore, 1989, 1998). While in the sahel there may be little evidence of environmental stability, there is certainly evidence of environmental resilience. Resistance and resilience are two possible effects of disturbance to ecosystems. Resistance is the ability of an ecosystem to resist disturbance while resilience may be defined as the ability of the system to return to its former state following a disturbance (Holling, 1973). It may also be defined as the measure of magnitude of disturbance which a system can absorb before it crosses a threshold and moves from one locally stable equilibrium position to another (Arrow et al., 1996). Biochemical buffer solutions may be a useful analogy in the explanation of environmental resilience.

It may be possible for environmental systems to absorb or mask change to a certain degree. The changes may be either physically induced or human induced and thus appear to be stable although changes are taking place within the system but are not apparent at a superficial level. Similarly, analyses of rangelands and also of moist forests (see below) at a small scale, reflect immense variation, far more than may be evident at a larger scale. If the disturbance in the case of the rangeland or forest persists, then a point will ultimately arrive when the system is forced to change significantly and may appear to stabilize at a new level.

Where disturbance levels or pressures on the system are low, the rate of change may be low, but where disturbance levels are high such as in the sahel where rainfall varies markedly from year to year, it can be argued that the biotic components of the system have the capacity to adapt to uncertainty and thus act as a buffer to change. Thus when the rains fail, the grass turns brown and fauna adapt in a variety of ways to conditions of moisture shortage. At this time, resistance of the environment to the disturbance may seem low, though the environment is far from being destroyed. It is this low resistance to perturbation that frequently earns the tropical environment the epithet 'fragile' though as Arrow et al. (1996) observe, low resistance may be related to a high level of resilience and the evidence suggests that resilient environments such as the sahel have a high capacity for recovery (Mortimore, 1989, 1998). The term 'fragile' is, arguably, inappropriate.

Resilience suggests environmental recovery, and the rains that followed the drought of 1973 saw the renewal of plant succession, though not necessarily following predictable paths. Recovery patterns of vegetation in a semi-arid grassland in North America over a period of 53 years following its disturbance through ploughing have been charted by Coffin and Lavenroth (1996). They reveal that recovery has followed paths very different from the Clementsian model. They confirm that semi-arid ecosystems may have low resistance to change in the face of severe disturbance, but that their capacity to recover is considerable. Because of the capacity of the environment to act as a buffer, resistance and resilience may be inversely related. Rarely are such concepts based on empirical evidence and as yet there is still considerable ignorance about the dynamic effects of change in ecosystem variables, about thresholds, about the capacity to absorb change and the loss of resilience (Arrow *et al.*, 1996). This is not due purely to negligence but because of the time element involved in studies which examine changes in species composition (Connell, 1987). In addition to natural factors, human management of such landscapes also plays a part in their capacity to recover. This may be influenced by competition for resources and in particular, the rights of access to land and water. Adaptation is the ability of plants and animals to cope with disturbance and one of the main aims of this book is to examine how people in West Africa manage non-equilibrial environments with limited access to technology.

## Non-equilibrium in the humid tropics

The semi-arid tropics readily lend themselves to discussions of non-equilibrium, but what of the humid zone? Here too, research suggests that ecological non-equilibrium rather than equilibrium is the norm, though evidence is limited in comparison with the semi-arid tropics. Classical views have cited moist forests of the tropics as some of the most species rich, complex, ancient, and stable ecosystems on earth but this is now much in doubt (Edwards *et al.*, 1994; Hubbell and Foster, 1987; Martin, 1991). As Sousa (1984) observes, 'For many communities, a self-reproducing climax state may only exist as an average condition on a relatively large spatial scale, and even this has yet to be rigorously demonstrated' (Sousa, 1984: 353). In the broadest of terms one could argue that succession, either primary or secondary, culminates in a climax of forest vegetation and the extent of forest in the humid tropics does give some credibility to the equilibrium viewpoint. On the other hand, Hopkins (1974) has identified patches in the moist forests of Ghana where no woody climax exists, and these are believed not to be the product of human activity. Whether existing conditions are at equilibrium or in a state of non-equilibrium in the moist formations of tropical forests would again appear to depend on the scale of study (Sousa, 1984; Connell, 1987; May, 1994), though Hubbell and Foster (1987) suggest that neither extreme position—equilibrium or non-equilibrium—is likely to be the case and that forest communities are likely to reveal evidence of equilibrial and non-equilibrial characteristics in different variables, in different areas, and at different spatial scales. The following study of a moist forest demonstrates that non-equilibrium exists in species composition in a forest at a small scale though at the larger scale such non-equilibrial characteristics may not be evident. These views are in line with the findings in semi-arid Africa, of Sousa (1984), Connell (1987), Fuls and Bosch (1992), Fuls (1993), and Sullivan (unpubl. Ph.D. thesis, 1998).

The question of what controls the mix of tree species at a local level in moist tropical forest where the number of rare species is high and where few species are common, has been the subject of much debate (Aubréville, 1971; He *et al.*, 1996; Hubbell and Foster, 1987). That forests are composed of a mosaic of patches of trees was identified by Aubréville (1971), but whether the forest as a whole is in equilibrium is debatable. At one extreme, proponents of the equilibrium hypothesis assume that tropical forests consist of assemblages of a wide variety of tree species which are stabilized through competition between the species. Because the assemblages consist of a wide variety of trees with few common species, in theory, trees should exhibit strong local self-inhibition in regeneration. When a species of tree becomes too common, processes influencing its density should reduce the *per capita* reproductive performance of individuals of the species relative to less common competing trees in the area (Hubbell and Foster, 1987). This idea is to some extent implicit in Aubréville's mosaic hypothesis which observes that the spatial variation of floristic composition in species-rich tropical forests is maintained by temporal variation in species composition at any given location. Aubréville (1971) goes on to state that most tree species in closed canopy forests of the Côte d'Ivoire appeared unable to regenerate beneath themselves or in the immediate vicinity. Thus a succession of different species would be expected to occupy a given site in the forest before the same species could re-establish itself there. According to Hubbell and Foster (1987), this is consistent with the equilibrium hypothesis because Aubréville is suggesting the possibility of predictable successional cycles of mature forest species, or of cylic or non-transitive tree replacement processes.

At the other extreme, one of several non-equilibrium hypotheses advocates that tropical tree communities are composed largely of generalist species which are either weakly stabilized by niche or not stabilized by niche at all. They thus coexist in spite of, and not because of, being functionally similar generalists. If this is true, the relationship between local tree abundance and *per capita* reproductive performance should be weak or absent (Hubbell and Foster, 1987).

Of the two extreme hypotheses, the non-equilibrium hypothesis was given greater support by Hubbell and Foster's study (1987) of the forest of Barro Colorado Island, Panama which concluded that the forest was not in equilibrium. In the main, results demonstrated that tree replacement was not homogenous in either spatial or temporal terms. However, there were some elements of the study which partially supported the equilibrium view of forests. Of the ten most common tree species which were the focus of the study, only two species, *Trichilia tuberculata* and *Alseis blackiana* gave strong evidence of self-inhibition of saplings in the vicinity of adults. These trees provided evidence of strong density dependence characteristics giving support to the equilibrium hypothesis. Of the remaining eight most common species studied, there was no evidence of self-inhibition and in two species there was significant positive association between saplings and adults. The authors have also shown that because of the low probabilities of self-replacment many different tree species are likely to occupy a site before it is reoccupied by the same species. For the trees studied, the results suggest that the probabilities of self-replacment are not statistically lower than the probability of replacement of any pair of species. The probability of any tree being replaced by another of the same type is so very low that local succession of tree species in any given place in the forest is almost certain to be unpredictable. The result that in the study area any tree species could replace any other tree species, also supported the hypothesis that many of the tree species in the forest were generalist rather than specialist species (Hubbell and Foster, 1987). These findings give more support to non-equilibrial rather than equilibrial hypotheses.

Detailed studies examining the equilibrium/non-equilibrium debate in tropical forests are still comparatively few but a study of species diversity in a tract of tropical rain forest in the Pasoh Reserve, Negeri Sembilan, Malaysia, further demonstrates the unpredictability of variables measured and adds to the weight of evidence that the forest community may not be in equilibrium (He *et al.*, 1996). In spite of such evidence for non-equilibrium in species composition, it is important to note that at a global, or even a regional level, the forest may appear to be in equilibrium—except where biotic factors such as forest clearance are taking their toll. Much more work needs to be done in the humid domain on non-equilibrium on a far wider range of variables both abiotic and biotic, in order to demonstrate more clearly whether or not the biota change in a predictable manner. The evidence so far from spatial studies suggests they do not. The next section considers changes over time.

## Non-equilibrium at a larger scale: change in forest area over time

Establishing the extent of change in the area under tropical forest is far from conclusive but there seems to be a general consensus that in West Africa human activity is reducing the area under forest. The main culprits are farmers who keep opening up land for cultivation, and to a much lesser extent, loggers and the mining industry (FAO, 1982; Myers, 1984, 1996; Rietbergen, 1989; World Resources Institute, 1992). There are also alternative arguments such as those of Morgan and Moss (1965) who, after years of observation on the forest-savanna boundary in Nigeria, concluded that smallholder farming did not necessarily lead to forest destruction. More recently, Fairhead and Leach (1996) have argued convincingly that smallholder farming does not necessarily degrade forest as in parts of Guinea the area under forest has actually increased in the forest-savanna boundary zone. Viewed at a larger scale and over a much longer time period, it can be demonstrated that the area under forest in the tropics has changed significantly in relation to climate change.

Until comparatively recently it was believed that tropical moist forests with their enormous complexity and apparent stability were some of the oldest vegetation formations on earth. This is now known not to be the case and there is growing evidence to support hypotheses for the expansion and contraction of the areas under tropical moist forest (Grove, 1992; Martin, 1991). The evidence has come from a variety of sources including palynology, the study of lake sediments, carbon dating, the study of species diversity and possible refugia. These suggest that on a timescale exceeding 20,000 years, non-equilibrium has prevailed, driven by abiotic factors and in particular by climate change. While the nature of climate instability and unpredictability, be it short or long term, is more evident in the savannas, it does not mean that conditions in the humid tropics are stable and arguing from the long-term perspective, environmental non-equilibrium could also be considered to prevail in the humid tropics. As early as 1911, Cowles (cited in Miles, 1987) recognized that changes in vegetation caused by climatic, topographic, and biotic changes occur at different rates and concurrently, providing a strong argument against the ability of ecosystems to stabilize for any significant length of time.

## Rainfall variability in West Africa

The above examples from the literature refer to the way in which biotic elements in both the semi-arid and the humid tropics are greatly influenced by abiotic fluctuations. While investigating land management methods by smallholders in West Africa, one abiotic variable, variability in rainfall, emerged as the most critical to agriculture and an attempt was made to look more closely at the nature of variation within and between seasons. At a regional level, rainfall data for the western sahel clearly reflect a trend of declining precipitation since the late 1960s (Fig. 1.4).

Within the drought period there is considerable variation: in some years rainfall is better than others while there may be several consecutive years of poor rains (Sivakumar, 1989, 1991). Unpredictable fluctuation in rainfall has meant that smallholders never know what to expect and that their survival depends on their capacity to work with an unstable environment and to harvest a crop whether the rains are good or bad. Chapter 6, which examines human responses to drought on the borders of the wet and dry savannas in West Africa, reveals the significant changes in the physical environment that have resulted from protracted drought.

In an attempt to look more closely at how unpredictable the rainfall has been, monthly rainfall data covering the period 1978 to 1997 were obtained from the Met. Office for nine stations in West Africa. The unpredictability of rainfall is usually associated with the drylands of West Africa (Sivakumar,

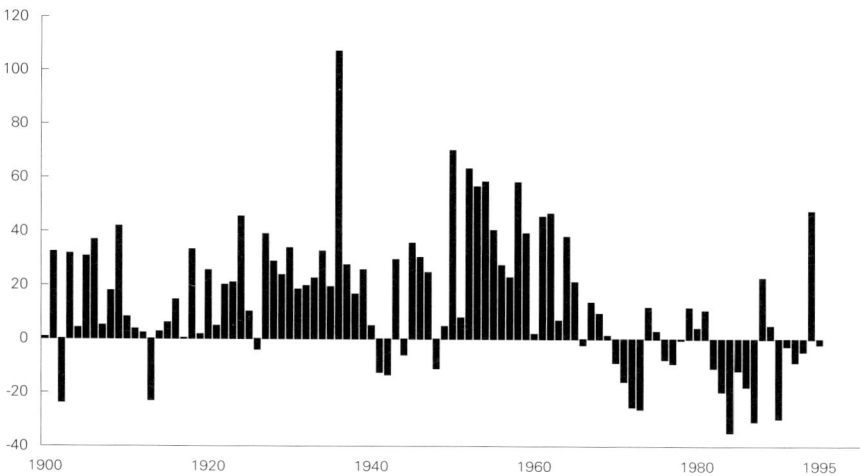

**Fig. 1.4.** Departures from the 1961–90 mean annual rainfall in the sahel, for the region 10–20 °N, 15–30 °E; June to September

*Source*: Derived from http://tao.atmos.washington.edu/data_sets/sahel/

1989, 1991) but it was believed that rainfall variability is also a problem for smallholders further south and an attempt was made to examine rainfall patterns in the humid zone as well. Data providing good coverage of the region are not easily available and efforts to obtain data for sites in the humid south, the Middle Belt, and the semi-arid north met only with partial success. Better representation of rainfall patterns throughout the region and in the south-west, in Liberia, Sierra Leone, and Guinea, for example, would have been valuable, but recent data were not available.

The data available are far from complete and in some stations such as Bissau, frequency of data is very poor. In most other stations the data are more frequent though for no station are they complete for the 19-year period. According to the Met. Office these are the best data available and in spite of their shortcomings the patterns they reveal, or the lack of them, further supports concepts of non-equilibrium. Owing to the poor quality of data, the analysis is inevitably qualitative. Fig. 1.5 shows the mean monthly rainfall for the nine stations in West Africa. Although the stations are not perfectly distributed across the region, they illustrate the variation in annual rainfall within the year, and also between the south and north of the region.

While we are conditioned to think in terms of means, in reality these can have little relevance to methods of land management in view of the considerable variation in rainfall each year. Fig. 1.6 shows both mean monthly rainfall and standard deviations about each monthly mean over the 19-year period for eight of the nine stations shown in Fig. 1.5. It was not possible to calculate standard deviations for several of the months for Bissau because the data were too

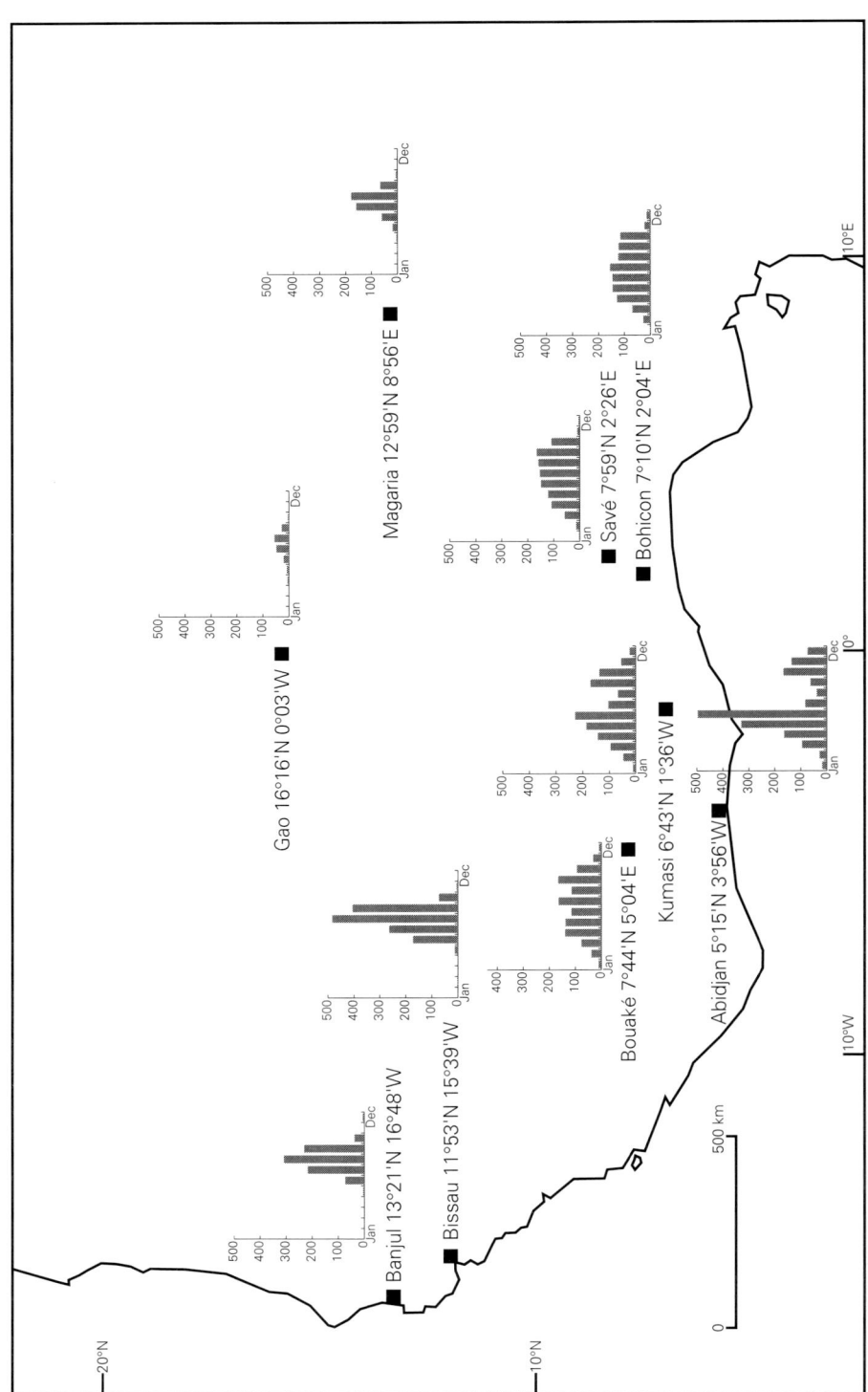

**Fig. 1.5.** Distribution of mean monthly rainfall at nine stations in West Africa, 1978–97 (mm)

*Source*: Rainfall data 1978–97 provided by the Met. Office, Bracknell.

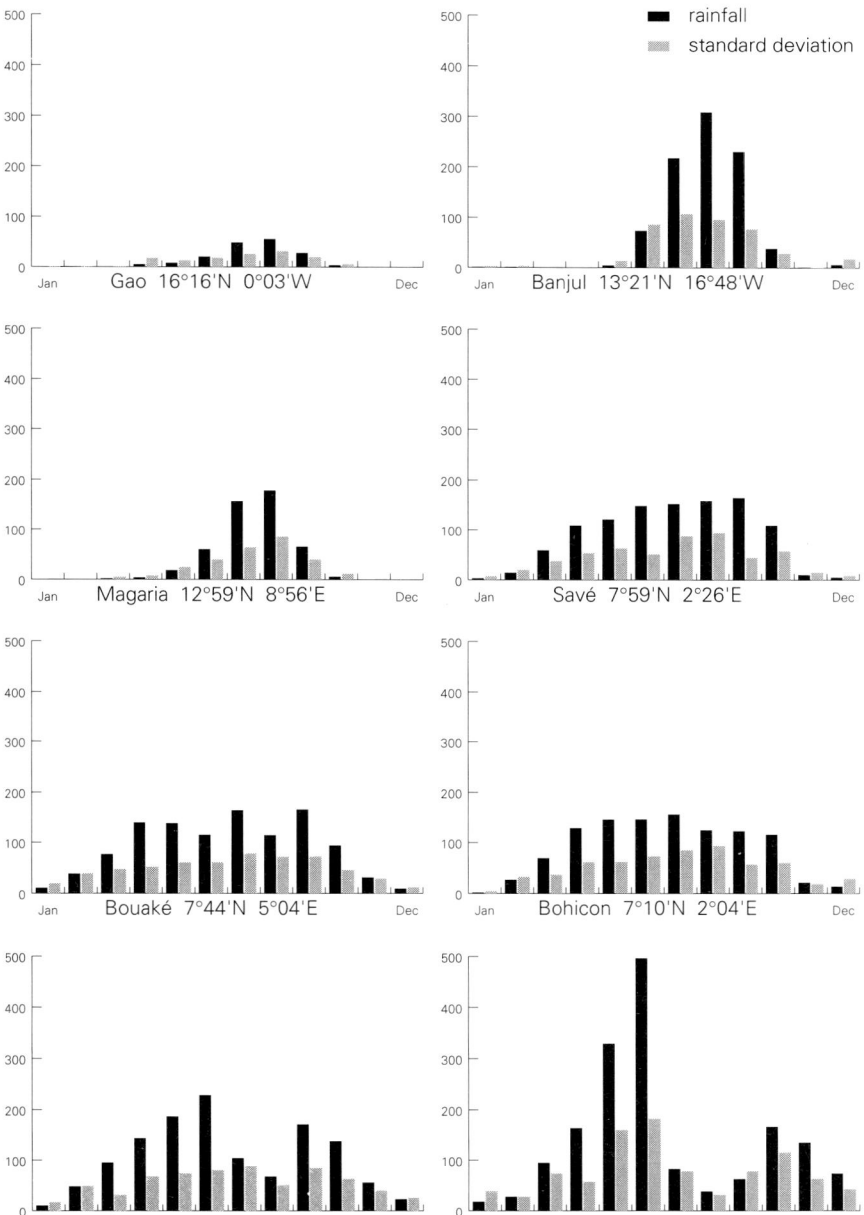

**Fig. 1.6.** Mean monthly rainfall and standard deviation for each of eight stations in West Africa, 1978–97 (mm)

*Source*: Rainfall data 1978–97 provided by the Met. Office, Bracknell.

few. What emerges from the data is that deviations about the mean are considerable not only in the semi-arid north, but in the humid south too. There is often a tendency to link environmental uncertainty with semi-arid areas and where rainfall is confined to so few months of the year, its failure can have dire consequences for local people who are heavily dependent on primary products. However, what the data emphasize is that uncertainty about the amount of rainfall in West Africa is no greater in the semi-arid north than it is in the humid south. Further south, variation in rainfall may pose serious problems for farmers as crops may be battered by heavy rains and hail storms, or washed away by floods. However, higher levels of rainfall in the more humid areas and the scope to harvest more than one crop in a year leave people less vulnerable than further north, and hence the effects of rainfall variability in the humid south have received less attention in the literature. Such findings, although revealing, are not new. Variability of rainfall throughout West Africa has long been known by climatologists and geographers. For example, Miller (1952) has shown the significant variability of rainfall about the mean at Zungeru, Nigeria over the period 1907–11. Harrison Church (1980) presented similar information on rainfall variability at more than thirty stations in Ghana from 1924 to 1933, and demonstrates variability of rainfall at Banjul, The Gambia, for the period 1901–19, and Sivakumar (1989) has shown the variability of rainfall in Niger during the drought that persisted from the early 1970s. He has extended this work (1991) to a review of rainfall variability throughout semi-arid West Africa. The relevant question, then, is why has such critical information been ignored by those involved in the theory and practice of development? Perhaps it is because interest in understanding the environment has never been afforded a sufficiently high priority.

In the more marginal rainfall months in the north, the standard deviation may be as great as the mean and in a very few cases, such as in April and May, in Gao, it is actually greater than the mean. The same is also true of January in Abidjan before the commencement of the main rainy season, and September, at the end of the little dry season. The tendency towards greater variability in the more marginal months particularly in the north, but to some extent in the south too, is apparent from Table 1.1, which charts the coefficients of variability (the ratio of the mean to the standard deviation). From these figures, variability may appear to be greater in the more marginal months because it is between zero and a very few millimetres of rain. However, variability is also considerable, and arguably greater in the wetter months but the ratio of mean to standard deviation is not so great and so this does not emerge from the table of coefficients of variability.

Fig. 1.7 lends weight to the non-equilibrial nature of rainfall by showing the range of precipitation in each month of the year for each of the nine stations in West Africa. A pattern of considerable variation about the mean in each month of the year is evident both in the semi-arid north and the more humid south. Although the data from Bissau are weak, they have been incorporated in the diagram. In an attempt to extract more detail from the data, Fig. 1.8 shows the

Table 1.1. Coefficients of variability of rainfall for stations in West Africa

|  |  | Jan. | Feb. | Mar. | Apr. | May | June | July | Aug. | Sept. | Oct. | Nov. | Dec. |
|---|---|---|---|---|---|---|---|---|---|---|---|---|---|
| Gao | 16.16°N | 0 | 0 | 1.84 | 3.35 | 1.58 | 0.86 | 0.51 | 0.55 | 0.69 | 1.54 | 0 | 0 |
| Banjul | 13.21°N | **2.71** | **2.88** | 0 | 0 | **2.77** | 1.13 | 0.47 | 1.05 | 0.32 | 0.70 | **2.59** | **2.66** |
| Magaria | 12.59°N | 0 | 0 | **2.65** | **2.03** | 1.32 | 0.63 | 0.39 | 0.47 | 0.59 | 1.96 | 0 | 0 |
| Bohicon | 7.10°N | **2.30** | 1.18 | 0.51 | 0.46 | 0.41 | 0.49 | 0.52 | 0.73 | 0.46 | 0.51 | 0.84 | **2.06** |
| Savé | 7.59°N | **2.25** | 1.35 | 0.62 | 0.47 | 0.50 | 0.33 | 0.56 | 0.58 | 0.26 | 0.51 | 1.39 | 1.68 |
| Bouaké | 7.44°N | 1.76 | 0.99 | 0.60 | 0.36 | 0.42 | 0.57 | 0.46 | 0.60 | 0.42 | 0.47 | 0.89 | 1.24 |
| Kumasi | 6.43°N | 1.59 | 0.97 | 0.31 | 0.45 | 0.37 | 0.33 | 0.80 | 0.72 | 0.47 | 0.44 | 0.67 | 1.04 |
| Abidjan | 5.15°N | **2.12** | 0.97 | 0.75 | 0.34 | 0.47 | 0.35 | 0.91 | 0.79 | 1.19 | 0.66 | 0.45 | 0.56 |

*Source:* Derived from data provided the Met.Office, Bracknell.
*Note:* Bold face figures are coefficients of 2.0 and above.

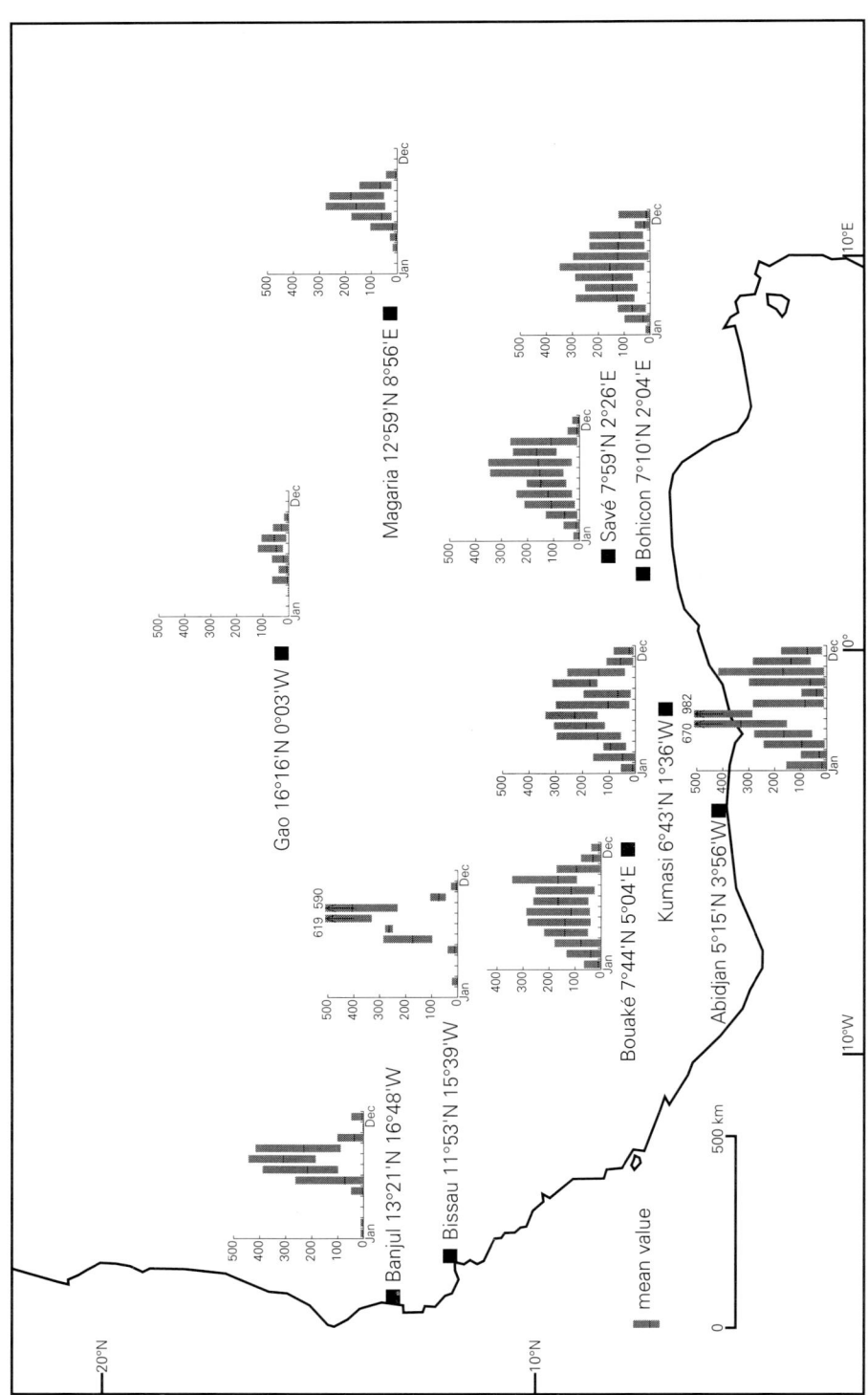

**Fig. 1.7.** Range of rainfall and mean in each month at nine stations in West Africa, 1978–97 in millimetres

*Source*: Rainfall data 1978–97 provided by the Met. Office, Bracknell.

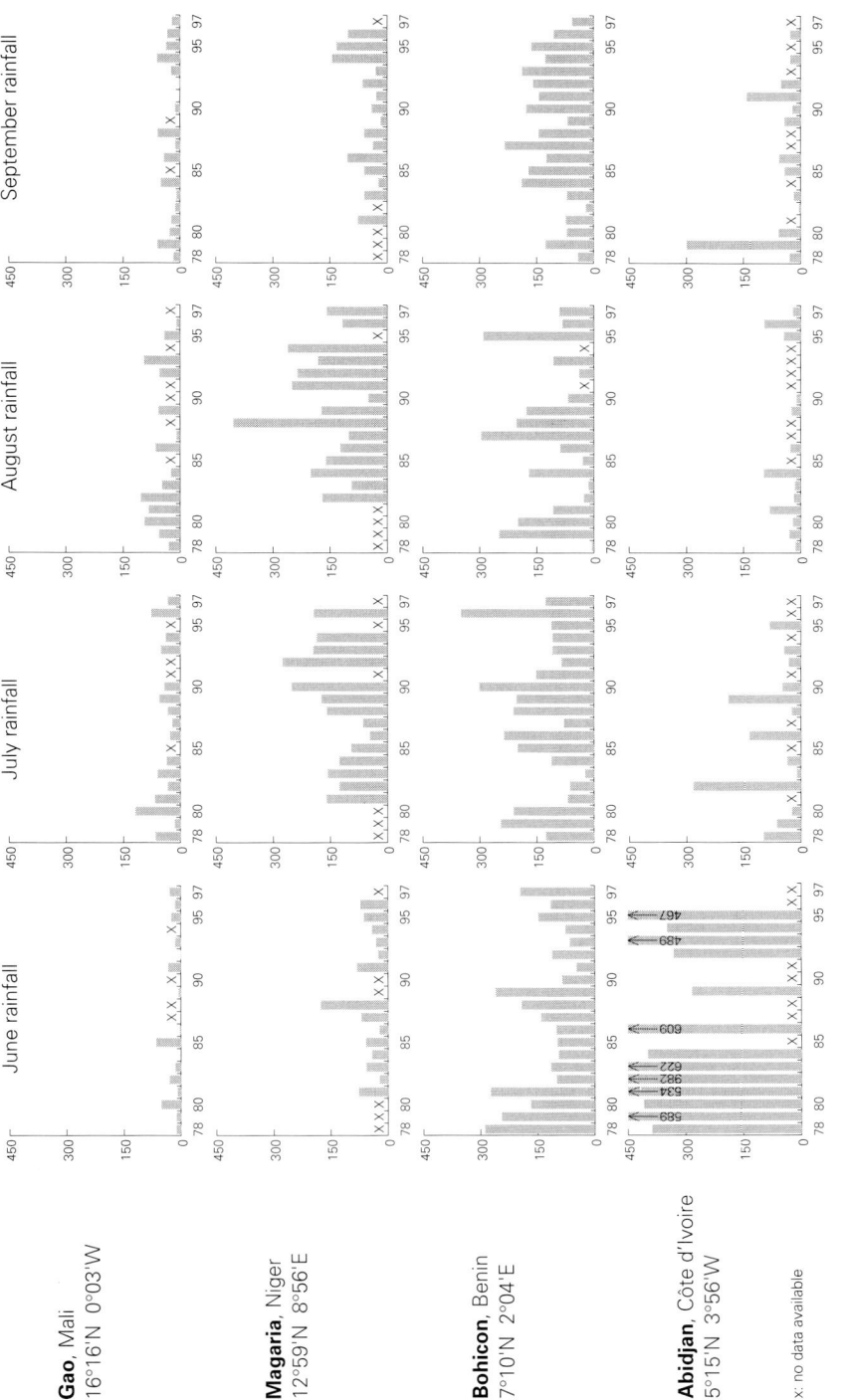

**Fig. 1.8.** Variations in rainfall during 4 months at four stations in West Africa (1978–97) in millimetres

*Source:* Rainfall data 1978–97 provided by the Met. Office, Bracknell.

range of rainfall for specific months of the year from 1978 to 1997. The first effect of this is to emphasize the inadequacy of the data by highlighting the months for which no data were available. That aside, the diagrams confirm the considerable variability of rainfall, in both timing and quantity, in Gao and Magaria in June, July, and August, the main growing months. In an environment where technology is limited, these data hint at some of the substantial problems facing smallholders—both cultivators and pastoralists.

Further south, the range within individual months is even greater. For example, the range of rainfall received in June, the wettest month in Abidjan, is very much greater than the range in August, the wettest month in Gao. Whilst this is no surprise as the average rainfall in June in Abidjan exceeds the annual total in Gao, it serves to strengthen the argument that uncertainty of rainfall is just as much a fact of life in the humid south as it is in the semi-arid north. In spite of their inadequacies the data do reflect the variability of rainfall in all nine stations and highlight the fact that as rainfall increases towards the humid south, a greater range occurs in more months of the year than in the north where the rainy season is restricted.

In addition to variability in both the timing and total quantity of rainfall, the intensity of climatic events create problems such as hurricanes, cyclones, and relentless heavy downpours. Such downpours may average 2.5 cm of rain per hour for 24 hours but within this time frame there can be periods of very intense rainfall, sufficient to trigger a cascade of changes. These may bring about instability and unpredictability in other aspects of the physical environment. In the semi-arid north where the environment is often more open, heavy rainfall can cause soil loss and erosion, particularly on slopes. In areas where there is dense vegetation cover and human disturbance of the environment is low, heavy rain storms may uproot trees and/or create land slips and because the soil cannot absorb all the rainfall, run-off is considerable and relentless gully formation can begin (Floyd, 1965). Morgan and Pugh (1969) describe erosion gully systems in south-eastern Nigeria exceeding 2.6 sq km in size and being 'hundreds of feet in depth' (Morgan and Pugh, 1969: 303).

Returning to concepts of equilibrium and non-equilibrium in the environment, Fig. 1.2(4) suggests that while physical factors alone may precipitate change such as described above, human intervention can give impetus to physical changes. For example, land which has been cleared of its vegetative cover for cultivation is usually very much more vulnerable to damage and periods of abundant rainfall in the humid tropics may be sufficient to cause flooding and wash away crops, particularly those near streams, lakes, and ponds. On slopes in south-eastern Nigeria and southern Côte d'Ivoire, for example, crops may be washed away and with them much of the top soil, bringing ruin not only to the areas losing their soil but also to the areas where the soil is deposited. Additionally, exposure of the soil can lead to the formation of hard pans (Grove, 1992; Goudie, 1973) and this is yet another uncertainty that smallholder farmers have to contend with.

In summary, fluctuations in the environment impinge on the farmer in many ways: the rains may not occur as and when expected. Their unpredictable nature is evident from Figs. 1.4 and 1.6–1.8. They may be excessively heavy or they may be light. The effects of heavy rain and wind can trigger other changes such as soil erosion which can lead to land degradation. It is accepted that all too frequently the loss of top soil may be overestimated. Soil erosion may be difficult to measure and it may be even more difficult to cure (Blaikie, 1985; Barrow, 1995; Stocking, 1996): nevertheless, it does exist and it is a major problem in parts of West Africa. The often visible change in the landscape that may be triggered by heavy rains may be perceived as environmental non-equilibrium or at least as a situation where the equilibrium of the physical environment (almost impossible to define), has altered or is altering, and if not irretrievably, then for a very long time in terms of a human working life.

## Conclusions

The evidence cited above suggests that non-equilibrial conditions are more likely to prevail than equilibrial conditions in tropical ecosystems. However, much depends on the scale of analysis, both in time and space, and on local environmental conditions. It has been observed that the equilibrium paradigm incorporates scope for change. Ecological equilibrium is rarely static, and on account of the problem of identifying what is occurring in the landscape, one cannot be certain whether ecological change represents non-equilibrium, disequilibrium, or dynamic equilibrium. But the question we have to ask is: does it matter whether ecological conditions are in a state of dynamic equilibrium or non-equilibrium or disequilibrium? Surely, what is of greatest importance to development practitioners is an awareness of the unpredictability of the environment and of its effects on human activity. It is this that must be incorporated into the thinking of development practitioners. Development failures, particularly in the rangelands, have been attributed to the long unchallenged equilibrium theory which in ecology was closely linked with Clements's concept of plant succession. While this may be so, one has to ask whether those interpreting Clements's ideas actually considered the implications of dynamic equilibrium or whether they considered equilibrium as static. There is little evidence that the dynamic element has ever been built into development projects. Perhaps creators of development schemes should accept some of the responsibility for failure and for not looking sufficiently closely at the underlying ecological assumptions or at their application at different scales.

If development initiatives are to be successful, at least at the production stage, surely much can be learnt by looking more closely at the autecological approaches of indigenous smallholders to crop and livestock production. Indigenous methods of land use have a history of success and for centuries have put into practice principles discovered relatively recently by theoretical ecologists. Indigenous methods thus merit careful examination for what they might usefully contribute to development. This theme is further developed in subsequent chapters.

Chapter 2 attempts to place West African agriculture in context and begins by considering a range of factors besides the physical environment which have influenced developments in the agricultural sector. One of the major strengths of this chapter is that it examines changes in agricultural data from the mid-1960s to the mid-1990s. It is rare to find data presented for such lengthy time periods in the literature. Although macro level statistics have many inadequacies and should be used only as indicators, nevertheless, it seems appropriate to examine statistical trends in agriculture as these are the basis of 'informed' decision and policy making.

Although the data in Chapter 2, with all their limitations, do reflect positive changes in agriculture over the region, the image of the agricultural sector becomes much more positive in Chapter 3 when the scale of investigation moves to field level. Here, enterprise and inventiveness abound. Chapters 3 and 4 focus on the humid domain and examine how producers have adapted an autecological approach to cope with different constraints such as high population densities in south-eastern Nigeria and insufficient people in parts of Sierra Leone. Chapter 4 examines how the commercial production of cocoa has dealt with a non-equilibrial environment, and also with a host of economic and political problems.

Chapters 5, 6, and 7 move to the savannas and sahel. The literature on savannas is relatively new and stimulating. It is explored in Chapter 5 which provides a background to the next two chapters on agriculture and pastoralism in the savannas. Chapter 6 focuses on smallholder adaptation to an environment dominated by seasonal rainfall and periodic drought. Chapter 7 reviews some of the characteristics of indigenous pastoralism in the savannas and sahel and demonstrates that indigenous approaches to rangeland livestock management are better suited to non-equilibrial environments than are many modern methods based on assumptions of equilibrium in the landscape. Having demonstrated the many merits of smallholder farming, Chapter 8 adopts a different approach, briefly reviewing some of the reasons why development schemes have not met with the success anticipated. One inevitable conclusion is that the non-equilibrial character of the environment has been under-represented in development planning. The ultimate conclusion is that attempts to achieve rural development must continue and that Africans must play a leading role in the development of their Continent.

## *References*

ADAMS, W. M. (1985), 'The downstream impacts of dam construction: a case study from Nigeria', *Transactions of the Institute of British Geographers*, 10 (3): 292–302.

ANDRAE, G. and BECKMAN, B. (1985), *The Wheat Trap: Bread and Underdevelopment in Nigeria* (London: Zed Books).

Arrow, K., Bolin, Bert, Costanza, Robert, Dasgupta, Partha, Folke, Carl, Holling, C. S., Jansson, Bengt-Owe, Levin, Simon, Mäler, Karl-Gorän, Perings, David, and Pimentel, David (1996), 'Economic growth, carrying capacity and the environment', *Ecological Applications*, 6 (1): 13–15.

Aubréville, A. (1971), 'Regenerative patterns in the closed forest of the Ivory Coast', in S. R. Eyre (ed.), *World Vegetation Types* (New York: Columbia University Press), 41–55.

Baker, Kathleen M. (1982), 'Structural change and managerial inefficiency in the development of rice cultivation in the Senegal River Region', *African Affairs*, 81 (325): 499–510.

—— (1985), 'The Chinese agricultural model in West Africa', *Pacific Viewpoint*, 2: 401–14.

Barrow, C. J. (1991), *Land Degradation: Development and Breakdown of Terrestrial Environments* (Cambridge: Cambridge University Press).

—— (1995), *Developing the Environment: Problems and Management* (London: Longman).

Bates, Robert H. (1981), *Markets and States in Tropical Africa: The Political Basis of Agricultural Policies* (Berkeley: University of California Press).

—— and Lofchie, Michael (eds.) (1980), *Agricultural Development in Africa* (Berkeley: University of California Press).

Behnke, Roy H. Jr., Scoones, Ian, and Kerven, Carol (eds.) (1993), *Range Ecology at Disequilibrium: New Models of Natural Variability and Pastoral Adaptation in African Savannas* (London: Overseas Development Institute).

Benneh, G. (1972), 'Systems of agriculture in Tropical Africa', *Economic Geography*, 40 (3): 244–57.

Berg, Elliot (1986), 'The World Bank's Strategy', ch. 2 in John Ravenhill (ed.), *Africa in Economic Crisis* (Basingstoke and London: Macmillan).

Blaikie, Piers (1985), *The Political Economy of Soil Erosion in Developing Countries* (London: Longman).

Bramwell, Anna (1989), *Ecology in the Twentieth Century: A History* (Newhaven and London: Yale University Press).

Bräutigam, Deborah (1998), *Chinese Aid and African Development: Exporting Green Revolution* (Basingstoke: Macmillan).

Chorley, R. J. and Kennedy, B. A. (1971), *Physical Geography: A Systems Approach* (London: Prentice Hall Inc.).

Clements, F. E. (1916), *Plant Succession: An Analysis of the Development of Vegetation* (Carnegie Institute of Washington Publication, 242).

—— (1936), 'Nature and structure of the climax', *Journal of Ecology*, 24: 252–84.

Coffin, Debra P. and Lavenroth, William K. (1996), 'Recovery of vegetation in a semi-arid grassland 53 years after disturbance', *Ecological Applications*, 6 (2): 538–55.

Connell, Joseph H. (1978), 'Diversity in tropical rainforests and coral reefs', *Science*, 199: 1302–10.

—— (1987), 'Change and persistence in some marine communities', ch. 16 in A. J. Gray, M. J. Crawley, and P. J. Edwards (eds.), *Colonization, Succession and Stability*. The 26th Symposium of the British Ecological Society held with the Linnaean Society of London (Oxford: Blackwell Scientific), 339–52.

Cooper, W. S. (1926), 'The fundamentals of vegetation change', *Ecology*, 7: 391–413.

Coppock, D. Layne (1993), 'Vegetation and pastoral dynamics in the southern

Ethiopian rangelands: Implications for theory and management', ch. 3 in R. H. Behnke Jr. *et al.*, *Range Ecology at Disequilibrium* (London: Overseas Development Institute), 42–61.

DAVIS, W. M. (1909), 'The geographical cycle', in D. W. Johnson (ed.), *Geographical Essays* (London: Grin & Co.), 254–6.

DE ANGELIS, D. L. and WATERHOUSE, J. C. (1987), 'Equilibrium and nonequilibrium concepts in ecological models', *Ecological Monographs*, 57: 1–21.

DE WILDE, JOHN (1967), *Experiences with Agricultural Development in Tropical Africa: Volume II, The Case Studies* (published for the International Bank for Reconstruction and Development; Baltimore, Maryland: Johns Hopkins Press).

DOYLE, C. J. (1990), 'Application of systems theory to farm planning and control: Modelling resource allocation', ch. 4 in J. G. W. Jones and P. R. Street (eds.), *Systems Theory Applied to Agriculture and the Food Chain* (Barling, Essex: Elsevier Science), 89–108.

EDWARDS, P. J., MAY, R. M., and WEBB N. R. (eds.) (1994), *Large Scale Ecology and Conservation Biology* (Oxford: Blackwell Scientific).

ELLIS, JAMES, COUGHENOUR, MICHAEL B., and SWIFT, DAVID M. (1993), 'Climate variability, ecosystem stability, and the implications for range and livestock development', ch. 2 in R. H. Behnke Jr. *et al.*, *Range Ecology at Disequilibrium* (London: Overseas Development Institute), 31–41.

ERLICH, PAUL R. (1989), 'Discussion: Ecology and Resource Management—Is ecological theory any good in practice?', ch. 21 in Jonathan Roughgarden, Robert M. May, and Simon A. Levin (eds.), *Perspectives in Ecological Theory* (Princeton, New Jersey: Princeton University Press), 306–18.

FAIRHEAD, JAMES and LEACH, MELISSA (1996), *Misreading the African Landscape: Society and Ecology in a Forest-Savanna Mosaic* (Cambridge: Cambridge University Press).

FAO (1982), *Tropical Forest Resources* (Rome: FAO, Forestry Papers, 30).

FLOYD, B. N. (1965), 'Soil erosion and deterioration in Eastern Nigeria: a geographical appraisal', *Nigerian Geographical Journal*, 8: 33–44.

FULS, E. R. (1992), 'Ecosystem modification created by patch over-grazing in semi-arid grassland', *Journal of Arid Environments*, 23 (1): 59–69.

—— and BOSCH, O. J. H. (1991), 'The Influence of below average rainfall on the regenerational traits of a patch-grazed semi-arid grassland', *Journal of Arid Environments*, 21 (1): 13–20.

GLEASON, H. A. (1917), 'The structure and development of the plant association', *Bulletin of the Torrey Botanical Club*, 44: 463–81.

—— (1926), 'The individual concept of the plant association', *Bulletin of the Torrey Botanical Club*, 53: 7–26.

—— (1927), 'Further views on the succession concept', *Ecology*, 8: 299–326.

GOLLEY, FRANK B. (ed.) (1977), *Ecological Succession* (Benchmark Papers in Ecology/5, Stroudsburg, Pennsylvania: Dowden, Hutchinson and Ross Inc.).

GOUDIE, A. S. (1973), *Duricrusts* (Oxford: Clarendon Press).

GROVE, A. T. (1992), *The Changing Geography of Africa*, (Oxford: Oxford University Press).

HAECKEL, ERNST (1866), *Generelle morphologie der Organismen* (Berlin: Reimer).

HARRISON CHURCH, R. J. (1980) (8th edn.), *West Africa* (London and New York: Longman).

He, Fangliang, Legendre, Pierre, and LaFrankie, James V. (1996), 'Spatial pattern of diversity in a tropical rain forest in Malaysia', *Journal of Biogeography*, 23: 57–74.

Helleiner, G. K. (1975), 'Smallholder decision making: Tropical African evidence', in L. G. Reynolds (ed.), *Agriculture in Development Theory* (New Haven: Yale University Press), 27–52.

Hinderink, J. and Sterkenburg, J. J. (1987), *Agricultural Commercialization and Government Policy in Africa*. Monographs from the African Studies Centre (Leiden, London and New York: KPI).

Holling, C. S. (1973), 'Resilience and stability in ecological systems', *Annual Review of Ecology and Systematics*, 4: 1–22.

Hopkins, Brian (1974) (2nd edn.), *Forest and Savanna: An Introduction to Tropical Terrestrial Ecology with Special Reference to West Africa* (Ibadan and London: Heinemann).

Hort, A. (ed.) (1916), *Theophrastus: an Enquiry into Plants. Book IV Of the Trees and Plants Special to a Particular District and Positions* (London: Heinemann).

Hubbell, Stephen P. and Foster, Robin B. (1987), 'The spatial context of regeneration in a neotropical forest', ch. 19 in A. J. Gray, M. J. Crawley, and P. J. Edwards (eds.), *Colonization, Succession and Stability*. The 26th Symposium of the British Ecological Society held with the Linnaean Society of London (Oxford: Blackwell Scientific), 395–412.

Iyegha, David A. (1988), *Agricultural Crisis in Africa: The Nigerian Experience* (Lanham, New York: University Press of America).

Jones, W. O. (1960), 'Economic man in Africa', *Food Research Institute Studies*, 1: 107–34.

Levin, Simon A. (1989), 'Challenges in the development of a theory of community and ecosystem structure and function', ch. 16 in Jonathan Roughgarden, Robert M. May, and Simon A. Levin (eds.), *Perspectives in Ecological Theory* (Princeton, New Jersey: Princeton University Press), 242–55.

McIntosh, Robert P. (1985), *The Background of Ecology: Concept and Theory* (Cambridge: Cambridge University Press).

Martin, Claude (1991), *The Rainforests of West Africa: Ecology, Threats, Conservation*. English translation by Linda Tsardakas (Basel: Birkhäuser Verlag).

May, R. M. (1976), *Theoretical Ecology: Principles and Applications* (Oxford: Blackwell Scientific).

——(1994), 'The effects of spatial scale on ecological questions', ch. 1 in P. J. Edwards, R. M. May, and N. R. Webb (eds.), *Large Scale Ecology and Conservation Biology* (Oxford: Blackwell Scientific), 11–17.

Miles, J. (1987), 'Vegetation succession: past and present perceptions', in A. J. Gray, M. J. Crawley, and P. J. Edwards (eds.), *Colonization, Succession and Stability*. The 26th Symposium of the British Ecological Society held with the Linnaean Society of London (Oxford: Blackwell Scientific), 1–30.

——(1979), *Vegetation Dynamics* (London: Chapman and Hall).

Miller, R. (1952), 'The Climate of Nigeria', *Geography*, 37 (178): 198–213.

Morgan, W. B. and Moss, R. P. (1965), 'Savanna and forest in Western Nigeria', *Africa*, 35 (3): 286–93.

Morgan, W. B. and Pugh, J. C. *West Africa* (London: Methuen).

Mortimore, M. (1989), *Adapting to Drought: Farmers, Famines and Desertification in West Africa* (Cambridge: Cambridge University Press).

MORTIMORE, M. (1998), *Roots in the African Dust: Sustaining the Drylands* (Cambridge: Cambridge University Press).
MOSS, R. P. (1992), 'Environmental constraints on development in tropical Africa', ch. 3 in M. B. Gleave (ed.), *Tropical African Development: Geographical Perspectives* (Harlow: Longman Scientific and Technical).
MYERS, NORMAN (1984), *The Primary Source: Tropical Forest and Our Future* (New York: Norton).
—— (1996), 'Biodiversity and depletion', ch. 20 in W. M. Adams, A. S. Goudie and A. R. Orme (eds.), *The Physical Geography of Africa* (Oxford: Oxford University Press), 356–66.
NETTING, ROBERT MCC. (1968), *Hill Farmers of Nigeria* (Seattle: University of Washington Press).
—— (1993), *Smallholders, Householders: Farm Families and the Ecology of Intensive, Sustainable Agriculture* (California: Stanford University Press).
NORMAN, D. W. (1972), *An Economic Survey of Three Villages in Zaria Province: Input/Output Study* (Samaru, Miscellaneous Papers, No. 37, Zaria, Nigeria).
NORMAN, D. W. (1974), 'Rationalising mixed cropping under indigenous conditions: The example of Northern Nigeria', *Journal of Development Studies*, 11: 3–21.
NOY-MEIR I. (1975), 'Stability of grazing system: an application of predator-prey graphs', *Journal of Ecology*, 63: 459–81.
ODUM, EUGENE P. (1971), *Fundamentals of Ecology* (Philadelphia, London, Toronto: W. B. Saunders Company).
O'NEILL, JOHN (1993), *Ecology, Policy and Politics: Human Well-Being and the Natural World* (London: Routledge).
O'NEILL, R. V. (1989), 'Perspectives in hierarchy and scale', ch. 10 in Jonathan Roughgarden, Robert M. May, and Simon A. Levin (eds.), *Perspectives in Ecological Theory* (Princeton: Princeton University Press), 140–56.
ONIMODE, BADE (1982), *Imperialism and Underdevelopment in Nigeria: The Dialectic of Mass Poverty* (London: Zed Books).
PÉLISSIER, PAUL (1966), *Les paysans du Sénégal: les civilisations agraires du Cayor à la Casamance* (Sainte-Yrieux, Haute Vienne: Imprimérie Fabriqué).
PORRITT, JONATHAN (1994), 'Translating ecological science into practical policy', ch. 16 in P. J. Edwards, R. M. May, and N. R. Webb (eds.), *Large Scale Ecology and Conservation Biology* (Oxford: Blackwell Scientific), 345–53.
POUND, R. and CLEMENTS F. E. (1898, 2nd edition 1900), *The phytogeography of Nebraska* (Lincoln: Nebraska) (Reprinted New York: Arno Press 1977).
REIJ, CHRIS, SCOONES, IAN, and TOULMIN, CAMILLA (1996), *Sustaining the Soil: Indigenous Soil and Water Conservation in Africa* (London: Earthscan).
RICHARDS, PAUL (1985), *Indigenous Agricultural Revolution* (London: Hutchinson).
—— (1986), *Coping with Hunger: Hazard and Experiment in an African Rice-Farming System* (London: Allen & Unwin).
RIETBERGEN, SIMON (1989), 'Africa', ch. 3 in Duncan Poore, Peter Burgess, John Palmer, Simon Rietbergen, and Timothy Synnott (eds.), *No Timber Without Trees: Sustainability in the Tropical Forest* (London: Earthscan).
ROWE, J. STAN (1997), 'From reductionism to holism in Ecology and Deep Ecology', *The Ecologist*, 27 (4): 147–51.

RUTHENBERG, HANS (1980) (3rd edn.), *Farming Systems in the Tropics* (Oxford: Clarendon Press).
——(JAHNKE, H., ed.) (1985), *Small Farmers in the Tropics: The Economics of Technical Innovations for Agricultural Development* (Oxford: Clarendon Press).
SCHIMPER, A. F. W. (1898), *Pflanzen Geographie auf Physiologiochen Grundlage* (Jena: Fischer).
——(1903), *Plant Geography Upon a Physiological Basis* (trans. W. R. Fisher) (Oxford: Clarendon Press).
SCOONES, IAN (1993), 'Why are there so many animals? Cattle population dynamics in the communal areas of Zimbabwe', ch. 4 in R. H. Behnke Jr. *et al.*, *Range Ecology at Disequilibrium*, 62–76.
SIMBERLOFF, D. (1983), 'When is an island community in equilibrium?' *Science*, 220: 1275–7.
SIMMONS, I. G. (1989), *Changing Face of the Earth: Culture; Environment; History* (Oxford: Basil Blackwell).
SIVAKUMAR, M. V. K. (1989), 'Agroclimatic aspects of rain-fed agriculture in the Sudano-Sahelian zone in ICRISAT', *Soil, Crop and Water Management in the Sudano-Sahelian Zone. Proceedings of an international workshop held at the ICRISAT Sahelian Center, Niamey, Niger, 11–16 Jan. 1987* (Patancheru Andhra Pradesh, India: ICRISAT), 17–38.
——(1991), *Drought Spells and Drought Frequencies in West Africa*. ICRISAT Research Bulletin no. 13 (Niamey, Niger: ICRISAT).
SMITH, ANDREW B. (1992), *Pastoralism in Africa: Origins and Development Ecology* (London: Hurst; Athens: Ohio University Press; Johannesburg: Witwatersrand University Press).
SOUSA, W. P. (1979), 'Disturbance in the marine intertidal boulder fields: The non-equilibrium maintenance of species diversity', *Ecology*, 60: 1225–39.
——(1984), 'The role of disturbance in natural communities', *Annual Review of Ecology and Systematics*, 15: 353–91.
SPEDDING, C. R. (1975), *The Biology of Agricultural Systems* (London: Academic Press).
STOCKING, M. A. (1996), 'Soil Erosion', ch. 18 in W. M. Adams, A. S. Goudie and A. R. Orme (eds.), *The Physical Geography of Africa* (Oxford: Oxford University Press).
STOTT, P. A. (1999), *'Tropical Rain Forest': The Political Ecology of a Hegemonic Myth* (London: Institute of Economic Affairs).
STRONG JR., DONALD, SIMBERLOFF, DANIEL, ABELE, LAWRENCE G., and THISTLE, ANNE B. (eds.) (1984), *Ecological Communities, Conceptual Issues and the Evidence* (Princeton: Princeton University Press).
SULLIVAN, SIAN (1996), 'Towards a non-equilibrium ecology: perspectives from an arid land', *Journal of Biogeography*, 23: 1–5.
——(1998), 'People, Plants and Practice in Drylands: Socio-Political and Ecological Dimensions of Resource-Use by Damara Farmers in North-West Namibia' (unpubl. Ph.D. thesis, Department of Anthropology, University College, London).
TANSLEY, A. G. (1916), 'The development of vegetation. A review of Clements' "Plant Succession" 1916', *Journal of Ecology*, 4: 198–204.
——(1920), 'The classification of vegetation and the concept of development', *Journal of Ecology*, 8: 118–49.

TOULMIN, C. (1992), *Cattle, Women and Wells: Managing Household Survival in the Sahel* (Oxford: Clarendon Press).

UPTON, MARTIN (1967), *Agriculture in South-Western Nigeria: A Study of the Relationship Between Production and Social Characteristics in Selected Villages* (Development Studies, no. 3, University of Reading, Department of Agricultural Economics).

VON BERTALANFFY, L. (1956), 'General Systems Theory', *General Systems Yearbook*, 1: 1–10.

—— (1962), 'General Systems Theory—A critical review', *General Systems Yearbook*, 1: 1–20.

WALKER, B. H. and NOY-MEIR, I. (1982), 'Aspects of stability and resilience of savanna ecosystems', in B. J. Huntley and B. H. Walker (eds.), *Ecology of Tropical Savannas*. Ecological Studies, 42 (New York: Springer Verlag), 556–90.

WALLACE, TINA (1979), 'Rural development through irrigation: Studies in a town on the Kano River Project' (Mimeo, Zaria, Nigeria: Centre for Social and Economic Research, Ahmadu Bello University).

WATT, A. S. (1919), 'On the causes of failure of natural regeneration in British oakwoods', *Journal of Ecology*, 7: 173–203.

—— (1947), 'Pattern and process in the plant community', *Journal of Ecology*, 35: 1–22.

WHITTAKER, R. H. (1953), 'A consideration of the climax theory: the climax as a population and pattern', *Ecological Monographs*, 23: 41–78.

WIENS, JOHN A. (1984), 'On understanding a non-equilibrium world. Myth and reality in community patterns and processes', ch. 25 in Donald Strong Jr. *et al.*, *Ecological Communities, Conceptual Issues and the Evidence* (Princeton: Princeton University Press), 439–57.

WILLIAMSON, MARK (1987), 'Are communities ever stable?', ch. 17 in A. J. Gray, M. J. Crawley, and P. J. Edwards (eds.), *Colonization, Succession and Stability*. The 26th Symposium of the British Ecological Society held with the Linnaean Society of London (Oxford: Blackwell Scientific), 353–70.

World Bank (1981), *Accelerated Development in Sub-Saharan Africa: An Agenda for Action* (Washington: World Bank).

World Resources Institute (1992), *World Resources 1992–1993: A Guide to the Global Environment* (New York and Oxford: Oxford University Press).

# 2

# Problems and Prognosis: Perspectives on Agriculture at a Regional Scale

Chapter 1 asserted that development strategies should be based on appropriate ecological models. It also made the claim (to be substantiated in subsequent chapters), that many of the methods used by smallholders were ecologically appropriate and might thus be incorporated in models for development. But if smallholder methods are so apt, one is forced to ask why is it that African agriculture is widely perceived as a sector in crisis, low in terms of productivity and largely stagnant? (World Bank, 1981, 1989; Berg, 1986; Eicher, 1986; Lowe, 1986; Gakou, 1987; Hinderink and Sterkenburg, 1987; Neimeijer, 1996). The answer, it would appear, depends very much on the scale of investigation. It is undeniable that there are major problems in the agricultural sector in West Africa, but these seem to be distorted at a regional scale. When viewed at a more local scale the image of the sector may be much more positive, as several of the subsequent chapters show. Inevitably, the interpretation of secondary source information is always open to debate.

The aim of this chapter is to review specific aspects of West African agriculture at a regional scale. This is an extremely difficult task as the very size of West Africa limits the coherence of any image of agriculture at a regional level. Furthermore, such regional images derived from international statistics seem to bear little relation to agriculture at village, farm, or field level. Nevertheless, the review conducted in this chapter is justified as aggregated statistics at a national, regional, and continental level are the usual bases of 'informed' decisions on African agriculture at a national level and in the international arena. Moreover, it is these data which influence government policies and actions towards the agricultural sector, so whatever one's views on the accuracy of macro level statistics, we ignore them at our peril.

The dubious quality of such data is debated in the literature (Raikes, 1988; Lipton and Longhurst, 1989). Lipton and Longhurst (1989) observe that 'only for three or four countries of sub-Saharan Africa, not including any of the four largest, can we be 95 per cent certain that food output in any given year lies within 40 per cent of the official figure' (Lipton and Longhurst, 1989: 353). Ever mindful of considerations such as these, the next section reviews selected agricultural statistics at a regional scale. The production, area, and yield of millet, sorghum, maize, paddy, groundnuts, and cassava are examined and some of the major trends that they reveal are analysed. (Tree crops, also of great importance in West Africa, are the focus of Chapter 4.) Following the discussion on crop statistics, government intervention in the agricultural sector prior to Structural Adjustment is considered to highlight the difficult environment in which smallholders have had to operate. Some of the effects of Structural Adjustment on the agricultural sector are then reviewed. Finally, in spite of long-held negative views towards smallholders, their capacity to survive is considered along with their skilful management of available resources.

For a region that extends the distance of London to Moscow, generalization is rarely satisfactory though this has never inhibited attempts to generalize, nor will it do so here. Lipton and Longhurst (1989) lament that although many authors caution against generalization at large scales in Africa, simply making the observation seems to absolve them of the responsibility that goes with it and too often, insufficiently critical generalization ensues. Doubtless, this chapter will be guilty of much the same charge though hopefully generalizations will not be too uncritical.

## Reviewing change in West African agriculture at a macro level: what the indicators say

### The area under agriculture

In terms of the area it covers, agriculture does not seem as important as the proportion of the population directly involved in the sector. According to the World Bank (1989), 14 per cent of the region was classified as 'cropland' in 1987, a term which also includes land in fallow (World Bank, 1989). Comparable data presented by The World Resources Institute (1992) suggest that cropland covered only 9.8 per cent of the region on average, from 1985 to 1987. A discrepancy of approximately 4 per cent over an area the size of West Africa is quite significant, particularly as both sources seem to depend heavily for their data on the FAO. This merely serves to confirm that aggregated statistics need to be treated with caution and considered as no more than indicators in any discussion.

What these data obscure is the very different nature and role of agriculture in the countries of the region. For example, in Mauritania which borders on the Sahara desert, less than 1 per cent of the total land area is devoted to agriculture while in Nigeria, both water and people are more abundant, and some 34 per cent of the country was designated 'cropland' in 1987–9 (World Resources Institute, 1992). Very broadly, the cultivated area appears to be directly related to both rainfall and population. Where one or both of these is limited, the area of cropland tends to be lower. While the Sahelian countries cover vast areas, much of this is marginal for cultivation, but in countries such as Guinea and parts of Côte d'Ivoire where rainfall is abundant and population comparatively low, the potential for the expansion of agriculture is considerable.

Agriculture has declined in importance since the mid-1960s. In 1965 it contributed just over 45 per cent of the Gross Domestic Product of West Africa; by 1987, this had fallen to 39 per cent without any significant restructuring of the economies of most states and in 1997 was just over 37% (IBRD/World Bank, 1999). Alarmists point to the increased food imports of the 1970s and early 1980s as a sign of a food production crisis but this may have been due to substantial rural–urban migration during this period and to the shift in urban diets to wheat and rice, neither of which is a traditional staple over most of the region. In addition, during the past 30 years, parts of West Africa have been hard hit by drought, by insect attacks, by political instability, by mismanagement within the region and by changes in the international situation. In spite of concern about food production levels, international statistics record an increase in agricultural production in the decade to 1996. The index of agricultural production which was equal to 100 in 1989–91 increased from 82.93 for West Africa in 1985 (the first year for which the index was calculated retrospectively), to 118.22 in 1996. Data for Liberia were unavailable and so could not be included in the West African statistics. Comparable figures for Africa as a whole were very similar. It is noteworthy that the increase has not been brought about largely by increases in cash crop production as the index of food crops closely mirrors the total index. In other words, there has been an increase in both cash crop and food crop production over the region. The situation becomes less encouraging when the *per caput* index of agricultural production is examined. This has hovered near 100 since 1985 and over the twelve years to 1996 was over 100 in only four years, and for only two years in the case of Africa as a whole. This suggests that agricultural production has barely kept pace with population growth though a major problem is that the precise size of the region's population is not known and the production of domestically consumed produce is estimated. In 1996 the *per caput* index of agricultural production for West Africa was 99.4, much the same as for Africa as a whole (99.9).

The increase in agricultural production in West Africa and throughout Africa has been achieved by a substantial increase in the area under cultivation (Morgan and Solarz, 1994). From 1965 to 1987 the area under cultivation increased by 16 per cent in West Africa compared with 22 per cent in Africa as a whole. As Tables 2.1–2.6 indicate there has been a substantial increase in the cultivated area under specific crops but in some cases, such as maize and paddy, a significant increase in yield has also contributed to increasing production. There is undoubtedly much more that could be achieved through the use of improved and appropriate technology (Macgregor, 1990). The

**Table 2.1.** Millet: Changes in production, area, and yield, 1965–7 to 1994–6

|  | Production | | Area | | Yield | |
| --- | --- | --- | --- | --- | --- | --- |
|  | Production 1994–6 ('000 tonnes) | % change in production (94–96/65–67) | Area 1994–6 ('000 ha) | % change in area (94–96/65–67) | Yield 1994–6 (kg ha$^{-1}$) | % change in yield (94–96/65–67) |
| Burkina | 783 | 135 | 1,230 | 60 | 637 | 47 |
| Mali | 793 | 2 | 1,353 | 48 | 586 | −32 |
| Niger | 1,858 | 228 | 4,812 | 240 | 386 | −4 |
| Nigeria | 5,130 | 118 | 5,139 | 18 | 997 | 86 |
| Senegal | 622 | 13 | 909 | −16 | 685 | 36 |
| West Africa | 9,671 | 93 | 14,068 | 50 | 705 | 11 |
| Africa | 12,308 | 36 | 19,432 | 35 | 634 | 1 |

*Note*: Burkina, Mali, Niger, Nigeria and Senegal produced 95 per cent of West Africa's millet in 1994–6.
All figures have been rounded to the nearest whole number.
*Source*: *FAO Production Yearbook* (1970), vol. 24; (1972), vol. 26; (1978), vol. 32; (1981), vol. 35; (1984), vol. 38; (1987), vol. 41; (1990), vol. 44; (1993), vol. 47; (1996), vol. 50.

following section focuses on some of the major field crops in West Africa and examines their performance over the past 20 to 30 years (depending on data availability).

## Variation in crop performance in West Africa

### Rainfed cereals: millet, sorghum, and maize

#### Millet

West Africa produces almost 80 per cent of the Continent's millet (Table 2.1). Indigenous to Africa, bulrush millet (*Pennisetum* spp.) is one of the most drought tolerant of the rainfed cereals and certain varieties can be grown where annual rainfall is as little as 300 mm. Sorghum is also drought resistant and can be harvested where rainfall is as little as 250 mm in a growing season (Moss, 1992), though it is less tolerant of poor soils than millet (Morgan and Pugh, 1969). By contrast, fieldwork by the author in The Lower Gambia (1992) suggested that when rainfall levels were low farmers believed that millet would provide a better harvest than sorghum. Both millet and sorghum produce a harvest with significantly less rainfall than maize though where annual rainfall is less than 588 mm risks of crop failure are always high (Mortimore, 1998).

The main millet-producing areas are the drier savannas and the sahel. In West Africa these include Burkina, Mali, Niger, Nigeria, and Senegal which together produce 95 per cent of the region's millet. Fig. 2.1 summarizes data from FAO Crop Production Statistics on levels of production, area cultivated and yield of millet in the major producing countries, in West Africa, and in Africa as a whole. It also reviews changes in these variables for the period

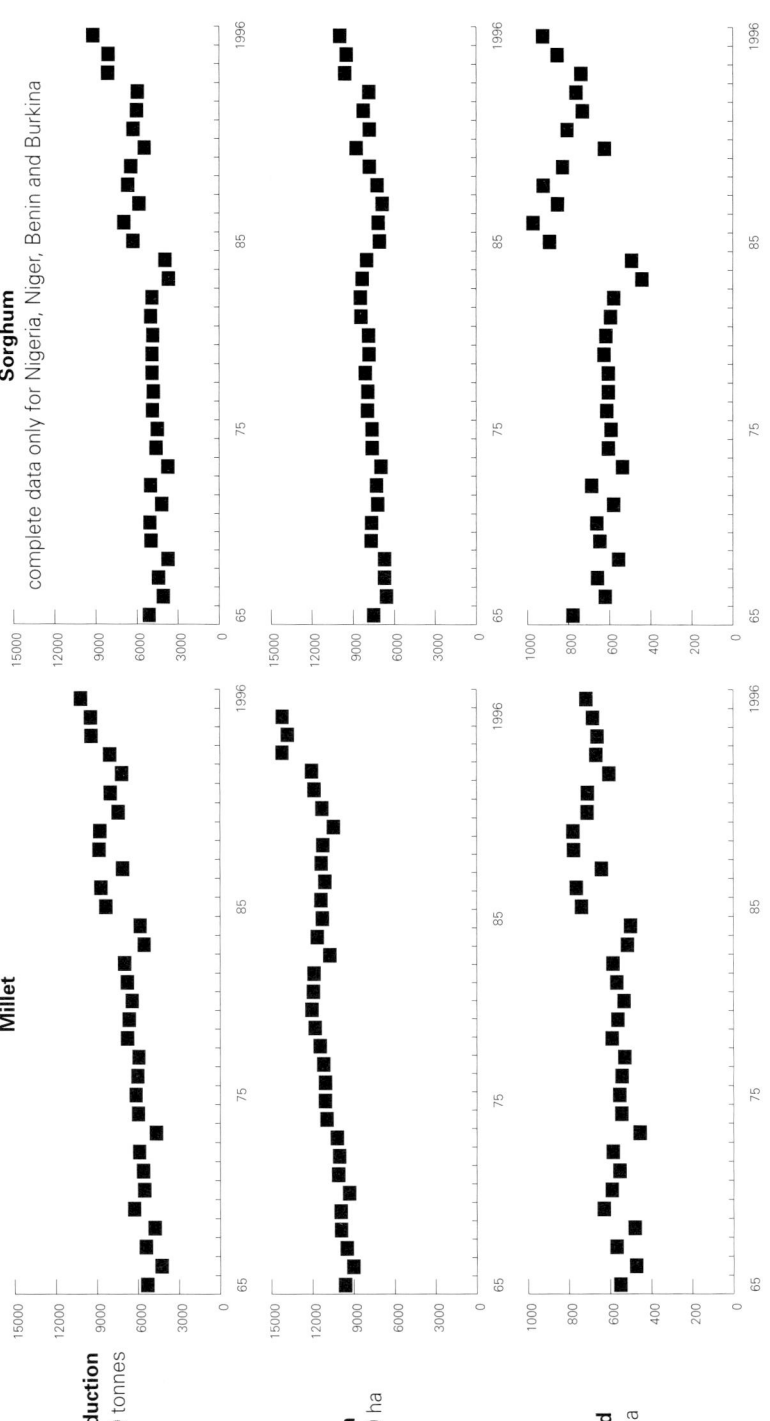

**Fig. 2.1.** Production, area, and yield of millet and sorghum in West Africa, 1965–96

*Sources:* FAO *Production Yearbook* (1970), vol. 24; (1972), vol. 26; (1978), vol. 32; (1981), vol. 35; (1984), vol. 38; (1987), vol. 41; (1993), vol. 47; (1996), vol. 50; FAO *Quarterly Bulletin of Statistics* (1995), vol. 8 1/2.

1965–7 and 1994–6. The accuracy of aggregated statistics such as these is always in doubt, not least because we have relatively little idea about the quality of data collection. Nevertheless, they are some of the most complete and easily accessible data available for the region for the past 30 years and although individual statistics should be treated with caution, the trends that they show are of interest.

Increases in production, area, and yield of millet have been quite significant in absolute terms, although the average annual increase has been very gradual over the past 30 years (Fig. 2.1). Millet production increased by over 93 per cent between 1965–7 and 1994–6 (Table 2.1). Much of this rise was due to an increase of almost 50 per cent in the area cultivated with millet, though a notable part of the increase has been achieved through improved yields which are now around 30 per cent higher in the region than they were in the mid-1960s. On Fig. 2.1 millet yields fluctuate the most and, to some extent, reflect variation in rainfall. The poor rains in 1973 and in 1983–4 stand out as years of low yield. Improved yields since 1985 may be due to increased uptake of higher yielding varieties or they may also reflect better rains over some of this period. Production for 1994–6 is the highest recorded for the region since the mid-1960s. Possible causes may be more favourable weather conditions, comparatively low levels of pest attack, and a trend towards increased domestic food production in response to Structural Adjustment.

Looking in more detail at the statistics of individual countries, it is evident that performance across the region is far from even. Nigeria remains the largest producer of millet, particularly in the semi-arid northern part of the country. Niger is also a significant producer, having shown the greatest percentage increase in production in the region (+228 per cent) and also the greatest percentage increase in the area under millet (+240 per cent) since 1965–7 (Table 2.1). When yields are examined there is again considerable variation. Yields in Nigeria are currently about one tonne per hectare and have increased by approximately 86 per cent since the mid-1960s. In The Gambia, Ghana, and Sierra Leone yields are also close to one tonne per hectare and well above the regional average, calculated by dividing total production for West Africa's fifteen states in 1994–6 by the area under millet over the same period. Yields in Burkina and Senegal are slightly below the regional figure while in Niger and Mali they have declined. In Niger, mean yields of $386 \text{kg ha}^{-1}$ for 1994–6 seem very low for a major producer. Grain crops are highly susceptible to insect pests and to birds. These may well influence yield statistics though their effects are not evident from the statistics.

The overall increase in yield for the region may be attributable to the increased uptake of improved varieties of millet. Some of these are available through official sources and several improved varieties have been developed by farmers through self-selection and exchange of seed. Fieldwork in The

Gambia in the 1980s and early 1990s revealed that the definition of an improved variety was far from clear. Perceptions included varieties which gave higher yields, varieties which matured quickly and were tolerant of drought, and varieties resistant to disease. Some of these definitions appeared to be paradoxical. Quicker maturing, drought resistant varieties of millet and groundnuts were widely grown in villages of the Lower Gambia in the early 1990s (Baker, 1992, 1995). These varieties which matured in 90 days yielded approximately one-third less than the slower maturing varieties which matured in 120 days. Improvement in this case meant lower yields but a greater chance of a harvest in drier years. Some of these varieties were new, but it should be noted that varieties with different maturation periods have existed for generations (Morgan and Pugh, 1969).

More recent field work in The Gambia (1999) revealed a significant change in the millet varieties used by farmers in the Western Division. They have now reverted to growing slower maturing varieties. The explanation offered was that birds consumed a higher proportion of the early millets than the later maturing varieties. The early harvests were particularly attractive to birds as these were some of the earliest grains to mature in the rainy season. Later in the season there was more food available and damage tended to be less severe. Birds were still a major problem later in the season and in order to minimize losses several villages decided that they would all plant one type of millet—a late variety, so that all the millet over a considerable area ripened at much the same time. Where the crop was ripening at different times the effect of bird attack was much more severe. Where one or two fields ripened at a time a farmer could lose up to 80 per cent of the harvest. Where all the fields matured at much the same time, the percentage loss overall was less. This had to be balanced against planting a slower maturing crop which might never mature if the rains failed. In 1998 they did.

Gambian millet yields have not changed significantly over the past 30 years but those in Nigeria have. Explanations for such differences are not evident from aggregated statistics. However, Nigerian farmers have had better access to improved varieties and international statistics confirm that availability and use of technology and other inputs such as fertilizer are greater in Nigeria than in other West African countries. This could be responsible for improved yields not only of millet but of other crops as well.

Niger's millet yields are comparatively low. Niger is drier than much of Nigeria, The Gambia, Ghana, and Sierra Leone and the broad pattern evident from the data suggests a direct relationship between rainfall and millet yields. Hodgkinson (1992) suggests that low rains have been the major cause of low yields in countries such as Niger. While this seems the most probable explanation for the variation in yields, it is also possible that seed is sown more thinly in drier areas. There was evidence of this occurring in northern Senegal in 1982 when farmers were battling against drought. Yield statistics may have been

**Table 2.2.** Sorghum: changes in production, area, and yield, 1965–7 to 1994–6

| | Production | | Area | | Yield | |
|---|---|---|---|---|---|---|
| | Production 1994–6 ('000 tonnes) | % change in production (94–96/65–67) | Area 1994–6 ('000 ha) | % change in area (94–96/65–67) | Yield 1994–6 (kg ha$^{-1}$) | % change in yield (94–96/65–67) |
| Burkina | 1,271 | 128 | 1,526 | 39 | 835 | 63 |
| Ghana | 368 | 272 | 311 | 96 | 1,182 | 89 |
| Mali* | 754 | | 770 | | 900 | |
| Niger | 364 | 23 | 1,939 | 277 | 186 | −68 |
| Nigeria | 6,482 | 80 | 5,904 | 14 | 1,098 | 58 |
| Major sorghum producers in West Africa** | 8,485 | 87 | 9,680 | 40 | 825 | 37 |
| Africa | 18,901 | 117 | 22,937 | 92 | 823 | 11 |

\* Production statistics based on mean of 1991–3 production, figures for 1994–6 not available. It was not possible to calculate the percentage increase for 1994–6/1965–7 for Mali because data were available only from 1988.
\*\* Statistics on sorghum for several West African countries were incomplete so figures summarizing production, area cultivated, and yield for the region could not be calculated.
*Note*: All figures have been rounded to the nearest whole number.
*Source*: *FAO Production Yearbook* (1970), vol. 24; (1972), vol. 26; (1978), vol. 32; (1981), vol. 35; (1984), vol. 38; (1987), vol. 41; (1990), vol. 44; (1993), vol. 47; (1996), vol. 50.

influenced by intercropping but there is no indication as to how this is accounted for in yield calculations.

In spite of all the limitations of macro-level statistics, the data do suggest that millet production has been increasing and that improved productivity of the land has played a part in some areas. Improvements in yields suggest the existence of potential for further production increases.

*Sorghum*

Data for sorghum or guinea corn are far from as complete as those for millet. Fig. 2.1 and Table 2.2 have been based on four major producers, Nigeria, Niger, Ghana, and Burkina. Mali too is a major producer, but statistics for sorghum production were not included in the FAO data until 1988 so have not been included in the graph. West Africa produces 40–50 per cent of the continent's sorghum, compared with approximately 53 per cent in the mid-1960s. Sorghum production has increased by approximately 87 per cent in West Africa, though the percentage increase is larger, over 117 per cent, for the Continent as a whole. How far one can rely on such statistics is doubtful. Since it is a subsistence crop, accurate records of production, consumption and other forms of crop disposal are rarely kept by farmers. Nigeria produces approximately 76 per cent of West Africa's sorghum and has seen a major increase in production, over 80 per cent since the mid-1960s. Ghana has experienced the highest relative production increase in the region (+272 per cent), though in absolute terms the quantity of sorghum produced in Ghana is minimal in comparison with that produced in Nigeria.

In all the major sorghum growing countries increases in production have been paralleled by an increase in the area cultivated. For the four countries Burkina, Ghana, Niger, and Nigeria, this has increased by just under 40 per cent between 1965–7 and 1994–6. For Burkina, Ghana, and Nigeria, percentage increases in area cultivated have been significantly lower than percentage increases in production, thereby revealing significant increases in yield over the 31-year period. Burkina's statistics suggest an increase in sorghum yields of over 60 per cent, yields in Nigeria have risen by a similar amount, over 58 per cent, and yields in Ghana have risen by almost 90 per cent over the same period. Not being a Green Revolution crop, sorghum is one of the cereals where much effort has been made on a local basis to improve the quality of seed. Fieldwork by the author in Senegal, The Gambia and in Mali suggests the widespread use of improved varieties.

By contrast, the increase in production of sorghum of just over 23 per cent in Niger has been achieved by a vast increase in area of 277 per cent and a decline in yields of around 68 per cent (Table 2.2). As with millet, explaining this decline is virtually impossible from the data alone and emphasizes the need for caution when drawing conclusions from highly aggregated statistics. There are a number of possible explanations: sorghum harvests in West Africa have suffered from drought, disease, and the effects of semi-parasitic weeds such as *Striga* spp. As the drought has continued since the late 1960s, farmers in Senegal have been found to seed cereals more thinly, reducing competition for very limited soil moisture. Prompted by low and erratic rainfall, similar practices in Niger may have contributed to lower sorghum yields. With the exception of Niger, yields in all the major sorghum producing countries in West Africa rose far more than yields at a Continental level. In spite of the limitations of macro level statistics, the steady and sustained increase in sorghum yields which is evident from Fig. 2.1 has undoubtedly contributed to increased production across the region. These developments cannot be ignored and suggest that potential exists for further increasing output from the numerous varieties of millet and sorghum that exist.

*Maize*

West Africa produced just under 25 per cent of the maize harvested in Africa in 1994–6 (Table 2.3), a significant increase on the 12 per cent produced in 1965–7. Maize production has leapt in every country in West Africa since 1965–7, though in both absolute and relative terms the increase has been by far the greatest in Nigeria (+541 per cent). As with the other crops, much of this increase has been achieved by an expansion of the area under maize by almost 170 per cent over the region. Côte d'Ivoire has shown the largest relative

**Table 2.3.** Maize: changes in production, area, and yield, 1965–7 to 1994–6

|  | Production | | Area | | Yield | |
|---|---|---|---|---|---|---|
|  | Production 1994–6 ('000 tonnes) | % change in production (94–96/65–67) | Area 1994–6 ('000 ha) | % change in area (94–96/65–67) | Yield 1994–6 (kg ha$^{-1}$) | % change in yield (94–96/65–67) |
| Benin | 531 | 138 | 490 | 16 | 1,083 | 106 |
| Burkina | 261 | 119 | 176 | −5 | 1,471 | 124 |
| Côte d'Ivoire | 555 | 457 | 722 | 429 | 769 | 5 |
| Ghana | 994 | 228 | 641 | 167 | 1,550 | 22 |
| Mali | 291 | 271 | 241 | 237 | 1,213 | 11 |
| Nigeria | 6,539 | 541 | 5,057 | 296 | 1,295 | 62 |
| Togo | 303 | 233 | 322 | 71 | 927 | 92 |
| West Africa | 9,714 | 326 | 7,887 | 169 | 1,233 | 43 |
| Africa | 41,450 | 119 | 26,152 | 64 | 1,584 | 34 |

*Notes*: In 1994–6, Benin, Burkina, Côte d'Ivoire, Ghana, Mali, Nigeria, and Togo produced 98 per cent of maize in West Africa.
All figures have been rounded to the nearest whole number.
*Source*: *FAO Production Yearbook* (1970), vol. 24; (1972), vol. 26; (1978), vol. 32; (1981), vol. 35; (1984), vol. 38; (1987), vol. 41; (1990), vol. 44; (1993), vol. 47; (1996), vol. 50.

increase in area (+429 per cent) but a comparatively modest increase in yields of around 5 per cent since 1965–7. Maize yields in Côte d'Ivoire remain relatively low at 769 kg ha$^{-1}$. Burkina presents an interesting contrast. Here, maize production has increased by almost 120 per cent, though the area cultivated has declined by nearly 5 per cent. Yield increases of around 124 per cent have thus been largely responsible for Burkina's increase in production. At 1,471 kg ha$^{-1}$ they are among the highest in West Africa but are still below the Continental average of 1,584 kg ha$^{-1}$.

Fig. 2.2 indicates that while there has been a broad trend of increasing production over the past 30 years, the greatest change has been in yields. Maize is a Green Revolution crop and there have been significant improvements in the quality of seed sown. High yielding varieties and improved varieties are widely used and success is greatest where farmers have access to inputs such as fertilizer, insecticide, pesticide, and irrigation, though in West Africa comparatively little maize is irrigated and use of other inorganic inputs still tends to be modest.

Fluctuations in yield closely mirror changes in rainfall over the region. The data on fertilizer consumption suggest that Nigeria alone uses almost 70 per cent of the fertilizer consumed in West Africa and the comparatively slow uptake of fertilizer in the region suggests that there is still considerable room for improvement in maize yields (International Fertilizer Industry Association, 1995; UNFAO Fertilizer Statistics, 1995). While West Africa is a long way behind the maize-growing states of Southern Africa, the marked upturn in production and yield is noteworthy. It may be temporary, or it may prove to be a sustained response to the need to increase domestic food production.

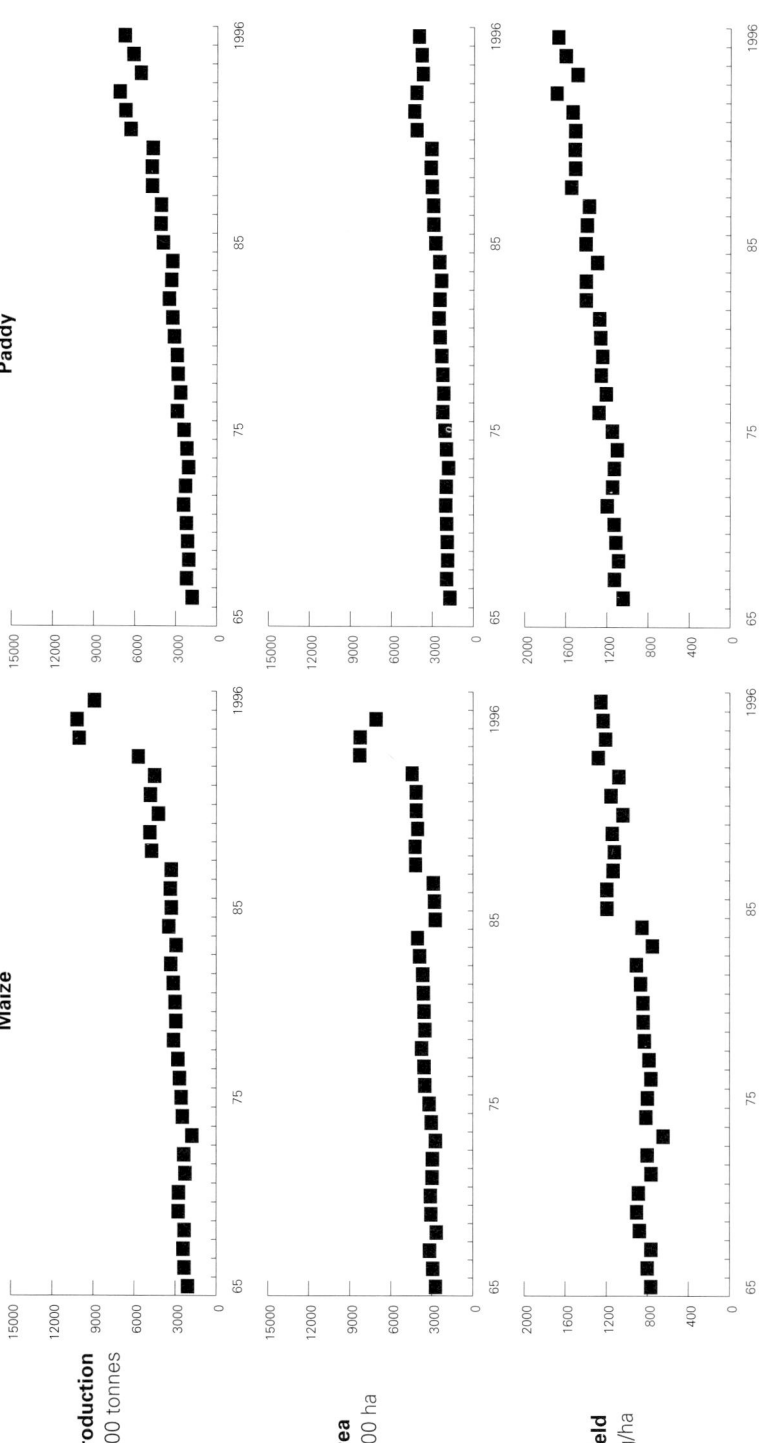

**Fig. 2.2.** Production, area, and yield of maize paddy in West Africa, 1965–96

*Sources:* FAO *Production Yearbook* (1970), vol. 24; (1972), vol. 26; (1978), vol. 32; (1981), vol. 35; (1984), vol. 38; (1987), vol. 41; (1993), vol. 47; (1996), vol. 50; FAO *Quarterly Bulletin of Statistics* (1995), vol. 8 1/2.

**Table 2.4.** Paddy: changes in production, area, and yield, 1966–8 to 1994–6

| | Production | | Area | | Yield | |
|---|---|---|---|---|---|---|
| | Production 1994–6 ('000 tonnes) | % change in production (94–96/66–68) | Area 1994–6 ('000 ha) | % change in area (94–96/66–68) | Yield 1994–6 (kg ha$^{-1}$) | % change in yield (94–96/66–68) |
| Côte d'Ivoire | 1,085 | 230 | 675 | 136 | 1,592 | 39 |
| Ghana | 189 | 285 | 91 | 135 | 2,069 | 66 |
| Guinea | 617 | 85 | 423 | 26 | 1,440 | 45 |
| Mali | 453 | 244 | 289 | 65 | 1,568 | 99 |
| Nigeria | 2,823 | 802 | 1,765 | 728 | 1,597 | 9 |
| Sierra Leone | 360 | –19 | 282 | –19 | 1,274 | –4 |
| West Africa | 6,165 | 206 | 3,869 | 109 | 1,593 | 45 |
| Africa | 15,152 | 132 | 6,986 | 94 | 2,168 | 17 |

*Notes*: In 1994–6, Côte d'Ivoire, Ghana, Guinea, Mali, Nigeria, and Sierra Leone produced 90 per cent of paddy in West Africa.
All figures have been rounded to the nearest whole number.
*Source*: *FAO Production Yearbook* (1970), vol. 24; (1972), vol. 26; (1978), vol. 32; (1981), vol. 35; (1984), vol. 38; (1987), vol. 41; (1990), vol. 44; (1993), vol. 47; (1996), vol. 50.

## Paddy

West Africa currently produces about 40 per cent of the Continent's paddy compared with approximately 30 per cent in 1966–8. There has been a major increase in paddy production across Africa over the past three decades (+132 per cent), and an even greater increase in West Africa (+205 per cent) (Table 2.4 and Fig. 2.2). All major paddy-producing countries have recorded significant increases in production, with the exception of Sierra Leone which has suffered a decline of around 19 per cent since 1965–7. Political unrest followed by civil war and managerial inefficiency in agricultural development initiatives may have contributed to this situation. Nigeria is by far the largest rice producer in West Africa but what regional statistics conceal is the extent of the increase in paddy production there. It has risen from 200,000 tonnes in 1966 to over 2.8 million tonnes in 1996 (Table 2.4).

As with other crops, statistics on rice are almost certain to conceal considerable error not least because rice is largely for domestic consumption and details of production, and disposal are rarely documented. Less than 1 per cent of the African rice crop enters international trade and over 90 per cent of this comes from Egypt (Brown, 1992). In Nigeria, there has been a major drive to increase domestic food production, a consequence of Structural Adjustment. The private sector has responded strongly to this with both smallholders and larger farmers producing for the domestic market. Cereals produced in the north are transported southwards to the large urban markets and increasingly, land in the south is being opened up for production of maize and rice in response to market demand. Large commercial farms (50 ha to 20,000 ha) producing food crops are a growth area in Nigeria (Lepper, 1998, pers. comm.; EIU, 1994).

Paddy yields ranged from just over $3,000\,kg\,h^{-1}$ in Mauritania, the result largely of irrigation projects on the Senegal River and its tributaries in the south of the country, to around $1,100\,kg\,h^{-1}$ in Liberia, a country with considerable potential for rice production but hindered by political unrest and warfare. On average, yields of around $1,600\,kg\,h^{-1}$ for the region for 1994–6 are still below the mean for Africa of $2,168\,kg\,h^{-1}$ for the same period (UNFAO, 1996), which in turn is well below the average for the major rice-producing areas of South-East and East Asia. Here, yields of 5 to 6 tonnes per hectare are not uncommon.

Most rice cultivated in West Africa is still low input and undoubtedly, this contributes to comparatively low yields. There are only two species of rice which are cultivated, *Oryza glaberrima* and *Oryza sativa*, and both these are to be found in West Africa. *Oryza glaberrima* is confined to West Africa, but *Oryza sativa* is found throughout the tropical world and is predominant in Asia (Grist, 1986; Oka, 1988). Morphologically, there is comparatively little difference between the two species though *O. glaberrima* always has a characteristic red pericarp. Grist (1986) notes that *O. glaberrima* has undergone less genetic diversification than *O. sativa* which, through its long history in Asia, has been subjected to intensive selection and dispersal. It is from the cultivars of *O. sativa* that the high yielding varieties of rice have been derived and within West Africa *O. glaberrima* is steadily being replaced by *O. sativa* (Grist, 1986). A wide range of types of rice are thus to be found in West Africa, each suited to the environment in which it is grown. Among these are upland rice which may be intercropped in rotational systems, flood rice, floating rice which can tolerate water level rises of 2 to 5 metres, such as at Mopti on the River Niger, and increasingly, where the essential package of inputs is available, higher yielding varieties. For the most part however, rice production is still dominated by low input methods, the success of higher yielding varieties being very much at the mercy of the rains. The success of flood rice is dependent on flood water levels, which in turn are influenced by fluctuations in rainfall and river regimes (indigenous methods of rice cultivation are discussed further in Chapter 6). The success of upland rice is similarly dependent on the annual rains and it is crucial that the crop receives adequate rain during the period from planting to flowering. The unpredictability of the annual rains, evident from the discussion in Chapter 1, reinforces the importance of the effects of environmental non-equilibrium on smallholder agriculture in Africa.

## Groundnuts

In contrast to the other crops being discussed, groundnut production, area harvested, and yield are currently at much the same level as they were about 30 years ago (Fig. 2.3). At a regional scale, groundnut production has been below current levels for much of the past 30 years, low points on both production and yield graphs in 1973 and 1983 being closely related to poor rainfall in those

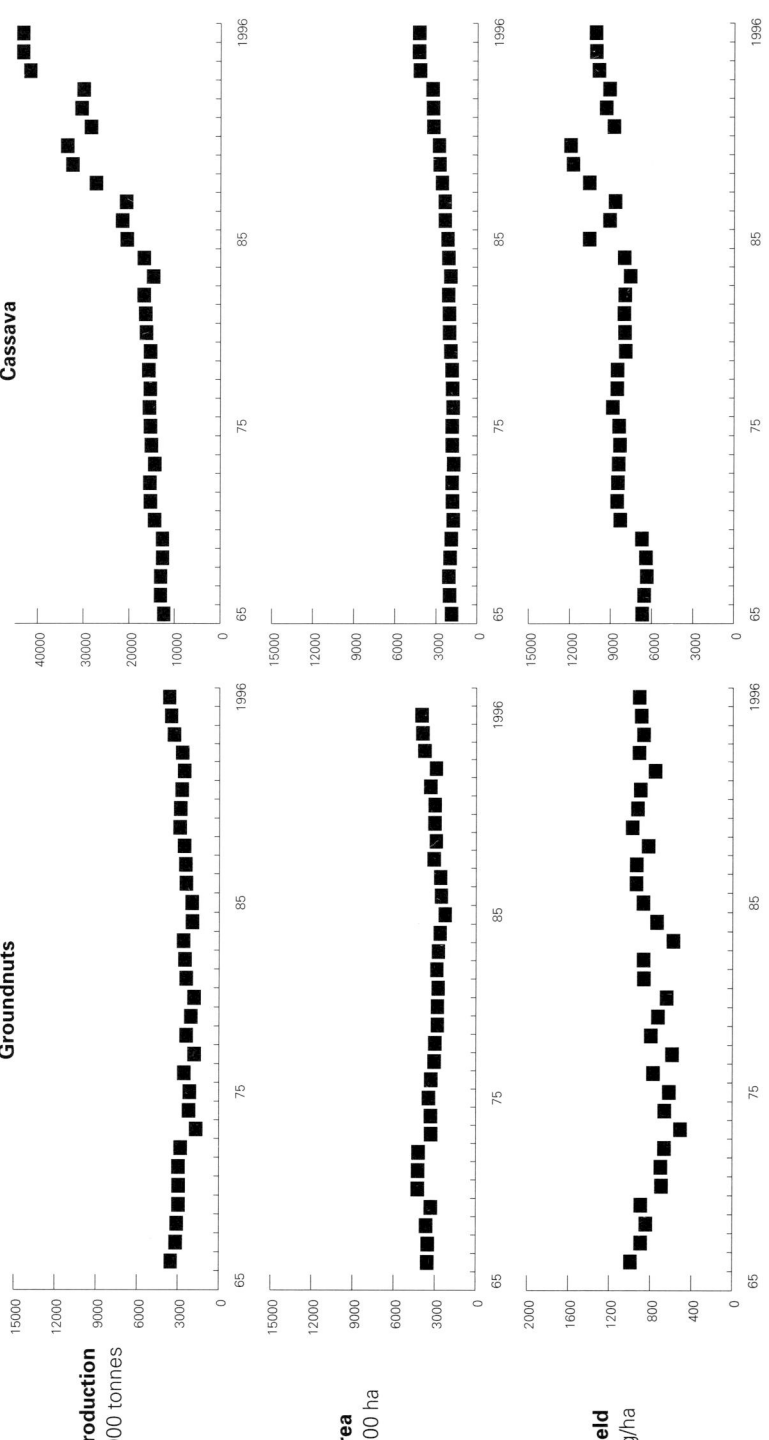

**Fig. 2.3.** Production, area, and yield of groundnuts and cassava in West Africa, 1965–96

*Sources*: FAO *Production Yearbook* (1970), vol. 24; (1972), vol. 26; (1978), vol. 32; (1981), vol. 35; (1984), vol. 38; (1987), vol. 41; (1993), vol. 47; (1996), vol. 50; FAO *Quarterly Bulletin of Statistics* (1995), vol. 8 1/2.

**Table 2.5.** Groundnuts: changes in production, area, and yield, 1966–8 to 1994–6

|  | Production | | Area | | Yield | |
| --- | --- | --- | --- | --- | --- | --- |
|  | Production 1994–6 ('000 tonnes) | % change in production (94–96/66–68) | Area 1994–6 ('000 ha) | % change in area (94–96/66–68) | Yield 1994–6 (kg ha$^{-1}$) | % change in yield (94–96/66–68) |
| Burkina | 210 | 87 | 254 | 6 | 826 | 77 |
| Mali | 173 | 37 | 196 | 74 | 890 | −20 |
| Nigeria | 1,585 | 7 | 1,721 | 46 | 921 | −27 |
| Senegal | 762 | −15 | 875 | −24 | 890 | 15 |
| West Africa | 3,495 | 8 | 3,916 | 10 | 892 | 21 |
| Africa | 5,999 | 22 | 7,516 | 29 | 798 | 9 |

*Notes*: In 1994–6, Burkina, Mali, Nigeria, and Senegal produced 73 per cent of groundnuts in West Africa. All figures have been rounded to the nearest whole number.
*Source*: *FAO Production Yearbook* (1970), vol. 24; (1972), vol. 26; (1978), vol. 32; (1981), vol. 35; (1984), vol. 38; (1987), vol. 41; (1990), vol. 44; (1993), vol. 47; (1996), vol. 50.

years. While the area under cultivation has followed a broadly similar trend, the lowest point on the area curve was reached in 1985, 2 years after production and yield were at their lowest. Groundnut production is important to Senegal, The Gambia, Burkina, Mali, Ghana, and Nigeria, though the Gambia cannot be counted among the major producers, production having fallen by 46 per cent from 1966. Nigeria is by far the largest producer (Table 2.5). Approximately one-third of West Africa's groundnuts and some 25 per cent of the entire groundnut crop of Africa are produced in Nigeria, though yields have fallen considerably here. Yields in West Africa range from nearly 1,800 kg h$^{-1}$ in Ghana to below 400 kg h$^{-1}$ in Niger but in most countries of the region they are slightly higher than the average yield in Africa of 798 kg h$^{-1}$.

Groundnut production has been severely damaged by the protracted drought in the drier savannas and the sahel and in some parts it has never quite recovered. In Niger, for example, groundnut production fell by over 70 per cent between 1966–8 and 1994–6. A sharp fall occurred with the drought of 1973 and when the rains improved, farmers moved into food crops at the expense of groundnuts. Some recovery has been achieved by government development programmes and external capital support but groundnut production is unlikely to increase further under present circumstances. Prices for cotton and grain crops are currently higher than for groundnuts and farmers have been converting their fields from the latter to cotton and cereals. Burkina has also suffered from drought but here groundnut production has increased by over 87 per cent between 1966–8 and 1994–6. An increase in the area cultivated of just under 6 per cent over this period points to an increase in groundnut yields of around 77 per cent. The effect of positive agricultural policies is evident. During the Sankara era in Burkina (1983–7), investment was significantly increased in the agricultural sector and particularly in cash-crop production. The focus on developing commercial agriculture has been sustained under the Compaoré regime. However, the notable increase in groundnut production may not be be sustained as here too cotton is currently a more profitable option (Hodgkinson, 1992).

In addition to the drought other factors have influenced groundnut production: groundnuts are susceptible to attack by insects such as millipedes and beetles and according to Gambian farmers these significantly reduce the size of the harvest when infestations are severe. Disease is also a problem and in Nigeria, the rosette virus has taken a major toll on groundnut production (Mortimore, 1989, 1998). Government policies and actions have not always stimulated groundnut production on a sustained basis and where production has been encouraged, as in Senegal, this has adversely affected neighbouring Gambia by encouraging cross-border trade in the crop. Demand on international markets has also affected production. The export of groundnut oil has suffered strong competition from soy bean, cotton seed, and sunflower oil production in the USA (Brown, 1992). In addition, the ability of the EU to satisfy almost all demand for vegetable oil from within the European Community has eroded the market potential for West African vegetable oils that once existed in Europe (Davenport, 1988). Furthermore, the EU has banned imports of groundnut oil cake and meal for use in animal feed where aflotoxin content is over $0.03\,\text{mg}\,\text{kg}^{-1}$ (Brown, 1992), and although the cake can be treated to reduce aflotoxin levels, Europe prefers to use alternative sources of animal feed. The market for edible groundnuts in Africa is also highly sensitive to production levels in the USA, which produces about 50 per cent of the world's import requirements.

## Root crops

Root crops are the dominant staple in the humid south of West Africa to the east of the Bandama River in Côte d'Ivoire. To the west of this cultural divide (Morgan and Pugh, 1969), rice becomes the dominant staple. Like all generalizations, this is broadly true although root crops are grown in virtually every country in West Africa.

The unreliability of macro-level statistics has been raised but in the case of root crops, statistics seem likely to be even more unreliable. Since root crops are grown largely for domestic consumption, records of production, area, and yields are prone to considerable inaccuracy. Errors are compounded because root crops are frequently intercropped and this causes complications in assessing the area they cover. They are not all harvested at once as the time of harvest for root crops is not as critical as it is for cereals. Cassava, for example, can be left in the ground and the roots harvested as and when required over an 18-month to 2-year period, depending on the variety. Whether production belongs to one year or the next can thus contribute to errors in data collection.

West Africa produces around 55 per cent of all roots and tubers in Africa (Table 2.6) though within the region yam (*Dioscorea* spp.) and cassava (*Manihot esculenta*) dominate. Some 95 per cent of Africa's yam crop is pro-

**Table 2.6.** Cassava: changes in production, area, and yield, 1965–7 to 1994–6

|  | Production | | Area | | Yield | |
| --- | --- | --- | --- | --- | --- | --- |
|  | Production 1994–6 ('000 tonnes) | % change in production (94–96/65–67) | Area 1994–6 ('000 ha) | % change in area (94–96/65–67) | Yield 1994–6 (kg ha$^{-1}$) | % change in yield (94–96/65–67) |
| Benin | 1,277 | 22 | 152 | −11 | 8,381 | 38 |
| Côte d'Ivoire | 1,564 | 206 | 310 | 121 | 5,045 | 102 |
| Ghana | 6,608 | 371 | 544 | 214 | 12,131 | 50 |
| Nigeria | 31,303 | 325 | 2,939 | 159 | 10,651 | 63 |
| West Africa | 42,725 | 233 | 4,261 | 117 | 10,026 | 54 |
| Africa | 83,699 | 335 | 9,957 | 330 | 8,438 | 1 |

*Notes*: In 1994–6, Benin, Côte d'Ivoire, Ghana, and Nigeria produced 95 per cent of cassava in West Africa. All figures have been rounded to the nearest whole number.
*Source*: FAO Production Yearbook (1970), vol. 24; (1972), vol. 26; (1978), vol. 32; (1981), vol. 35; (1984), vol. 38; (1987), vol. 41; (1990), vol. 44; (1993), vol. 47; (1996), vol. 50.

duced in West Africa. Cassava is much more widespread across the Continent and West Africa produces only 51 per cent of Africa's cassava. Within West Africa root crop production is dominated by Nigeria, which produces approximately 82 per cent of the yam harvest and over 73 per cent of the region's cassava. Côte d'Ivoire is a distant second with regard to yam production, accounting for 10 per cent of West Africa's harvest, and Ghana, the second largest producer of cassava, accounts for only 15 per cent of the region's production. The statistics clearly indicate that in absolute terms, production of both cassava and yam has increased significantly over the past 20 years (Table 2.7) but that a significant change in pattern has occurred. In absolute terms cassava production has increased by a factor of three, whereas yam production has not quite doubled (Table 2.7). While yam dominated root crop production in West Africa in 1973, cassava has now taken over the dominant position and the proportion of yam relative to cassava has declined.

The reversal of the relative importance of yam and cassava raises interesting issues. The first point is that although the two crops are frequently discussed together, their environmental demands for successful cultivation are very different and management needs are also dissimilar. Cassava is grown over a much wider area within West Africa and can be grown where annual rainfall is as little as 500 mm, as long as it is well distributed. Cassava can tolerate fairly poor soils though it does not necessarily impoverish soils (Moss, 1992). Yam on the other hand prospers between latitudes 4 °N and 10 °N. Its range is thus more limited than cassava, but it does require more fertile soils. It would appear from Fig. 2.3 that cassava production has grown more rapidly since the mid-1980s. There are several possible explanations for this: according to Bennison (1987) this may be in response to population growth. According to farmers in the Western Division of The Gambia, the protracted drought has resulted in a shortfall of grain crops and cassava production has been increased to meet the demand for food. Furthermore, the drought has provoked considerable rural–urban migration which has left farms short of labour. Cassava cultivation, because of its com-

**Table 2.7.** Changes in the relative importance of yam and cassava in West Africa

| Total production ('000 tonnes) | 1973 | 1996 |
| --- | --- | --- |
| Roots and tubers | 35,319 | 76,921 |
| Cassava | 14,421 (40.83%)* | 43,246 (56.22%) |
| Yam | 18,704 (52.96%) | 30,135 (39.18%) |

\* Figs. in parentheses are percentages of total root and tuber production.
*Source*: *FAO Production Yearbook* (1978), vol. 32; (1996), vol. 50.

paratively limited labour requirements and market potential, has thereby gained popularity. Farmers frequently sell their cassava to contractors who are then responsible for having the crop harvested, yet another labour-saving mechanism. The crop is then exported to major markets at Dakar and Kaolack in Senegal where prices are higher than they are in the Banjul market. A further factor stimulating cassava production may be that as the austerity programmes adopted throughout West Africa in the early and mid-1980s started to bite, demand for cheap, domestically produced food increased. Cassava has proved a suitable crop as it is both easier and cheaper to produce than yam (Upton, 1967).

To the north of the yam growing zone cassava production may well have increased at the expense of crops such as groundnuts, but in the yam growing zone cassava may have replaced yam to some extent because yam is a more labour-intensive crop (Shimada, 1993). Although yam is preferred as part of the diet, the cultivation of cassava rather than yam releases labour which may be put to a range of other uses either within the farming system, or outside it. Table 2.7 therefore has far more subtle implications than simply the replacement of yam by cassava as the dominant root crop.

Cassava yields in West Africa have increased by around 54 per cent between 1965–7 and 1994–6, much of the increase having occurred over the past decade (Fig. 2.6). Some of this rise may be attributed to the increased use of varieties which are both higher yielding and apparently more tolerant of disease. As with other crops, there is little suggestion that smallholder farmers apply inorganic fertilizer to root crops though in the case of yam, organic matter is beneficial and improves harvests.

## Summarizing the changes

Trends of production and yield in most of the crops reviewed are more positive than was predicted by the 'ecodoomsters' of the 1980s and there is evidence

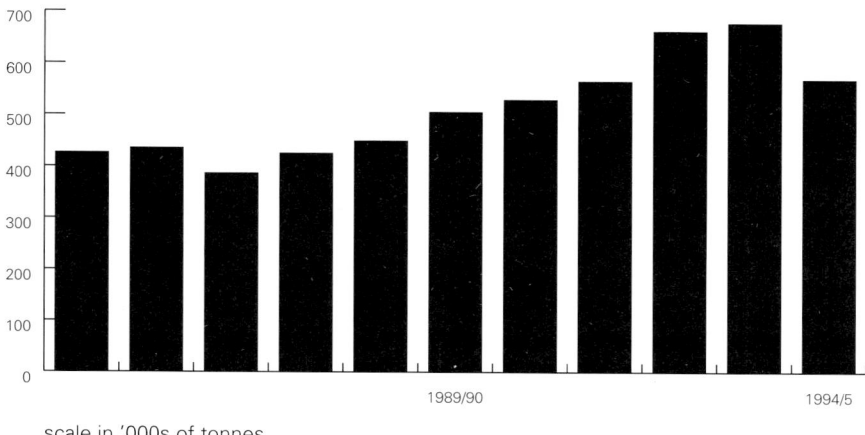

**Fig. 2.4.** Fertilizer consumption in West Africa, 1983/4–1994/5
Source: FAO *Fertilizer Statistics*.

that development is taking place in other aspects of agriculture, albeit slowly. The irrigated area is increasing and will continue to do so (UNFAO, 1996). The use of mechanization is increasing. There has been a gradual increase in the number of tractors in service since the early 1970s, although some 50 per cent of the existing machines are in Nigeria. West Africa is still well behind the rest of Africa as the region possesses under 4 per cent of tractors in service on the Continent but the potential for improvement is there. Consumption of inorganic fertilizer has also increased since the mid-1980s (Fig. 2.4), Nigeria consuming almost 70 per cent (UNFAO *Fertilizer Yearbook*, 1995). In relation to the entire Continent in 1983–4, West Africa consumed around 11 per cent of fertilizer used in Africa. By 1994–5, this had risen to over 16 per cent (UNFAO *Fertilizer Yearbook*, 1995). One concern is the fall in fertilizer consumption in 1994–5, the first time since 1986–7. Gambian farmers in the Western Division complained bitterly about the rising cost of fertilizer. All but a very few in the villages visited had ceased to use fertilizer because of this. There are suggestions that the rising cost of fertilizer subsidy has also had a negative effect in other parts of Africa (Richardson, 1996). In a Continent where agricultural development is of prime importance, removal of the fertilizer subsidy as directed by Structural Adjustment does appear shortsighted.

Although regional statistics tend to mask patterns, overall trends suggest that agriculture in West Africa is far from stagnant. The question then, is why has productivity not increased to any greater degree? The technology is in existence to achieve this. Several factors may be considered at a regional level: first, the drought. There can be little doubt that 30 years of rainfall largely below the mean for the previous 30 years (mid-1930s to mid-1960s) (Fig. 1.4), has taken a toll on agriculture. Although statistical analyses at a regional scale cannot be used to confirm the link between rainfall and crop yields as the data are not suitable, nevertheless, it would seem that the most bleak years in terms of agricultural production were those when rainfall levels in semi-arid West Africa were particularly low.

Since the mid-1980s the statisics reveal an increase in field-crop production. They do not, however, point to the causes. Improved rainfall may well be an important contributory factor as may higher producer prices for certain crops following the implementation of Structural Adjustment Programmes. There have also been improvements in seed quality and where rainfall has been adequate these new seeds may have contributed to increased production. There are likely to be many more reasons than these, most of which are unpredictable. In The Gambia, for example, crop production is very much influenced by rainfall, disease, and pest attack and while price is important, the first three factors appear to be the major determinants of production. In years where the rains are favourable and disease and pest attack is low, production tends to be higher. It is difficult to check such hypotheses, there being few official estimates of crop loss due to pest attack. In the author's experience the most precise information about influences on production is from the farmers—but inevitably, the sample size is very limited.

While factors such as rainfall, price, and pest attack focus on current influences on production, the effects of history cannot be ignored. Investment in smallholder farming has been consistently low and government policies and actions over decades have increased uncertainty and added to the constraints on agriculture. The effects of some of these are considered in the following section.

## Government intervention in agriculture prior to Structural Adjustment

In addition to uncertainty in the physical environment, decisions taken at national and international levels have done little to promote confidence and stability within the agricultural sector. Prior to the advent of Structural Adjustment, government intervention in the agricultural sector was considerable throughout the region. Since Colonial times governments have controlled agricultural markets and prices particularly for cash crops such as cocoa, coffee, and groundnuts. Although farmers were paid a comparatively low proportion of the world market price (Bates, 1981), nevertheless, producer prices were stable and they were relatively assured. The benefits to the smallholder farmer could and should have been greater, however, they appear to have been sufficient to stimulate major increases in the production of crops such as cocoa. After Independence, which was around 1960 for much of West Africa, commodity prices on the world market began to fluctuate. By this time marketing board funds, accumulated predominantly for stabilization of producer prices, had been diverted to other uses and farmers had to face price fluctuations. Where guaranteed prices existed they were held down and agricultural exports were often taxed heavily (Bates, 1981; Hinderink and Sterkenburg, 1987). In the Independence era attempts have been made to control food prices though this has proved less successful than the control of cash crop prices.

There seems to have been some perception at government level that farmers were compensated for low producer prices by subsidies on many agricultural inputs. However, there was never any guarantee that essential inputs would be accessible. Fieldwork by the author in Senegal, The Gambia, Mali, and southern Mauritania between 1980 and 1990 revealed a range of problems confronting farmers in their attempts to obtain agricultural inputs. These included a lack of response to orders for inputs, receiving the wrong orders and receiving orders for fertilizer, insecticide, and pesticide too late, though they were still charged for them (Baker, 1982). In the Senegal Delta and in The Gambia many farmers could not afford to use even subsidized fertilizer on the occasions when it was available thus suggesting that subsidized inputs were no real compensation for low producer prices. Furthermore, while governments were attempting to limit producer prices (and to a lesser extent producer costs through subsidized inputs), the cost of farm labour was rising with inflation particularly in countries with sovereign currencies. There was also a lack of private sector credit (Killick, 1992) and farmers had no alternative but to buy consumer goods and labour in inflated markets while producer prices were depressed (Bates, 1981). Depression in the agricultural sector intensified: labour migration increased to supplement declining incomes; private sector credit was not (and is still not) readily accessible to the majority. Government investment in agriculture was focused at a larger scale and largely neglected the smallholder sector. As a consequence, the capacity of smallholders to invest in their land declined. The more negative effects of government intervention in agricultural markets and prices have been documented extensively in the literature (World Bank, 1981; Bates, 1981; Andrae and Beckman, 1985; Berg, 1986; Hinderink and Sterkenburg, 1987; Mikell, 1989).

The problems created by inflation and low agricultural prices were aggravated by manipulation of exchange rates during the 1970s and early 1980s particularly in countries with sovereign currencies. Most of these countries experienced higher rates of inflation than did their trading partners, a factor which contributed to a major appreciation of real exchange rates. Ghana, Nigeria, and Sierra Leone experienced some of the worst effects of overvalued currencies. Where currencies became overvalued, imports of all kinds became comparatively cheap and in the agricultural sector cheap food imports had the effect of undercutting domestic producers. Other distortions also resulted. Imported foodstuffs such as maize, rice, and wheat flooded into West Africa making it unprofitable for African farmers to produce these crops (Bates, 1981; Andrae and Beckman, 1987). With domestic inflation running high in countries such as Ghana and Nigeria, desperate attempts were made to smuggle produce across the border to Franc Zone countries where they could be exchanged for hard currency. This was damaging not only to exporting countries which lost the benefits of their domestic produce, but also to importing countries which were flooded with agricultural produce from their neighbours. For example, cocoa production figures in both Ghana and Côte d'Ivoire were distorted because of such cross-border trade. In the savannas, groundnuts were moved back and forth across the Senegambian border by farmers wishing to earn the highest possible prices for their produce. Problems caused by overvaluation are documented extensively by the World Bank (1981), and by authors such as Pellow and Chazan (1986), Berg (1986), and Killick (1992).

Members of the Franc Zone (mainly former members of French West Africa whose common currency, the CFA franc, is pegged to the French franc) suffered less from inflation, overvalued exchange rates and from real exchange rate volatility during the 1970s and much of the 1980s than did most of their neighbours with sovereign currencies (Lane and Page, 1991). However, in the latter part of the 1980s the situation changed and the competitive position of CFA members compared with non-CFA members weakened in terms of exports, investments and savings (Elbadawi and Majd, 1996). Inflation increased in the Franc Zone as the French franc strengthened, a result of its effective linkage with the Deutschmark through membership of the European Monetary System (Killick, 1992). According to Lane and Page (1991), much of the success of the Franc Zone may be attributed to specific institutional arrangements between African members and France. Statutory restraints upon domestic credit to West African governments appear to have been a major factor in lower rates of monetary growth in comparison with non-Franc Zone countries. However, as Killick (1992) notes it has become apparent that some Franc Zone member governments were able to circumvent regulations of regional central banks designed to control government spending and found alternative means of increasing the latter. Accumulating large external debts and using public enterprises for the non-payment of bills were among the many measures used to evade expenditure limits.

Franc Zone countries have also suffered adversely from competition with non-Franc Zone neighbours who, by the latter part of the 1980s were following IMF and World Bank Structural Adjustment Programmes. These required, among other things, that currencies be devalued. As imports became more expensive for the non-Franc Zone countries, so their exports became more competitive and exports into neighbouring Franc Zone states increased further, stimulated by the overvalued CFA franc. Recipient countries were flooded with produce which was exchanged for convertible CFA francs, much in demand in West Africa. The artificial nature of this situation caused distortions for both 'exporter' and 'importer' nations and not until January 1994 was the CFA franc devalued. This was intended to assist the Franc Zone countries in Africa and to reduce the cost of the linkage to the French monetary authorities which guarantee the convertibility of the CFA franc. Devaluation was not achieved without pressure from the IMF which made new loans and debt arrangements dependent on devaluation. France also applied pressure, tying its future aid to an agreement between the IMF and the CFA countries. As foreseen by the Franc Zone countries, devaluation of the CFA has had serious social and economic consequences and in several countries riots broke out as the price of basic commodities rose sharply. In response, France agreed to write off the debt of the ten poorest countries, some of which are in West Africa, and 50 per cent of the debt of the four richest countries which included Cameroon and Côte d'Ivoire (Lensink, 1996). The full impact of devaluation on the farming community is still emerging.

## The impact of Structural Adjustment Programmes

Structural Adjustment Programmes were prescribed by the IMF and World Bank for most African countries as the means of reducing their debts and regenerating agricultural production for domestic markets and for export. The principal elements of Structural Adustment concern international trade, exchange rate, fiscal and agricultural policy reforms and the nature of governance (Lensink, 1996). Although concern here is primarily with agriculture, other aspects of Structural Adjustment impinge on the agricultural sector.

Structural Adjustment has positive and negative effects on the smallholder. According to Greenaway *et al.* (1997) a comparison of the economic performance of African (and other) economies pre- and post-adjustment has shown mixed results. They conclude that success is very much dependent on the nature and timing of those development programmes which are implemented. Such views are also reflected in Barrett's work (1998). This goes back to one of the fundamental assertions in this book, that if indigenous knowledge were utilized to a greater extent, development initiatives might prove to be more pertinent. According to Berry (1997) working in Ghana, 'for many of the nation's farmers, the gains from a decade of structural adjustment have been painfully small' (Berry, 1997: 1226). Along similar lines, Ihonbvere (1993) argues that Structural Adjustment is not leading to economic recovery and is making life almost impossible for the poor majority. Oyejide (1991) considers that the social costs of Adjustment have been grossly underestimated and in Ghana attempts are being made to improve conditions through PAMSCAD, an externally funded programme designed to ameliorate the social costs of adjustment. Potts (1995) has shown that Adjustment has put such pressures on urban dwellers that it has stimulated a trend of urban–rural migration in some areas. Thus Structural Adjustment is affecting the agricultural sector both directly and indirectly.

In an attempt to stimulate production of both food and cash crops, African governments have had to undertake to pay farmers 'realistic prices' for their crops. What constitutes a realistic price is open to debate and although fixed prices have increased, very frequently they have done so more slowly than the rate of inflation (derived from EIU, 1998). Liberalization of markets and prices has left farmers free to get the best prices possible but these may be no improvement on pre-liberalization prices, particularly for farmers in more remote areas and where transport to market is a problem (Lepper, pers. comm., 1998). A specific example of where theory and practice divide relates to groundnut marketing in The Gambia. Here groundnut prices have increased steadily since the mid-1980s when the then government adopted a Structural Adjustment Programme. Clearly, increased producer prices have influenced farmers' decisions to extend the area under groundnuts though the benefits of higher prices are perceived by the farmers to have been eroded by the elimination of the fertilizer subsidy. In 1998/9 chaos in both the purchasing of groundnuts from the farmers and in the now privatized Gambia Groundnut Corporation (GGC) resulted in many farmers selling their groundnuts in Senegal or receiving less than half the groundnut price announced by the government, if they sold their crop in The Gambia. On paper the price has continued to creep up. In reality many farmers have lost badly.

Berry (1997), working in the Asante region of Ghana, argues that while liberalization may ease access to markets, it may also increase exposure to market fluctuations. Among tomato growers of Asante risk is high and prices obtained for the crop can be good though when there is a glut, tomato prices plummet. This would seem to be the point at which the colonial governments introduced marketing boards in an alleged attempt to reduce fluctuations in market prices and to obtain an assured output. Structural Adjustment Programmes may have stimulated increased production through increased producer prices and liberalization of the market. However, efforts to reorganize local credit supplies along commercial lines and to attract foreign investment at a local level in order to develop agriculture have not been widespread (Berry, 1997).

Prompted by the demand by the donor agencies for a reduction in government involvement in agriculture, marketing organizations have seen considerable change. In Nigeria, for example, all marketing organizations have been discontinued though some are now re-forming in response to demand. Services such as quality control which were offered by the former marketing organizations have not been undertaken by the private sector. This has created substantial problems in the cocoa sector, for example, as low-quality produce has been rejected by the world market. In most West African countries, however, it is only marketing organizations for certain products that have been closed down and for the most part these organizations have been re-established and reformed. The aims of reforms are to reduce the costs of these organizations, to improve their efficiency, and to improve the consistency of prices offered to producers. In Côte d'Ivoire, for example, the private sector has been allowed to compete with the marketing parastatals with regard to domestic purchase of cocoa and coffee, though government control over exports has been maintained (Lensink, 1996). Changes in marketing systems which were often inflicted with the minimum of notice have not occurred without significant confusion. In many cases the private sector was not fully prepared for takeover and in consequence, the collection of agricultural produce from more remote areas has been neglected. This was particularly true of cocoa produced in the more remote regions of Côte d'Ivoire (Lepper, pers. comm., 1996) where initially, private traders were not sufficiently well organized. In Nigeria, producers within easy reach of markets have been receiving acceptable prices for their cocoa while those without easy access to transport routes have been finding disposal of their produce at comparable prices much more difficult (Lepper, pers. comm., 1998). Women in some rural areas have been particularly disadvantaged. Working in Naro Moro, Kenya, Goodwin (unpubl. MA dissertation, 1999) has shown that women do not have the time to seek the best prices for their produce and so sell at comparatively low prices to buyers coming to the villages.

Development of the private sector has become the medium- to long-term objective of Adjustment Programmes in Africa. Work by Bennell (1997) on sub-Saharan Africa shows that privatization is on the increase though there are considerable bottlenecks because government bureaucracies have not been prepared to cope with it. The privatization of land, for example, has been responsible for generating much uncertainty among farmers in parts of West Africa (Firmin-Sellers and Sellers, 1999). Governments have title to much of the land in the region, though in practice tenure has remained in communal hands. Prompted by the conclusions of writers such as Hardin (1968), a belief has developed that investment in land held under communal tenure is not as great as on land that is privately owned. To some extent this may be true but writers such as Cornia (1994), Sjaastad and Bromley (1997), and de Zeeuw (1997) argue that indigenous land tenure systems have not been investigated in sufficient depth and that there is much merit in them.

Where attempts have been made to replace communal tenure with private tenure, tension and uncertainty have been generated. One example involves land in the Senegal River valley which is suitable for irrigation. The floodplain has long been used as dry season grazing by semi-nomadic pastoralists (Park, 1993). In recent years the implementation of private tenure rules has denied pastoralists access to some of these lands, with the result that traditional land-use systems have been severely disrupted. Conflict resulted in bloodshed because indigenous people were not consulted about change nor were any alternative arrangements made to cater for their needs. Similar problems of access due to changing rules of tenure are evident in the vicinity of Lake Kainji in Nigeria, where a national park now traverses the traditional migration route of pastoralists (Ayeni, 1983). Berry's (1997) experience of privatization in Ghana suggests that it can aggravate rather than allay uncertainty over access to productive resources and certainly this is true in the case of women. According to Lastarria-Cornhiel (1997), poor rural women in parts of Africa are losing access to land as privatization occurs. Nowhere are bureaucratic problems associated with land privatization more evident than in Nigeria where the backlog of applications to register title to land is causing severe delays to its privatization and has added to uncertainty about rights of access (Lepper, pers. comm. 1998). This must inevitably affect management decisions at field level. It is important to recognize, nevertheless, that over much of West Africa, particularly in the more remote regions, communal rules of tenure still continue to operate. However, as population continues to increase and if urban to rural migration continues in the wake of public sector retrenchment, as demanded by Structural Adjustment, many rural areas may suffer. Not only will there be growing pressure on land and on the nature of tenure systems in which land is held, but increasing social problems seem sure to follow as the poorer and less powerful members of society are further marginalized and the numbers of landless people grow.

In certain aspects of Structural Adjustment the transition to a set of circumstances which would lead to growth of the agricultural sector has not occurred because processes of change have been impeded. The pattern is apparent with regard to currency devaluation. Devaluation is thought to have stimulated exports, but structural bottlenecks have arisen where, for example, the production of exports involves the use of imported inputs, the prices of which have increased as a result of devaluation (Yiheyis, 1997). Similarly, the decline in fertilizer use over the region in the mid-1990s may be an indication that fertilizer is too expensive, particularly now that subsidies have been reduced or have been removed altogether (UNFAO *Fertilizer Yearbook*, 1995). A recent research project by the International Food Policy Research Institute (IFPRI), on agricultural market reforms in several countries of Africa[1] has revealed that fertilizer use has atrophied with the removal of subsidies and with reduced government participation in the distribution of fertilizer. The decline in use is particularly evident on small farms in remote locations (IFPRI, 1998). Fieldwork by the author in the Western Division of The Gambia in 1999 confirmed that the increase in fertilizer prices following the removal of the subsidy has resulted in its reduced use. These findings, however, relate to only a small sample area though they are confirmed by Gambia's National Environment Agency (1997). A comparison of statistics produced by the National Agricultural Sample Survey for the Gambia for 1992 and 1997 (1993, 1998) suggests a decline in fertilizer use for groundnuts and for coarse grains in Gambia's Western Division. It is difficult to be precise about the nature of change as figures are given for the percentage area treated with fertilizer under different crops. No indication is given as to the intensity of fertilizer use nor are there any data on quantities of fertiliser used. Hutchful (1996) working in Ghana has documented the negative effects of removal of the fertilizer subsidy on fertilizer use. Richardson (1996) has confirmed that fertilizer use in part of Kenya has declined because of the weakness of the Kenyan shilling in recent years. Thus currency manipulation may have increased uncertainty in the agricul-

---

[1] Benin, Ghana, Madagascar, Malawi, and Senegal.

tural sector, and may have affected decisions about choice of crops. The World Bank believes that many African currencies are still overvalued despite massive adjustments (Lensink, 1996), and this could signal further retrenchment of essential imports.

On a more positive note it would seem that Structural Adjustment may be stimulating increased agricultural production in parts of West Africa. In Nigeria, for example, constraints on imports and the shortage of foreign exchange that followed devaluation of the naira encouraged smuggling into neighbouring Franc Zone countries before the devaluation of the CFA. Since then, the incentive to smuggle has decreased and Northern Nigerian farmers are turning to the production of crops for Southern Nigerian markets. There is also evidence that food production is increasing in the south and east. Extensive areas have been bought by private individuals, some of whom are believed to have connections with government, and the land is being planted with rice and maize destined for the domestic market. International statistics cited above confirm both an increase in area cultivated with crops in Nigeria and in production but whether this is a result of Structural Adjustment or other factors is impossible to say without further empirical evidence. In the north, private money is also purchasing large areas for cattle ranches with the aim of producing meat for urban markets. Whereas the government, as part of the austerity measures, is reducing the amount of machinery imported for the agricultural sector, mechanization is high on the list of demands by the private sector and money derived from oil is now being invested in the land as prospects in urban areas decline (Lepper, pers. comm., 1998). Nigeria, with its large markets is far better placed to respond to changes brought about by Structural Adjustment than countries with smaller populations.

Although under Structural Adjustment there is greater emphasis on the development of agriculture, farmers are afforded relatively little protection. Policies being implemented do not seem to have reduced their vulnerability and farmers remain under pressure. In times of stress this may appear to induce the adoption of ecologically unsound techniques which can add to problems of environmental non-equilibrium (Figs. 1.1–1.4 above).

## The impact of external factors on agriculture

West African countries are heavily dependent on the export of primary products and in consequence, smallholder farmers are susceptible to fluctuations in world market prices for these commodities, and linked to price is demand. Most West African countries are heavily dependent on one or two exports such as cocoa and coffee in the humid south and groundnuts, cotton, and animal products in the north. A major problem confronting African producers is that in terms of real prices, primary products have suffered a long-run trend of decline. They are also prone to greater world market price instability (Killick,

1992). Terms of trade for most African countries deteriorated between 1970 and 1981, stabilized in the late 1980s and by 1988 stood at 60 per cent of the 1970 to 1973 level. While terms of trade remain poor, they are not expected to deteriorate to the levels of the 1980s (Sparks, 1993, 1998). The purchasing power of Africa's exports has fallen by 22 per cent since 1987, owing principally to a decline in world petroleum prices and Africa's share of the world market for primary (non-petroleum) exports has nearly halved since 1970 (Sparks, 1998). Arguably, pressure on West African countries through international financial institutions such as the World Bank to increase production of export crops could be counter-productive and further depress world market prices, particularly of those crops for which world demand has not been rising. While such possibilities give cause for concern (Koester, et al., 1989), Killick (1992) argues that African governments have themselves made the minimum of effort to limit the effects of such price movements by diversifying their economies and reducing their dependence on exports facing weak global demand. While West African governments must accept some of the guilt for this, their economies have not been helped by poor terms of trade over much of the past 30 years (World Bank, 1989).

A second concern is whether agricultural protectionism by OECD countries has depressed prices received by West African exporters (Olagbaju and Adeseun, 1992). According to Killick (1992) the main commodities affected by EU agricultural protectionism are sugar, edible oils and fats, rice, and certain other grains. As West Africa is a major importer of cereals, it benefits from OECD's policies of protectionism. According to Ng and Yeats (1997), OECD trade preferences have made market access for Africa more favourable than for many exporters. They argue that Africa's trade barriers, which are higher than in most other developing countries, are a major cause of Africa's declining importance in world trade. This view is shared by Onafowora and Owoye (1998), who also argue that African governments should pursue more liberal trade policies. Nevertheless, the EU's commitment to supply as much as possible of its own market with domestically produced goods—oil seeds for example for the vegetable oils market—has meant that West Africa's potential to supply oils and fats to Europe has been reduced (Davenport, 1988). Such external factors have inevitably limited the capacity of West African countries to diversify their agricultural sectors.

This section has shown that human actions at different levels have at worst disadvantaged agriculture and at the very least added to the confusion and uncertainty within which farmers operate. However, in spite of minimal support, and in spite of difficult environments, smallholder producers remain the most productive element of West African agriculture. Before moving on to Chapter 3 which begins the focus on smallholder methods of land management, a brief review follows of the literature on changing attitudes to smallholder farming and to indigenous knowledge.

## Changing perceptions of indigenous knowledge

There has long been a negative attitude towards methods of land management by farmers in developing countries, though over the past 20 to 30 years respect for indigenous knowledge has been growing. Recognition of the appropriateness of many indigenous skills is still far from universal and there is as yet little evidence of indigenous expertise being put to use in rural development schemes. Some NGOs do incorporate local people actively in decision-making but for the most part rural Africans still play a marginal role in efforts to develop their Continent. Negative attitudes towards rural Africans date back to the early days of European presence in West Africa when Africans were considered backward in terms of their development (Perham, 1961: 852) and their methods of land management were perceived as wasteful and environmentally destructive (Dumont, 1966, cited in Iyegha, 1988; Stebbing, 1935, 1938). Such Western views were driven by ignorance about the reasons behind indigenous methods and more particularly, by concern for the potential loss of highly valued timber and forest products which Europeans perceived to be disappearing as land was cleared for indigenous farming (Moloney, 1887). But not all Europeans were derisory about African methods of land management. For example, O. T. Faulkner in Nigeria took time to investigate and to understand the merits of some indigenous farming practices (Richards, 1985). African governments have perpetuated negative attitudes towards smallholders and Iyegha (1988) argues that during the Second Development Plan in Nigeria (1970-4), the Nigerian government perceived peasant farmers as a serious hindrance to agricultural development, despite their significant contribution to the economy. Similar conclusions have been reached by Onimode (1982) for Nigeria, by Saris and Shams (1991) for Ghana and as Bates (1981) has shown, policies have been discriminatory towards smallholder producers throughout the region.

Negative attitudes towards farmers throughout the developing world continued through the 1950s and 1960s (Jewitt, 1996), but at the same time the work of Hill (1963, 1970, 1972) in Ghana and Berry (1975) in Nigeria aptly demonstrated the skill of indigenous cultivators in adapting their farming systems to produce cash crops such as cocoa. The complexity and rationality of indigenous farming systems, both the commercial and the non-commercial elements, were clearly demonstrated by these authors and by others such as Benneh (1972), Haswell (1953, 1963), Jones (1960), Upton (1967), Pélissier (1966) and Udo (1978). Awareness of the skill of indigenous farmers, not just in Africa, is also recorded in the work of Cohen (1967) and of Schultz (1964) who claimed that given the opportunity, farmers could turn sand into gold. However, such attitudes, based frequently on field evidence, failed to significantly alter enduring views, particularly among the aid givers, of the ignorance and incompetence of the peasant farmer.

The development of Green Revolution Technology in the late 1960s is yet another reason why smallholder agriculture in West Africa has been marginalized. Although more successful in Asia, many large-scale, capital-intensive schemes were established in Africa. The role of indigenous farmers on these schemes was mainly as labour, rarely as decision-makers. Farmers were provided with the land which was prepared by the parastatal in charge, seed was usually provided, farmers were instructed when to weed, apply fertilizer, insecticide, and pesticide, and they were told when to harvest. The government was the main market for the produce and while this reduced farmers' problems of marketing their crops, they had no alternative but to accept the price fixed by the government (Baker, 1982; Idachaba, 1980; Wallace, 1979, 1980). Such projects aimed to provide food security while virtually ignoring peasant production but this has proved to be impossible. The many causes of failure of such schemes have prompted a smaller-scale approach to agricultural development throughout the region. In Nigeria, for example, smaller-scale Agricultural Development Projects (ADPs) escalated in the 1980s, replacing earlier, larger-scale initiatives (Iyegha, 1988). But development schemes have not really focused on the smallholder. It is more a case that large-scale, capital-intensive agricultural schemes have failed to achieve their objectives and in consequence, smallholder production remains the major source of supply of agricultural produce. The opinion that Third World farmers were 'ignorant and needed to be taught how to farm' (Hecht, 1987: 18-19) has thus been shown to be erroneous in the case of West Africa.

The failure of large-scale capital-intensive agriculture saw the growth of the 'bottom-up' ideologies (Brokensha *et al.*, 1980; Chambers, 1978, 1979, 1981, 1983, 1992, 1994; Richards, 1985, 1986) where the fundamental argument was that indigenous farmers behaved rationally and that the methods they used were neither wasteful nor nonsensical, but were very appropriate to the environment. Associated with a change in paradigm on agricultural development from 'top-down' to 'bottom-up', was the work on how to extract accurate information from farmers. The approaches of Chambers which advocated talking to farmers and encouraging them to lead the conversation and thus direct the researcher seems second nature to field research now, but in the early 1980s they contrasted markedly with methods adopted by the Social Sciences in the 1960s and 1970s. This was a time when disciplines such as Geography were attempting to approach field research in as scientific and objective a manner as possible in the belief that this would maximize accuracy and minimize bias. Producing data with the minimum bias, often in considerable volume, was a means of offsetting the losses incurred through apparent objectivity. The accuracy of any field results is always open to question, but it would appear that such methods which characterized the scientific 'revolution' in the Social Sciences may well have been the cause of many erroneous judgements on smallholder agriculture in the tropics. A smaller-scale approach which attempts to minimize as far as possible the preconceptions, and hence misconceptions, of the researcher may well be what is necessary to redress the balance. Over the past two decades many seemingly more appropriate methodologies have been developed (Farrington and Martin, 1988; Biggs and Farrington, 1991). One of these has been the identification of the wealth of indigenous knowledge that exists and the increasing confirmation that the transfer of technology model for development is wholly inappropriate in view of the different cultures and different environments of the temperate zone and of the tropics.

Two models introduced to assist the understanding of ecological and socio-economic issues in research in developing countries were farming systems research (FSR) and rapid rural appraisal (RRA) (Chambers, 1981). These approaches have been criticized because of the emphasis they tended to place on questioning local people rather than on listening to them (Chambers, 1983, 1992, 1994; Chambers, Pacey, and Thrupp, 1989). This, it was argued, tended to dissuade local communities from participating fully in the development of projects and from voicing their own concerns and requirements (Jewitt, 1996). To compensate for this a number of 'farmer first', 'farmer participatory research', and 'participatory rural appraisal' models have been developed to encourage local people to play a more active part in focusing development projects. Criticisms of these have led to the development of 'beyond farmer first' models to refine approaches to development yet further.

Prompted by the desire to obtain a more accurate understanding of conditions in rural Africa, methodologies have burgeoned and with them an increased appreciation of the complexity of African rural life and of the appropriateness of land management techniques. Farming systems research, however much criticized, emphasized the need for a holistic approach (Spedding, 1975, 1979). In other words, FSR studies stressed that farming was only part of a much wider and more complex way of life for most rural Africans and that one had to look beyond farming itself, to the socio-economic and to the physical environment, in order to begin to understand the nature of indigenous African agriculture. The work of Norman (1972, 1974, 1980) and of Swindell (1978, 1982, 1985) add weight to this view. In the move to find new and more appropriate methods of collecting information in the field, Spedding (1975, 1979) argued that less emphasis had come to be placed on the research itself, that is, on the crucial question of how land productivity might be increased. This argument continues to be relevant today.

Prompted by the growth of 'bottom-up' methodologies for rural development there has been greater consideration of access to resources and interest has developed in community-based forms of property ownership and resource management. According to Jewitt (1996) interest in common property resources can be traced back to the 1960s when publications such as Rachel Carson's 'Silent Spring' (1965), Garrett Hardin's 'The Tragedy of the Commons' (1968), Ehrlich's *The Population Bomb* (1966), and The Club of Rome's *Limits to Growth* (Meadows *et al.*, 1972) generated concern about the existing and future relationship between population and resources. Utopian visions of community-oriented and decentralized societies appeared in Goldsmith *et al.*'s 'Blueprint for Survival' (1972) and Schumacher's *Small is Beautiful* (1973). However, the gloomy predictions of such literature have been countered to some extent by the more critical and measured analysis of relationships between population growth and environmental conditions and between poverty and environmental degradation (Blaikie, 1985; Blaikie and Brookfield, 1987; Eckholm, 1984; Joeckes *et al.*, 1994; Redclift, 1987; Tiffen *et al.*, 1994). Hardin's model of the tragedy of the commons has received considerable criticism and much of this, which is particularly relevant in an African context, is considered further in Chapter 7. Proponents of community-based resource management have demonstrated that Hardin's 'commons' were not common property resources, but open access resources (Berkes, 1989; Berkes and Taghi Farvar, 1989; Stevenson, 1991). As Jewitt (1996) observes, 'The distinction between common property and open access resources is an extremely important one as community access to true commons is retricted, whereas that to open access resources is not' (Jewitt, 1996: 9).

## Conclusions

A review of aspects of the agricultural sector in West Africa at a regional scale has shown that in spite of the difficulties farmers have faced, the production of most crops *has* increased over the past 30 years. This may have been achieved largely by an increase in the cultivated area, but there is also evidence of yields increasing and this suggests that potential exists for raising land productivity further. If this is to take place development planning must begin, as Moss (1992) suggests, with the autecology of the crops concerned. When crops are matched to local environmental conditions, this ecological unit may be fitted into the larger picture of agricultural development which involves a wide range of interdisciplinary links. Such an approach places the focus on crop plants and the people who grow them. The literature review above confirms that no one is better placed to be the focal point of development than indigenous farmers, who are innovators and skilled managers of the African environment. Chapter 3 now moves to a much smaller scale and examines how smallholder farmers in humid environments have adapted autecological land management strategies to cope with different constraints.

### *References*

AHN, PETER M. (1970), *West African Soils* (Oxford: Oxford University Press).
ALKALI, AHMED (1991), 'The World Bank: Financing rural development and the politics of debt in Nigeria', *Africa Development*, xvi (3/4): 163–80.
AMANOR, KOJO SEBASTIAN (1994), *The New Frontier: Farmers' Response to Land Degradation: A West African Study* (London: Zed).

ANDRAE, G. and BECKMAN, B. (1987), *The Wheat Trap: Bread and Underdevelopment in Nigeria* (London: Zed).
AYENI, J. S. O. (1983), 'Rangeland problems of the Kainji Lake Basin Area of Nigeria', *Environmental Conservation*, 10 (3): 239–45.
BAKER, K. M. (1982), 'Structural change and managerial inefficiency in the development of rice cultivation in the Senegal River Region', *African Affairs*, 81 (325): 499–510.
——(1992), 'Traditional farming practices and environmental decline with special reference to The Gambia', ch. 9 in K. Hoggart (ed.), *Agricultural Change, Environment and Economy: Essays in Honour of W. B. Morgan* (London: Mansell), 180–202.
——(1995), 'Drought, agriculture and environment: a case study from The Gambia, West Africa', *African Affairs*, 94 (374): 67–86.
BARBIER, EDWARD B. (1987), *Cash Crops, Food Crops and Agricultural Sustainability*. Gatekeeper series no. SA2, Sustainable agriculture programme (London: IIED).
BARRETT, CHRISTOPHER (1998), 'Immiserized growth in liberalized agriculture', *World Development*, 26 (5): 743–53.
BATES, R. H. (1981), *Markets and States in Tropical Africa: the Political Basis of Agricultural Policies* (Berkeley: University of California Press).
—— and LOFCHIE, MICHAEL (eds.) (1980), *Agricultural Development in Africa* (Berkeley: University of California Press).
BENNEH, GEORGE (1972), 'Systems of agriculture in Tropical Africa', *Economic Geography*, 40 (3): 244–57.
BENNELL, PAUL (1997), 'Privatization in Sub-Saharan Africa: progress and prospects during the 1990s', *World Development*, 25 (11): 1785–1803.
BENNISON, HUGH (1987), 'Cassava, its developing importance', *The Courier*, 101: 69–71.
BERG, ELLIOT (1986), 'The World Bank's strategy', ch. 2 in John Ravenhill (ed.) *Africa in Economic Crisis* (Basingstoke: Macmillan Press).
BERKES, F. (ed.) (1989), *Common Property Resources: Ecology and Community-Based Sustainable Development* (London: Belhaven Press).
—— and TAGHI FARVAR, M. (1989), Introduction and Overview, pp. 1–17 in F. Berkes (ed.) (1989), *Common Property Resources: Ecology and Community-Based Sustainable Development* (London: Belhaven Press).
BERRY, SARA (1975), *Cocoa, Custom and Socio-Economic Change in Rural Western Nigeria* (Oxford: Clarendon Press).
——(1997), 'Tomatoes, Land and Hearsay: Property and History in Asante in the Time of Structural Adjustment', *World Development*, 25 (8): 1225–41.
BIGGS, STEPHEN and FARRINGTON, JOHN (1991), *Agricultural Research and the Rural Poor* (Ottawa, ON, Canada: International Development Research Centre).
BLAIKIE, PIERS (1985), *The Political Economy of Soil Erosion in Developing Countries* (London: Longman).
BLAIKIE, PIERS and BROOKFIELD, HAROLD (eds.) (1987), Land Degradation and Erosion (London: Methuea).
BRÄUTIGAM, DEBORAH (1998), *Chinese Aid and African Development: Exporting Green Revolution* (Basingstoke: Macmillan Press).
BROKENSHA, D., WARREN, D. M. and WERNER, O. (eds.) (1980), *Indigenous Knowledge Systems and Development* (Lanham, New York and London: Academic Press).
BROWN, RICHARD (1992), 'Major commodities of Africa', in *Africa South of the Sahara 1993* (London: Europa Publications), 21–53.

CARSON, RACHEL (1965), *Silent Spring* (London: Penguin Books).
CHAMBERS, R. (1978), 'Towards Rural Futures', IDS Discussion Paper No. 134 (London: Institute of Development Studies/International Institute of Environment and Development).
——(1979), 'Rural Development: Whose Knowledge Counts?', Special Issue, *IDS Bulletin*, vol. 10: 2 (Institute of Development Studies, University of Sussex).
——(1981), 'Rapid Rural Approach: Rationale and Repertoire', *Public Administration and Development*, 1: 95–106.
——(1983), *Rural Development: Putting the Last First* (New York: Longman).
——(1992), 'Rural Appraisal: Rapid, Relaxed and Participatory', IDS Discussion Paper No. 311 (London: Institute of Development Studies/International Institute of Environment and Development).
——(1994), 'Paradigm Shifts and the Practice of Participatory Development', IDS Working Paper No. 2 (London: Institute of Development Studies/International Institute of Environment and Development).
——PACEY, A., and THRUPP, L. A. (eds.) (1989), *Farmer First. Farmer Innovation and Agricultural Research* (London: Intermediate Technology Publications).
COHEN, P. (1967), 'Economic Analysis and Economic Man'. Pp. 91–118, in R. Firth (ed.) (1967), *Themes in Economic Anthropology* (London: Tavistock).
COMMANDER, S. (ed.) (1989), *Structural Adjustment and Agriculture: Theory and Practice in Africa and Latin America* ODI, in collaboration with James Currey (London and Portsmouth: Heinemann).
CORNIA, GIOVANNI ANDREA (1994), 'Neglected issues in the decline of Africa's agriculture: Land tenure, land distribution and research and development constraints', in G. A. Cornia and G. K. Helleiner (eds.), *From Adjustment to Development in Africa* (London: Macmillan), 217–47.
CROOK, RICHARD C. (1990), 'Politics, the cocoa crisis and administration in Côte d'Ivoire', *Journal of Modern African Studies*, 28 (4): 649–69.
DAVENPORT, M. (1988), *European Community Trade Barriers to Tropical Agricultural Products* (ODI Working Paper no. 27, London: ODI).
DE ZEEUW, FONS (1997), 'Borrowing of land, security of tenure and sustainable land use in Burkina', *Development and Change*, 28: 583–95.
DIBUA, J. I. (1989), 'Government's agricultural policy and rural development in Nigeria: The case of Bendel state, 1964–88', *The African Review*, 16 (1): 40–53.
DUMONT, R. (1966), *False Start in Africa* (New York: Praeger).
DUNN, JUSTINE (1996), 'The role of indigenous woody species in "farmer-led" agricultural change in south east Nigeria, West Africa' (unpubl. Ph.D. thesis, University of London).
ECKHOLM, E. (1984), *Fuelwood: The Energy Crisis that Won't Go Away* (London: Earthscan).
EICHER, CARL K. (1986), 'Facing up to Africa's food crisis', ch. 7 in John Ravenhill (ed.), *Africa in Economic Crisis* (Basingstoke: Macmillan Press).
EHRLICH, P. (1966), *The Population Bomb* (New York: Ballantine).
EYOH, DICKSON L. (1992*a*), 'Structures of intermediation and change in African agriculture: A Nigerian case study', *African Studies Review*, 35 (1): 17–39.
——(1992*b*), 'Reforming peasant production in Africa: Power and technological change in two Nigerian villages', *Development and Change*, 23 (2): 37–66.
Economist Intelligence Unit (1994 and 1998), *Nigeria Country Profile* (London: Economist).

ELBADAWI, IBRAHIM and MAJD, NADER (1996), 'Adjustment and economic performance under a fixed exchange rate: a comparative analysis of the CFA zone', *World Development*, 24 (5): 939–51.
FARRINGTON, JOHN and MARTIN, ADRIENNE (1988), *Farmer Participation in Agricultural Research: A Review of Concepts and Practices* (London: Overseas Development Institute).
FIRMIN-SELLERS, KATHRYN and SELLERS, PATRICK (1999), 'Expected failures and unexpected successes of land titling in Africa', *World Development*, 27 (7): 1115–28.
GAKOU, MOHAMED LAMINE (1987), *The Crisis in African Agriculture* (London: Zed).
GOLDSMITH, E. R. D., ALLEN, R., ALLABY, M., DAVOLL, J., and LAWRENCE, S. (1972), 'A Blueprint for Survival', *The Ecologist*, 2 (1): 1–43.
GOODWIN, ELIZABETH (1999), 'Gender issues in Naro Moro location, Kenya: Women's aspirations and constraints' (unpubl. MA dissertation, Department of Geography, University of London).
GORSE, JEAN EUGENE and STEEDS, DAVID R. (1987), *Desertification in the Sahelian and Sudanian Zones of West Africa* (Washington: World Bank).
Government of The Gambia, National Agricultural Data Centre, Department of Planning, Ministry of Agriculture (1993), *Statistical Yearbook of The Gambia Agriculture: 1992; 1992/93 National Agricultural Sample Survey* (Banjul, The Gambia: Department of Planning, Ministry of Agriculture).
Government of The Gambia, National Agricultural Data Centre, Department of Planning, Department of State for Agriculture (March 1998), *Statistical Yearbook of Gambian Agriculture: 1997; 1997/98 National Agricultural Sample Survey* (Banjul, The Gambia: Department of Planning, Department of State for Agriculture).
GREENAWAY, DAVID, MORGAN, WYN and WRIGHT, PETER (1997), 'Trade liberalization and growth in developing countries: Some new evidence', *World Development*, 25 (11): 1885–92.
GRIST, D. H. (1986) (Sixth Edition), *Rice* (London: Longman).
GROVE, A. T. (1992), *The Changing Geography of Africa* (Oxford: Oxford University Press).
HARDIN, G. (1968), 'Tragedy of the the Commons', *Science*, 162: 1243–8.
HASWELL, M. R. (1953), *Economics of Agriculture in a Savannah Village* (London: Her Majesty's Stationery Office).
——(1963), *The Changing Patterns of Activity in a Gambian Village* (London: Her Majesty's Stationery Office).
——(1975), *The Nature of Poverty* (London: Macmillan).
HECHT, S. B. (1987), 'The Evolution of Agroecological Thought', pp. 1–20 in M. A. Altieri, with contributions by R. B. Norgaard, S. B. Hecht, J. G. Farell, and M. Liebman (1987), *Agroecology. The Scientific Basis of Alternative Agriculture* (Boulder: Westview Press. London: Intermediate Technology Publications).
HILL, P. (1963), *The Migrant Cocoa Farmers of Southern Ghana: A Study in Rural Capitalism* (Cambridge: Cambridge University Press).
——(1970), *Studies in Rural Capitalism in West Africa* (Cambridge: Cambridge University Press).
——(1972), *Rural Hausa: A Village and a Setting* (Cambridge: Cambridge University Press).
HINDERINK, J. and STERKENBURG, J. J. (1987), *Agricultural Commercialization and Government Policy in Africa* (Monographs from the African Studies Centre, Leiden, London and New York: KPI).

HODGKINSON, E. (1992), Sections on 'Economy', in various 'Country Surveys', in *Africa South of the Sahara 1993* (London: Europa Publications).

HULME, MIKE (1994), 'Historic records and recent climatic change', ch. 4 in Neil Roberts (ed.), *The Changing Global Environment* (Oxford: Basil Blackwell), 69–98.

HUTCHFUL, EBOE (1996), 'Ghana' in Poul Engberg-Pedersen, Peter Gibbon, Phil Raikes and Lars Udsholt (eds.) *Limits of Adjustment in Africa* (Centre for Development Research Copenhagen, in association with Oxford; Portsmouth, New Hampshire USA: James Currey and Heinemann).

IBRD and World Bank (1999), *World Development Report* (New York: Oxford University Press).

IDACHABA, FRANCIS SULEMANN (1980), *Agricultural Research Policy in Nigeria* (Washington DC: International Food Policy Research Institute).

IHONBVERE, JULIUS O. (1993), 'Economic crisis, Structural Adjustment and social crisis in Nigeria', *World Development*, 21 (1): 141–53.

International Fertilizer Industry Association (IFIA) (1995), *Fertilizer Indicators: Graphs and Diagrams* (Information and Market Research Service, Paris: IFIA).

International Food Policy Research Institute (1998), 'Agricultural market reform in Sub-Saharan Africa', Commentary, in *IFPRI Report*, vol. 20 (1), February, 1 and 4.

IYEGHA, DAVID A. (1988), *Agricultural Crisis in Africa: The Nigerian Experience* (Lanham, New York: University Press of America).

JEWITT, S. L. (1995), *Europe's 'Others'? Forestry Policy and Practices in Colonial and Post-Colonial India* (Environment and Planning: Society and Space), 13: 67–90.

—— (1996), 'Agro-ecological Knowledges and Forest Management in the Jharkhand, India: Tribal Development or Populist Impasse?' (unpubl. Ph.D. thesis, University of Cambridge).

JOECKES, S. with HEYZER, N., ONIANG'O, R., and SALLES, V. (1994), 'Gender, Environment and Population', *Development and Change*, 25: 137–65.

JONES, W. O. (1960), 'Economic man in Africa', *Food Research Institute Studies*, 1: 107–34.

KILLICK, TONY (1990), *Exchange Rates and Structural Adaptation* (ODI Working Paper no. 33, London: ODI).

—— (1992), *Explaining Africa's Post Independence Experiences* (ODI Working Paper no. 60, London: ODI).

KOESTER, ULRICH, SCHAFER, HARTWIG, and VALDÉS, ALBERTO (1989), 'External demand constraints on agricultural exports', *Food Policy*, 14 (3): 243–54.

LANE, CHRISTOPHER and PAGE, SHEILA (1991), *Differences in Economic Performance Between Franc Zone and other Sub-Saharan Countries* (ODI Working Paper no. 43, London: ODI).

LASTARRIA-CORNHIEL, SUSANA (1997), 'Impact of privatization on gender and property rights in Africa', *World Development*, 25 (8): 1317–33.

LENSINK, ROBERT (1996), *Structural Adjustment in Sub-Saharan Africa* (Harlow, Essex: Longman).

LIPTON, M. and LONGHURST, R. (1989), *New Seeds and Poor People* (London: Unwin Hyman).

LOWE, R. G. (1986), *Agricultural Revolution in Africa* (London and Basingstoke: Macmillan).

MACGREGOR, J. (1990), 'The crisis in African agriculture', *African Insight*, 20 (1): 4–16.

MAXWELL, S. (1988), Editorial, 'Cash crops in developing countries', *Institute of Development Studies Bulletin*, 19 (2): 1–4.

MEADOWS, D. H., MEADOWS, D. L., and RANDERS, J. (1972), *The Limits to Growth* (New York: Universe Books).
MIKELL, G. (1989), *Cocoa and Chaos in Ghana* (New York: Paragon House).
MOLONEY, ALFRED (1887), *Sketch of the Forestry of West Africa with Principal Reference to its Present Principal Commercial Products* (London: Sampson Low, Maarston Searle and Rivington).
MORGAN, W. B. and SOLARZ, JERZY A. (1994), 'Agricultural crisis in sub-Saharan Africa: Development constraints and policy problems', *Geographical Journal*, 160 (1): 57–93.
—— and PUGH, J. C. (1969), *West Africa* (London: Methuen).
MORTIMORE, M. (1989), *Adapting to Drought: Farmers, Famines and Desertification in West Africa* (Cambridge: Cambridge University Press).
—— (1992), 'The intensification of peri-urban agriculture: The Kano close-settled zone, 1964–86', ch. 11 in B. L. Turner II, Goran Hyden, and Robert W. Kates (eds.), *Population Growth and Agricultural Change in Africa* (Gainesville: University Press of Florida), 358–400.
—— (1998), *Roots in the African Dust* (Cambridge: Cambridge University Press).
MOSLEY, PAUL, HARRIGAN, JANE, and TOYE, JOHN (1991), *Aid and Power: The World Bank and Policy Based Lending* (2 vols.) (London: Routledge).
MOSS, R. P. (1992), 'Environmental constraints on development in tropical Africa', ch. 3 in M. B. Gleave (ed.), *Tropical African Development: Geographical Perspectives* (Harlow: Longman Scientific and Technical), 50–92.
NATIONAL, ENVIRONMENT AGENCY (N.E.A.) (1997), *State of the Environment Report—The Gambia* (Banjul: N.E.A.).
NEIMEIJER, DAVID (1996), 'The dynamics of African agricultural history: Is it time for a new paradigm?' *Development and Change*, 27 (1): 87–110.
NELSON, H. D., DOBERT, M., MCDONALD, G. C., MCLAUGHLIN, J., MARVIN, B., and MOELLER, P. W. (1974), *Area Handbook for Senegal* (Washington: American University).
NG, F. and YEATS, A. (1997), 'Open economies work better! Do Africa's protectionist policies cause its marginalization in world trade?' *World Development*, 25 (6): 889–904.
NICHOLSON, SHARON and GLOHN, HERMANN (1980), 'African environmental and climatic changes and the general atmospheric circulation in Late Pleistocene and Holocene', *Climatic Change*, 2: 313–48.
NORMAN, D. (1972), *An Economic Study of Three Villages in Zaria Province*, vols. 1 and 2 (Samaru, Nigeria: Institute of Agricultural Research).
—— (1974), 'Rationalizing mixed cropping under indigenous conditions: The example of Northern Nigeria', *Journal of Development Studies*, 11: 3–21.
—— (1980), 'The farming systems approach: Relevancy for the small farmer' (East Lansing: Michigan State University, Department of Agricultural Economics, MSU Rural Development Paper No. 5).
O'CONNOR, ANTHONY M. (1991), *Poverty in Africa* (London: Belhaven).
OKA, H. I. (1988), *Origins of Cultivated Rice*, Developments in Crop Science 14. (Tokyo: Japan Scientific Societies Press, and Amsterdam: Elsevier).
OLADIPO, E. O. and KYARI, J. D. (1993), 'Fluctuations in the onset, termination and length of the growing season in Northern Nigeria', *Theoretical and Applied Climatology*, 47 (3): 241–50.
OLAGBAJU, J. O. and ADESEUN, G. O. (1992), 'The problem of declining agricul-

tural export products', ch. 3 in S. A. Olanrewaju and Toyin Falola (eds.), *Rural Development Problems in Nigeria* (Aldershot: Avebury), 43–55.

OLOMOLA, A. S. (1989), 'Dimensions of institutional and policy deficiencies in the Nigerian agricultural credit system', *Development Policy Review*, 7: 171–83.

ONAFOWORA, OLUGBENGA and OWOYE, OLUWOLE (1998), 'Can trade liberalization stimulate economic growth in Africa?' *World Development*, 26 (3): 497–506.

ONIMODE, BADE (1982), *Imperialism and Underdevelopment in Nigeria: The Dialectic of Mass Poverty* (London: Zed Books).

OSHIN, OLASIJI (1992), 'The historical roots of the rural development problem', ch. 2 in S. A. Olanrewaju and Toyin Falola (eds.), *Rural Development Problems in Nigeria* (Aldershot: Avebury), 19–42.

OYEJIDE, T. ADEMOLA (1991), 'Adjustment with growth: Nigerian experience with Structural Adjustment policy reform', *Journal of International Development*, 3 (5): 485–98.

PARK, THOMAS K. (ed.) (1993), *Risk and Tenure in Arid Lands* (Tuscon and London: University of Arizona Press).

PÉLISSIER, PAUL (1966), *Les paysans du Sénégal: Les civilisations agraires du Cayor à la Casamance* (Sainte-Yrieux, Haute Vienne: Imprimérie Fabriqué).

PELLOW, DEBORAH and CHAZAN, NAOMI (1986), *Ghana: Coping with Uncertainty* (Boulder, Colorado: Westview).

PERHAM, M. (1961), 'African Nationalism', *The Listener*, 66 (1704): 851–5.

PICKETT, J. (1990), 'The low income economies of sub-Saharan Africa: Problems and prospects', in J. Pickett and H. Singer (eds.), *Towards Economic Recovery in Sub-Saharan Africa*.

—— and SINGER, H. (eds.) (1990), *Towards Economic Recovery in Sub-Saharan Africa* (London: Routledge).

POTTS, DEBORAH (1995), 'Shall we go home? Increasing urban poverty in African cities and migration processes', *Geographical Journal*, 161 (3): 245–64.

RAIKES, PHILIP (1988), *Modernising Hunger, Famine, Food Surplus and Farm Policy in the EEC and Africa*. Catholic Institute for International Relations in collaboration with James Currey (London: Heinemann).

REDCLIFT, MICHAEL (1987), *Sustainable Development: Exploring the Contradictions* (London and New York: Methuen).

RICHARDS, P. (1985), *Indigenous Agricultural Revolution* (London: Hutchinson).

—— (1986), *Coping with Hunger* (London: Allen & Unwin).

RICHARDSON, JULIE (1996), *Structural Adjustment and Environmental Linkages: a Case Study of Kenya* (London: Overseas Development Institute).

RIMMER, D. (1984), *The Economies of West African States* (London: Weidenfeld & Nicolson).

—— (1991), *Africa 30 Years On* (London: Royal African Society and James Currey).

ROY, SUMIT R. (1989), 'Agrarian crisis and technology in Nigeria', *Africa Quarterly*, xxv (1–2): 1–10.

SARRIS, A. and SHAMS, H. (1991), *Ghana under Structural Adjustment: the Impact on Agriculture and the Rural Poor* (New York: New York University Press for IFAD).

SCHULTZ, T. (1964), *Transforming Traditional Agriculture* (New Haven: Yale University Press).

SCHUMACHER, R. (1973), *Small is Beautiful* (London: Abacus).

SHIMADA, SHUHEI (1993), 'Migration and change in agricultural production systems

in rural Nigeria: A case study', *Science Reports of the Tohoku University*, 7th ser. (Geography), 43 (2), Dec.: 63–90.
SJAASTAD, ESPEN and BROMLEY, DANIEL (1997), 'Indigenous land rights in Sub-Saharan Africa: appropriation, security and investment demand', *World Development*, 25 (4): 549–62.
SPARKS, DONALD, L. (1993), 'Economic trends in Africa South of the Sahara' in *Africa South of the Sahara* (London: Europa Publications), 8–12.
——(1998), 'Economic trends in Africa South of the Sahara' in *Africa South of the Sahara* (London: Europa Publications), 10–17.
SPEDDING, C. R. W. (1975), *The biology of Agricultural Systems* (London: Academic Press).
——(1979), *An Introduction to Agricultural Systems* (London: Applied Science).
STEBBING, E. P. (1935), 'The encroaching Sahara: the threat to the West African colonies', *Geographical Journal*, 88: 506–24.
——(1938), 'The man-made desert in Africa—erosion and drought', Supplement to the *Journal of the Royal African Society*, vol. xxxvii, no. cxlvi: 40.
STEVENSON, G. G. (1991), *Common Property Economics: A General Theory and Land Use Applications* (Cambridge: Cambridge University Press).
SWINDELL, K. (1978), 'Family farms and migrant labour: The strange farmers of The Gambia', *Canadian Journal of African Studies*, 12 (1): 3–17.
——(1982), 'From migrant farmer to permanent settler: The strange farmers of The Gambia', in J. I. Clarke and L. A. Losinski (eds.), *Redistribution of Population in Tropical Africa* (London: Heinemann).
——(1985), *Farm Labour* (Cambridge: Cambridge University Press).
TIFFEN, M., MORTIMORE, M., and GIKUCHI, M. (1994), *More People Less Erosion: Environmental Recovery in Kenya* (Chichester: John Wiley & Sons).
TOSH, J. (1980), 'The cash crop revolution in tropical Africa: An agricultural reappraisal', *African Affairs*, 79: 79–94.
UDO, R. (1978), *A Comprehensive Geography of West Africa* (Ibadan: Heinemann Educational).
UNFAO (1995), *Fertilizer Yearbook* (Rome: FAO).
UNFAO(i) (1995), *Quarterly Bulletin of Statistics*, vol. 8, 1/2 (Rome: FAO).
UNFAO (1965–96), *Production Yearbooks* (Rome: FAO).
UPTON, MARTIN (1967), *Agriculture in South-Western Nigeria: A Study of the Relationship Between Production and Social Characteristics in Selected Villages* (Development Studies no. 3, University of Reading, Department of Agricultural Economics).
WALLACE, TINA (1979), 'Rural development through irrigation: Studies in a town on the Kano River Project' (Mimeo, Zaria, Nigeria: Centre for Social and Economic Research, Ahmadu Bello University).
——(1980), 'Agricultural projects and land in Northern Nigeria', *Review of African Political Economy*, 17: 59–69.
WARREN, ANDREW and KHOGALI, MUSTAFA (1992), *Assessment of Desertification and Drought in the Sudano-Sahelian Region, 1985–91* (New York: UN Sudano-Sahelian Office (UNSO).
World Bank (1981), *Accelerated Development in Sub-Saharan Africa: an Agenda for Action* (Washington: World Bank).
——(1989), *Sub-Saharan Africa: From Crisis to Sustainable Growth* (Washington: World Bank).

World Bank (1994/5), *World Development Report* (Washington: World Bank).
World Resources Institute (1992), *World Resources, 1992–93* (Oxford: Oxford University Press).
YIHEYIS, ZELEALEM (1997), 'Export Adjustment to Currency Depreciation in the Presence of Parallel Markets for Foreign Exchange: The Experience of Selected sub-Saharan African Countries in the 1980s', *The Journal of Development Studies*, 34 (1) Oct., 111–30.

# 3

# Smallholder Adaptation: The Humid Domain

The non-equilibrial character of the environment in both the semi-arid and the humid tropics was discussed in Chapter 1. Environmental non-equilibrium is most frequently associated with semi-arid areas where swings in environmental variables, particularly rainfall, make human existence precarious. However, the analysis of rainfall patterns over the past 19 years in West Africa (Chapter 1), has shown that unpredictability of rainfall is a characteristic common to both the semi-arid and the humid zone. Although environmental non-equilibrium in the latter is of a different nature from that of the semi-arid realm, it can pose equally serious problems for farmers. This chapter investigates how indigenous smallholder farmers, in different locations in the humid zone, cope with an uncertain environment and also with a range of other constraints. The location-specific, autecological methods of land management which are considered support the argument that development strategies need to be appropriate to the farmers and to the physical environment. This is emphasized by Moss (1992) who argues that environmental information, which is essential for the effective planning and implementation of practical agricultural development, is neglected.

Environmental unpredictability is a characteristic that smallholder farmers in the humid tropics have learnt to cope with. There is no doubt that the environment both causes and sustains damage, but physical destruction of the environment is minimized by cultivators who understand local ecological conditions and who over the centuries have developed low technology, autecological systems which avoid the worst effects of the climate. Surely any form of rural development strategy should capitalize on such indigenous knowledge. In spite of this, human activity can also be ecologically unsound and has done much to aggravate environmental problems and to provoke environmental instability. Forest land damaged by loggers, for example, can initiate a process which leads to land degradation, as can cultivators whose approach to the environment is not sensitive. Migrants from the semi-arid north to the humid zone who are unfamiliar with local ecological conditions have been associated with such types of ecologically unsound management of the environment (Wiersum *et al.*, 1985*b*). Climate change may also play a part as suggested by Fig. 1.2, but at present, human activity is arguably the dominant cause of non-equilibrium in the humid tropics (Boers, 1990; Ofomata, 1975).

A difficult physical environment is not the only determinant of the characteristics of indigenous farming. Fig. 3.1 shows that the nature of farming is influenced by a great many variables, access to land, labour and capital being of major significance. The influence of each of these is far from equal and their effects may differ with changes in time, location, and scale. However, as access to capital is so limited in much of West Africa and as farming is commonly characterized as 'low input', so much depends on the skill and initiative of cultivators—both men and women—and their ability to manage the physical environment to greatest effect to meet the needs of crops and animals. We thus return to the critical importance of the physical environment. In high input, capital intensive systems, problems caused by variations in soil type, for example, can in large measure be overcome by the use of technology: soils which are particularly difficult to dig by hand provide little resistance to farm machinery; infestation of land with noxious weeds may be eliminated with chemicals, and low levels of soil fertility may be compensated by treatment with inorganic fertilizer. In low input systems the farmer does not have the technology to reduce variation in the physical environment significantly, but instead is forced to work with it.

## Focusing on farming at the local scale

That agriculture is not stagnant was evident from Chapter 2, but the extent of the dynamism and the adaptation of farming systems to a range of different influences are more readily apparent at the local scale. Examining the varied and vigorous responses of indigenous farming to different circumstances, examples of farming systems have been drawn predominantly from two ecological zones in the humid tropics. The first is south-eastern Nigeria where total annual rainfall is around 2,500 mm and where most rain falls from April/May to October/November, where there are no truly dry months and where humidity levels are persistently high (Idachaba, 1985; Harrison Church, 1980; Agboola, 1979). The second is from central Sierra Leone where similarly heavy total annual rainfall (2,750 mm at Njala) is also distributed seasonally but where the dry season is much more marked and where there is a moisture deficit from November/December to April/May (Richards, 1986). Examples of farming systems where seasonality of rainfall is pronounced are supplemented from the Bassila region of Benin. The locations of these examples of indigenous farming systems are shown on Fig. 3.2 together with profiles of mean annual rainfall. These data were not always available for specific locations so rainfall profiles for sites nearby have been included.

Within south-eastern Nigeria, the focus will be placed on the characteristics of farming in eastern Cross River State where population densities are com-

# Smallholder Adaptation

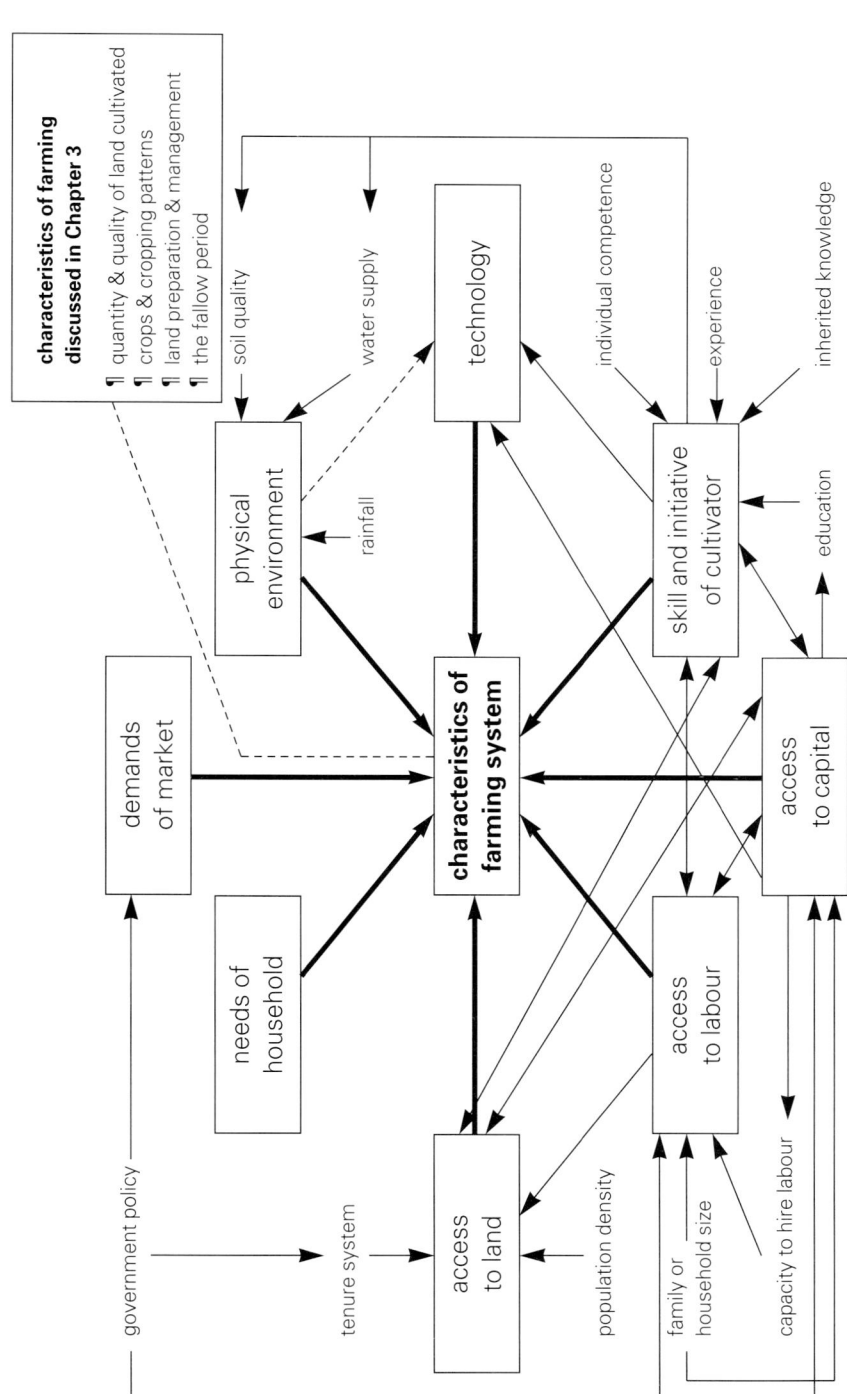

**Fig. 3.1.** Factors influencing characteristics of indigenous farming

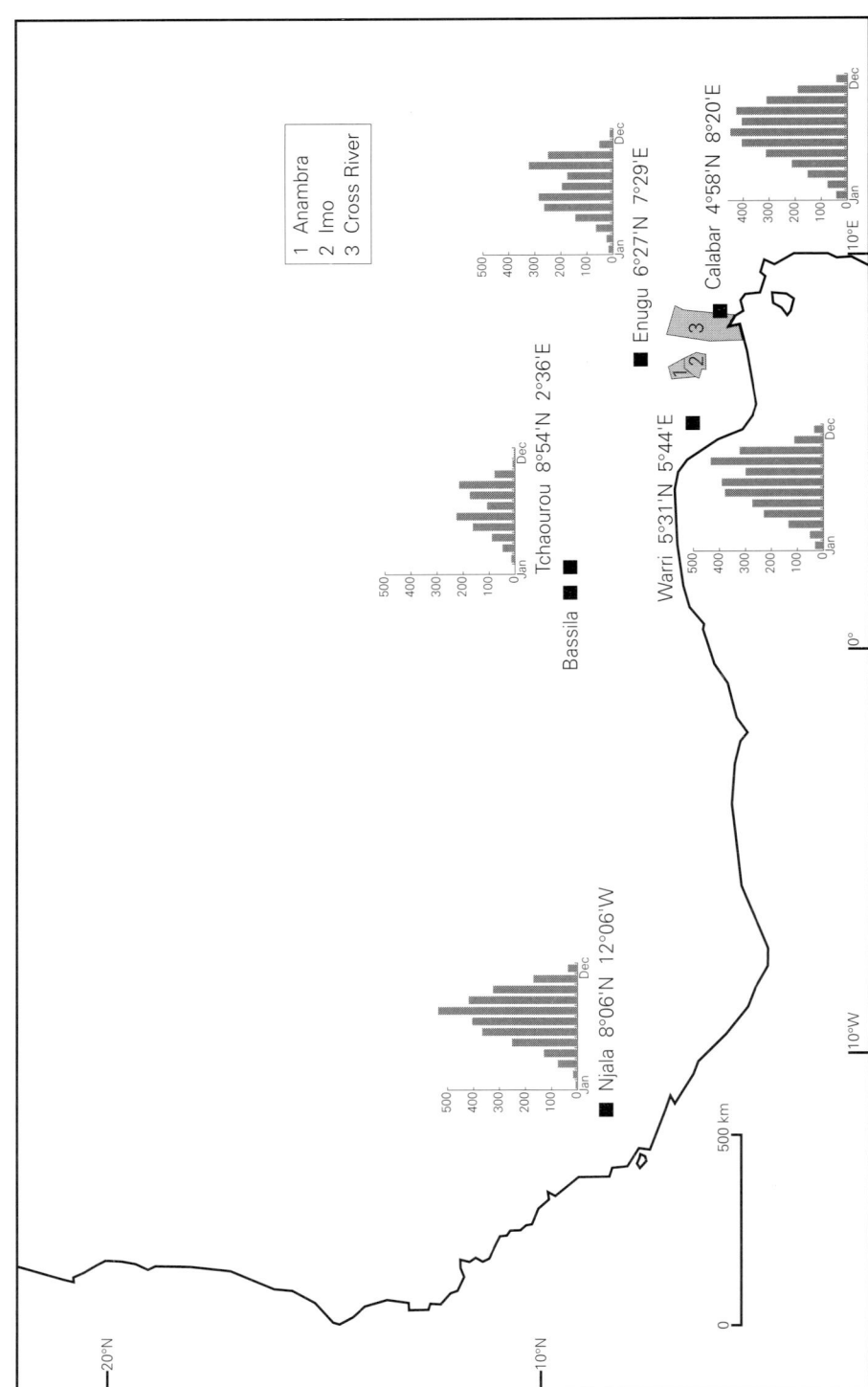

**Fig. 3.2.** Location of sites under discussion in the humid zone

*Source:* Rainfall data: Harrison Church (1980).

paratively low for the region (around 400 per square mile, 156 km$^{-2}$) and where there is still forest land to be cleared for farming. Indigenous farming will also be considered in the more densely peopled Anambra and Imo States where population densities are believed to rise to 1,900 km$^{-2}$ in the Awka-Nnewi region of Anambra (Okafor, 1992) and to around 1,400 km$^{-2}$ just over the boundary in Imo State. While population levels are undeniably high, Goldman (1992) observes that it is almost impossible to know precisely how high population densities are, not least because of the history of problems associated with the collection of census data in Nigeria. Soils of these areas are inherently poor. They are heavily leached, acidic, and have a tendency to lose minerals and organic matter rapidly (Okafor, 1992; Goldman, 1992; Martin, 1992; Udo, 1982). Typical of soils of the humid tropics, they are poor stores of nutrients, the bulk of these being held in the biomass (Nye and Greenland, 1960). Successful cultivation of these soils in the virtual absence of external inputs is thus dependent on careful management.

By contrast with south-eastern Nigeria, population levels in central Sierra Leone are far lower and in the 1974 census, population densities of 10–20 km$^{-2}$ in parts of central Sierra Leone were not uncommon (Richards, 1986). With low rates of rural population growth, these areas still remain comparatively thinly peopled. In both these sample areas of Cross River and central Sierra Leone smallholders use bush fallow systems of farming. These are perceived to be the main type of indigenous farming in West Africa (Udo, 1982). A plot of land is cultivated for a period and then left to fallow. This is followed by the clearance and cultivation of another plot while the first is in fallow. Several plots may be in cultivation at any one time and at different stages in the process. Although broadly similar, there are many differences between bush fallow farming systems. Those selected for discussion contrast markedly with the permanently cultivated compound gardens of Imo and Anambra where population densities are very high, though these may well have resulted from the intensification of rotational bush fallow (Okafor, 1992).

The aim of this chapter is to examine how indigenous farming with limited external inputs has adapted to cope with an unpredictable physical environment in the face of a range of different circumstances. To examine how farmers cope, selected aspects of indigenous farming are investigated. These include among other variables, the quantity and quality of land cultivated; choice of crops and cropping patterns; preparation of land for rainfed cultivation and finally, the role of fallow and bush land in dryland farming. Although other variables could be chosen, these are sufficient to demonstrate the skill of indigenous cultivators who, through their knowledge of the physical ecology, have developed autecological systems which are closely adapted to different environmental socio-economic and political circumstances. Not all aspects of farming selected receive equal treatment in different farming systems. For example, methods of land clearance in eastern Cross River State and in Sierra Leone are paid more attention than are methods of land clearance on the compound gardens of Imo and Anambra where permanent cultivation prevails. Similarly, management of burning is of greater importance where there is a marked dry season. This highlights the nature of smallholder adaptation to differing circumstances. While adaptation to physical conditions is fundamental, the importance of socio-economic and political factors is also examined and the relative importance of each of these on farmer decision-making is considered.

## Making the most use of available land

Availability of land is largely related to density of population (Okafor, 1992; Goldman, 1992; Martin, 1992; Cornia, 1994). In eastern Cross River State where land is still relatively plentiful and in keeping with the traditional bush fallow model, farmers still cultivate several different plots, each normally being under different ecological conditions. According to Dunn and Agom (1993) in Cross River a man, his wife and three to four children, cultivate approximately one hectare of land per season and may have permanent access to between two and four plots of equivalent size. While one plot, or approximately one hectare is being cultivated, the others lie fallow. The number of plots under cultivation is thus linked directly to the length of the fallow, which in turn may be related to population density and demand for land. Cultivating several plots of land increases the range of crops that can be grown and helps to limit the spread of disease. In the eastern part of Cross River State where land is still available, fieldwork by Dunn and Agom identifies land cultivated in four different environments (Dunn and Agom, 1993). First, there is land where rainfed crops, mainly cassava, yam, and maize are intercropped. Staple crops are usually produced on rainfed land though the patterns of crops grown may vary with micro level changes in soil conditions. Secondly, on land in more moist areas such as stream valleys where the quality of the soil is good and land floods regularly, rice is cultivated. Beyond the flooded area a range of speciality crops including sugarcane are grown. It is noteworthy that although riverine land is highly prized, environmental conditions can be unpredictable, the floods being either inadequate or extending beyond expected bounds and destroying crops intolerant of flooding. Thirdly, land for vegetable gardens is usually near a water source, as a regular water supply is vital. Fourth is compound land, which may also be the site of the vegetable garden.

Throughout the area the oil palm is ubiquitous. This is of major economic importance but in contrast to planted crops, generates naturally in forest fallows and so is rarely planted (Agboola, 1979). The benefits of cultivating land under different ecological conditions can be considerable: it enables the production of a wide range of crops at different times of the year, some of which are for subsistence, and some for sale. Speciality crops such as rice or tobacco may be grown on the wetlands and crops requiring close supervision can be grown in the compound. Cultivating land under different ecological conditions staggers demand for labour and with different environments producing different crops, it reduces the risk of crop loss from insect pests, from disease and from environmental hazards such as shortage of water, or more particularly excessively heavy rain, which is common in the humid south of West Africa. This is one example of adaptation to an unpredictable environment.

As population density increases and with it demand for land, so the intensity with which land is farmed increases and in parts of Anambra and Imo States increased intensity of cultivation compensates for the reduction in the number of plots, and the diversity of habitats which are cultivated. Three stages can be identified in a transition of farming techniques between eastern Cross River State which is a comparatively land rich area, and the very densely peopled areas of Imo and Anambra. First, as we have seen above, in Cross River State land under different ecological conditions may still be cultivated by one farm family. Fallows are around four years, not sufficient according to Lagemann (1977) and Nye and Greenland (1960) for recovery of the soil but nevertheless, quite a luxury in comparison with other areas. Secondly, in parts of Anambra State, Nigeria, where there is greater pressure on land, both compound land and rainfed dryland fields continue to be farmed. However, the dryland fields receive less attention than in Cross River and are fallowed for a shorter period. Greater focus is placed on the cultivation of compound land. Finally, where population densities are very high as in the Awka-Nnewi region of Anambra and in parts of Imo, where land is scarce, the luxury of cultivating more than one plot has been lost; the luxury of a fallow period has been lost and farmers are confined to intensive cultivation of compound land (Goldman, 1992; Okafor, 1992; Martin, 1992). This is a trend continuing throughout the region. Work by Grove (1951) in Awka Nnewi classified some 35 per cent of the farmland as compound land and according to Okafor (1992), the proportion of compound land has since risen to 63 per cent of farmland in the same area, a probable reflection of local adaptation to growing pressure on land. Arguably, increased intensity in cultivation as land resources diminish does not necessarily suggest any progression of the sort described by Boserup (1965). It reflects adaptation, rather than the technological advancement of farming with increasing population density.

In stark contrast to parts of Imo and Anambra, land is available sufficient for the cultivation of several plots in central Sierra Leone and in the case of upland rice, cultivation frequently follows the valley profile (Richards, 1986; Gwynne-Jones *et al.*, 1978). The upper parts of the slope support rice that is entirely rainfed while the lower slopes are watered by a combination of rainfall, runoff and the rising water-table (Richards, 1986). The lower slopes which receive both rainfall and runoff are suitable locations for early rice, particularly if the soils have a comparatively high silt fraction and can retain moisture. But where these lower slope soils have a high gravel fraction, water retention is poor and rice can only be planted once the rainy season is well progressed. Soils of the upper slopes which depend more on direct rainfall than on runoff are sown latest in the season but produce the bulk of the harvest. Here too, variation in soils is important and those with a higher silt fraction, normally better at retaining moisture and nutrients, are preferred to those with a high gravel content. The link between rice cultivation and local ecological conditions is evident.

The motivation behind the cultivation of a range of different ecological environments is economic, though the physical environment initially defines the range of environments available for cultivation. Where prize locations such as wetlands are limited in area, local politics can be critical in determining who has access to how much land, and to land of what quality. Much depends on the prevailing land tenure. Generalization about tenure systems is extremely difficult and the following paragraphs demonstrate the complexity of tenurial relations and highlights their importance in farm decision-making.

In the areas under discussion, that is south-eastern Nigeria, central Sierra Leone, and Bassila in Benin, communal tenure frequently dominates in rural areas, though since the Land Use Decree of 1978 in Nigeria, all rights over land have been vested in the Offices of the State Governors (Williams, 1992). Implemenation of the Decree has not been smooth and while nearly 25,000 applications for certificates of occupancy had been made by 1989, less than 55 per cent of these had been processed because of the enormous burden this placed on the administration. Bottlenecks were inevitable. It is not uncommon for formal tenure to be found side by side with communal tenure which is still poorly understood (Bräutigam, 1992), and which, according to Cornia (1994), allows the exchange of land to take place remarkably smoothly.

Many types of tenure exist in West Africa and the following examples do no more than scratch the surface of a highly complex system: at one end of the spectrum there may be freehold land where people have title and with this a right to sell as they choose. Some land is leasehold and of this some may be rented. Some land may be pledged in return for a loan and here the right of cultivation, though not the development of the land, remains with the person who provided the loan until it is redeemed. However, land which is leased or pledged does not necessarily give the occupier the right to the products of trees which may continue to be harvested by the original occupier. Failure to redeem the loan over a long period may result in a transfer of land rights. Community land held on trust may be available from the community with exclusive rights only during the period of occupancy. Such usufruct rights are what many people have over the fields they cultivate, though compound land is more commonly freehold. Ususfruct rights are particularly convenient as they allow land to be cultivated for a period, then to be fallowed, a time when the community has access to the products. It is also a way of ensuring that different soil types are shared between members of a village. There are, however, many rules governing access to communal land. Intense population pressure in areas where compound farms dominate the environment such as in the Awka-Nnewi region of Anambra, has led to fragmentation of land as inheritance laws divide the plot between the sons of a household (Okafor, 1992). Different types of land holding may exist side by side and the same family may hold land in different types of tenure. Investment in land is thought to be greatest where access rights are assured (Cornia, 1994) and where this is so, applications of limited supplies of manure are most likely to be made to freehold fields than to leasehold plots or to communal land (Morgan and Pugh, 1969; Okafor, 1992; Olaloku et al., 1979). However, a great deal more empirical work needs to be done on communal systems of land tenure before firm conclusions can be drawn on the impact these have on agricultural development (Barrows and Roth, 1990).

Socio-economic factors such as access to labour and capital also influence the number of plots cultivated. Assuming there is access to land, decisions about the number and ecological characteristics of plots to be cultivated are defined by the needs of the household and this, in large measure, is decided by the family elders and the head of the household. This is clearly demonstrated by Richards (1986) and by Dunn and Agom (1993). There may be friction over the proportion of food crops grown relative to commercial crops. Younger people who are trying to accumulate money frequently wish to focus more on cash crops, though food security is usually the prime objective of heads of households (Richards, 1986; Baker fieldwork, 1990, 1999). A large family or household is more likely to have the labour resources to farm a larger area. Where labour resources are limited, access to capital is critical in order to hire the necessary help. Without adequate access to labour or capital, the potential of the farm family to cultivate several plots is limited and inevitably income is lower and risks of loss, greater.

As Fig. 3.1 shows, the skill and initiative of those cultivating the land are also of critical importance in determining the success of indigenous farming. Nevertheless, skill and initiative cannot be exercised *in vacuo*. More often than not, farm decision-making is channelled by a range of socio-economic and political factors, at village or household level, and these may be subject to influences from a national and global level (Fig. 3.6).

## Crops and cropping patterns

The cultivation period in the humid south extends from one calendar year to the next without a clear break, as in south-eastern Nigeria, whereas in central Sierra Leone seasonality of rainfall is more marked and there is a clear break. Most planting begins with the rains in April/May, and harvesting takes place towards the end of the rainy season in August, September, and October (Gwynne-Jones et al., 1978; Morgan and Pugh, 1969; Osemeobo, 1987a).

The choice of crops cultivated is determined by a wide range of factors: the preferences and needs of the family; the proportions for consumption and for the market; the amount and type of land available and the availability of labour and capital. The ways in which crops are grown are very much related to environmental conditions, the techniques involved being mainly to meet the needs of the crop while at the same time minimizing soil deterioration. This is demonstrated with regard to cultivation of dryland fields.

In the Cross River region of south-eastern Nigeria, the dryland fields support mainly food crops such as cassava, yams, cocoyams, maize, pumpkins, okra, and melons. Cash crops are rice, cocoa, and bananas most of which are cultivated mainly on or near the wetlands or in the compounds. The oil palm, also an important cash crop, is ubiquitous though rarely planted, growing readily in forest fallows (Agboola, 1979). Plants are intercropped though Dunn and Agom (1993) note the absence of any particular pattern. Trees which contribute a harvest may also be found on an intercropped plot. Formal rotations among field crops are rare, a characteristic common to other parts of the region too (Upton, 1967). Visually, the effect is one of disorder, but plants sown together, which have different physical forms, different nutrient requirements (particularly micronutrients), different rooting depths, and different rates of development are able to capitalize on nutrients and water at different depths in the soil. They do not compete for either light or space and intercropping can also help guard against pest attack and against soil erosion which is a problem in Cross River and is frequently evident between yam mounds (Dunn and Agom, 1993; Richards, 1985). Autecological techniques are thus tailored specifically to the environment. Higher soil productivity, higher labour productivity, and lower variablity of crop yields are all characteristics of intercropping, a practice which was largely ignored during the colonial era and the merits of which have been appreciated mainly since the 1960s (Peter and Runge-Metzger, 1994).

In parts of Imo and Anambra States, where population density is higher than in eastern Cross River, but where there is still sufficient land for households to cultivate more than one plot, the choice of crops cultivated is similar to those in eastern Cross River, though changes are apparent in dryland cultivation. Higher levels of population pressure have reduced the amount of land available and in consequence the number of plots to be cultivated. This in turn has reduced the proportion of income derived from the land, has stimulated migration and increased dependence on non-farm income and has reduced the availability of labour for agriculture. This is a serious constraint on farming. Okafor (1992) notes that although there is sufficient land for households to cultivate an outfield as well as compound land, shortage of labour has resulted in more intensive cultivation of the compound than the outfield. Crops such as pumpkins, melons, and okra are cultivated in what Okafor (1992) terms the outfields in Cross River, but in Anambra, these are cultivated in the compounds. These crops require more attention than the staples yam, cassava, and maize, the latter being intercropped in the outfields where they receive less attention and their lower than potential yield supplements the output from the compound. The tradition of intercropping is retained because it increases the total yield from the plot and also because the combination of crops is a guard against disease.

The focus on crops in the compound is largely determined by the availability of labour though an additional reason may be related to the tenurial situation. Compound land is frequently freehold while access to land further away may not be so assured. In consequence, the bulk of any inputs that are available, limited amounts of manure for example, are used on compound rather than on usufruct land (Morgan and Pugh, 1969). As a result, the quality of soil on compound land is better than it is further away and greater demands are made on it to support the main economic trees such as oil palm, coconut palm, kola, citrus and bread fruit (Okafor, 1992), in addition to a wide range of field crops including root crops, vegetables, and herbs. Compound land is more productive in terms of output per unit area than most other types of cultivated land because of higher labour inputs and enhanced soil quality. Because it is highly productive, it is given priority in terms of labour allocation.

In the most densely peopled parts of Imo and Anambra, where people do not have the luxury of cultivating more than one plot of land and where they do not have the luxury of a fallow period, compound gardens produce most of the needs of the household from a single, permanently cultivated plot. They support a higher number of species than the compounds in the example described above, where an outfield was still cultivated, and consist of a planned mixture of trees, staple field crops, vegetables, speciality crops, and herbs. Niñez (1987) describes the arrangement of trees and crops in compound gardens as a multistoreyed farming ecology capable of mobilizing nutrients, water, and sunlight from different levels of the soil and the atmosphere. Such an arrangement of crops can modify the micro-climate and micro-crop environment in a beneficial way (Beets, 1990). A further benefit of intercropping is that it staggers labour use, a scarce resource in south-eastern Nigeria. Not only do the planting dates of crops differ, but so do the times for weeding and harvesting. In the struggle to get the most out of their land farmers grow crops in mixtures and also sequentially so dietary needs can be met. Interspersed with the major staples, cocoyams, cassava, and vegetables are trees of economic value such as oil palm, coconut palm, kola, citrus, and breadfruit (Netting, 1993; Lagemann, 1977). In some areas compound gardens are largely in the hands of women (Goldman, 1992), but in others work is more equitably shared among men, women and children (Okafor, 1992).

Continuous cultivation is possible only through a substantial energy subsidy, the provision of which is time consuming and costly in terms of labour. Okafor notes that inorganic fertilizers are used on compound gardens in Anambra when farmers have the resources to buy them. Igbokwe (1996) states that inorganic fertilizer used to be widely used on compound farms in the Maku region of Enugu State but the removal of subsidies on agricultural inputs has resulted in reduced usage. Fertilizer is used relatively efficiently in compound gardens because the dense cover of crops with different physical forms reduces nutrient loss from the soil. Higher yields per unit area from compound gardens reflect the sensitive management of the soil.

In spite of the density of population in these parts of south-eastern Nigeria, labour is probably even more scarce here than elsewhere. Fragmentation of compound farms as a result of the prevailing tenurial system has resulted in these small, intensively cultivated areas providing an ever decreasing share of family needs. Economic pressures have forced migration and higher incomes have been both sought and found in activities other than farming. A starvation of compound farms of labour has inevitably followed (Okafor, 1992). The success of compound gardens has thus depended on the ability of indigenous cultivators to manipulate the physical environment and to make the best use of labour and capital resources to meet the demands of the crops they cultivate. This, surely, is an autecological approach to farming working at its best.

Schreckenberg further confirms that in the moist savannas of the Bassila area of Benin crops cultivated are a reflection of resource availability and of the needs and preferences of the farm family (Schreckenberg, pers. comm., 1996). For example, the decision as to whether to plant yam or cassava is influenced by soil type, labour availability, cost of production, and market demand. This is of critical importance as yams are a comparatively expensive crop to produce and it has to be worthwhile for the farmer to do so (Osemeobo, 1987*a* and 1987*b*). According to Shimada (1993), working in Nigeria, tasks associated with the production of crops such as yam, maize, cocoyam, and melon are much harder than tasks associated with cassava production. Ridging, mounding, and the cutting of yam stakes are all very hard work, they are also seasonal and can be avoided by the cultivation of cassava. However, decisions about which crop varieties to grow and whether or not the household should plant improved varieties of cassava are made by the head of the household, individual family members having very little say in such decisions (Poulson and Spencer, 1991). Socio-economic and political factors are thus of significance in influencing patterns in farming.

In central Sierra Leone rice is the dominant crop and grows in two different types of environment, upland rice and swamp or wet rice, though upland rice dominates smallholder production. It is generally preferred to swamp rice for its flavour, even though the latter is higher yielding. This is an interesting contrast with rice grown in The Gambia (see Chapter 6) where it is the swamp rice which is preferred. Several varieties of rice may be sown. Rice may be sown as a monocrop but it is often mixed with a small amount of other seed such as sorghum, millet, okra, maize, pigeon-pea, benniseed, and cotton. These are broadcast together and each is harvested as and when it matures, providing variety in the diet (Gwynne-Jones *et al.*, 1978; Richards, 1986). While Richards (1985) explains that intercropping reflects the indigenous cultivator's deep knowledge of the environment, it is also evident that intercropping is influenced by resource availability, not least the availability of land and labour.

Thus explanations of cropping patterns can be invoked from both physical and from political ecology.

Unlike southern Nigeria where a patch of rainfed land may be cultivated for several years, in central Sierra Leone it is cultivated for no more than two years before being left to fallow. Rice is cultivated in the first year and this allegedly causes such a depletion of nutrients that another patch of bush land is cleared for the next rice crop. In the second year crops of lesser value are cultivated. These may include cassava, cocoyam, groundnuts—where the soil does not have too high a clay content—ginger or tobacco, the latter two being export crops (Jalabi, 1978; Gwynne-Jones *et al.*, 1978). While the success of the household rice farm dominates family farming, there are nevertheless some fields that are privately cultivated with groundnuts, sweet potatoes, and vegetables (Wiersum *et al.*, 1985*a*). These are usually in the dry swamps and are commercial ventures. With the permission of the head of the household, an individual may also plant a crop of cowpeas on rice land immediately after the harvest (Richards, 1986). A common problem in all the areas discussed is a shortage of labour, though in Sierra Leone the abundance of land relative to labour has not meant that soil degradation is the problem it is in south-eastern Nigeria.

Work on field crops is divided between men and women and their roles vary considerably throughout Africa. In the Cross River region men tend to be in charge of cash crops while women are more involved with food crops, in particular the vegetable gardens and maize fields. Yams are harvested by men for whom they are a prestige crop (Agboola, 1979), though women do help with harvesting cassava and dominate the processing of oil palm. With regard to bananas, traders from the urban areas come to the villages about once a week to buy the crop. While the division of labour is often described in very precise terms in the literature, field experience indicates that in reality there is flexibility and when labour is in short supply for some good reason such as illness or migration, there is cooperation between women and men in order to to ensure that household needs are met (Dunn, pers. comm., 1993; Richards, 1985; Baker fieldwork, 1990/1). In Sierra Leone land clearance is the responsibility of the men, the task of women and children during this period being to keep the men well supplied with food. Women become involved in the rice farm once the seed is sown. They tidy the fields ensuring that all the seed is covered, taking out the most prominent weeds at this early stage and continue to maintain the fields through weeding. Finally, both men and women work together to harvest the crop (Richards, 1986). While tasks undertaken may differ, men, women and children work together with a common cause of reaping a successful harvest.

## Dynamism in cropping patterns

In spite of perceptions of indigenous farming as stagnant and resistant to change, farmers are constantly readjusting farming systems in response to changing circumstances. In south-eastern Nigeria, and indeed in much of West Africa, cassava (*Manihot esculenta*) cultivation has increased at the expense of yam (*Dioscorea* spp.) which requires fertile soils and which is costly in terms of labour, planting material and staking requirements (Agboola, 1979; Martin, 1992; Okafor, 1992; Shimada, 1993). It is mainly for these reasons that the yam has been displaced by cassava, which has become the most important food crop in the region and also the main income-earning crop, exceeding both oil palm and yam (Goldman, 1992; Martin, 1992). Okafor (1992) reached similar conclusions in Anambra state, Shimada (1993) demonstrates the growing importance of cassava throughout southern Nigeria, and FAO statistics reveal that these findings extend throughout West Africa, as cassava production has increased both in absolute terms and in relation to yams (Chapter 2). Odemerho and Avwunudiogba (1993) have observed that in Nigeria traditionally, cassava was intercropped but this is giving way to monoculture. They note that traditional practices for intercropping cassava have limited soil erosion, but where it is monocropped, problems of nutrient depletion are being aggravated by soil erosion. Cassava has several advantages over yam: it can tolerate poor soils, has low labour requirements for its production though not necessarily for processing, is cheap and easy to propagate from stem cuttings, and is easily 'stored' in the soil until required allowing flexibility as to its time of harvest. When processed into cassava meal (*garri*), it is easily transported to market. The direct relationship between increasing land pressure and cassava cultivation has been noted by many authors including Oluwasanmi *et al.* (1966) in Uboma village, Imo State, and in other parts of West Africa by Morgan (1955) and Forde (1937). Its history has also been traced in the Ngwa region, south-eastern Nigeria, by Martin (1992). Introduced to the region in 1907 by Christian pastors, cassava was being interplanted with yam by the early 1950s. By the early 1960s, cassava had risen to dominance as the staple Ngwa foodstuff and by 1980–1, it had replaced palm produce as the main cash crop. Cassava continues to remain an attractive proposition as there are now many improved varieties (Akoroda *et al.*, 1987) a development which has not been paralleled by improved varieties of yam (Igbokwe, 1996). The widespread adoption of cassava reflects the readiness of farmers to make the best use of opportunities. With limited access to labour and capital and with ever growing population pressure, farmers have adopted cassava because in economic terms it has enabled their continued survival when so few other opportunities exist.

Cassava is not the only change in smallholder farming. Goldman (1992) notes that the number of crops declining in importance in the cultivation system in parts of Imo State is greater than those that have been adopted over the past 20 to 30 years. Two species less in evidence are the cocoyams (*Colocasia esculenta* and *Xanthosoma sagittifolium*). At one time these were some of the most lucrative crops grown and even exceeded yams in importance (Lagemann, 1977). However, the spread of soil-borne disease and the physical effort involved in their cultivation (Shimada, 1993) have reduced the area under these crops. Other crops which have declined in importance include groundnuts (*Arachis hypogaea*), also a victim of disease, several varieties of bean, coconuts (*Cocos nucifera*), and several species of yam (*Dioscorea* spp.). Crops introduced have been far fewer and have included new varieties of maize and cocoyam, the latter being disease resistant. Similar types of innovation are noted by Amanor (1992, 1994) in south-eastern Ghana where dynamism is equally apparent in rice cultivation. Indigenous farmers keep numerous varieties of seed and plant, only those which they anticipate will thrive. For example, if the rains have been comparatively poor or arrive late, appropriate varieties may be sown. Farmers interchange their seed regularly, thus maintaining the genetic strength of stocks and introducing new varieties.

Dynamism is also apparent in compound gardens which have modified farming techniques to improve their efficiency. For example, yams, where they are still cultivated, are not always grown on mounds but in deep compost filled trenches from which yields are significantly higher (Okafor, 1992). In parts of humid Cameroon where population density forces cultivation of erosion prone hillsides, local farmers have developed broad ridges along the slope, covered the whole year by a high density mix of crops such as cassava, maize, and sweet potatoes. In heavily mulched furrows which run parallel with the ridge rice, bananas, and coffee are grown. The high crop density on the ridges reduces the impact of rain storms and ridges and furrows serve to reduce run-off and soil erosion (Beets, 1990). Similarly, appropriate methods of terrace cultivation have been developed in southern Anambra. On steeper slopes unique forms of terrace cultivation slow the flow of water while on lower slopes of hillsides fields are surrounded by low walls about a metre high or are surrounded by fences both of which encourage rain water to percolate into the soil. They are also a deterrent to foraging animals. Compound gardens thus reflect major emphasis on conservation of nutrients and energy and the heavy investment of labour resources in soil conservation techniques.

The skilful arrangement of species, refined over generations on compound farms, and the high levels of organic inputs have been a substitute for fertilizers and other technical improvements characteristic of farming in the West. While compound fields are highly successful at increasing output from a limited area with limited access to modern technology, the question is how much more can they achieve? Nigeria has seen a sustained increase in fertilizer use since the early 1970s (FAO 1995) though this trend has been reversed over the past 2 or 3 years, possibly in response to increased fertilizer prices resulting from the removal of subsidies. Although compound gardens reflect considerable success in terms of land management, this part of south-eastern Nigeria is still a food deficit area where most families derive a significant and increasing part of their income from non-farm activities. Any further increase in land productivity will be predicated on further improvements in soil management either from the increased use of inorganic fertilizers, now increasingly costly, or from the skilful combination of organic and inorganic inputs, an area where development money and expertise could prove useful.

Concluding this section, it is evident that the choice of crops and cropping patterns is determined by a wide range of socio-political, economic, and environmental factors. Of fundamental importance is the land available—both the amount and the quality—as are the needs and preferences of the family, the labour supply, local market conditions, and domestic views on food security. While these and many more variables may influence the choice of crops grown, ultimately, successful crop cultivation which involves planting the right crops in the right place, in the right combinations, at the right time and adjusting cropping patterns to keep abreast of changes be they political, social, economic or environmental, is a reflection of the cultivators' expertise in understanding the nuances of the environment and coping with its unpredictability. Certain practices are perceived as being far from sound in ecological terms, the decline in the length of fallow being one example (World Bank, 1981). While this is detrimental if the energy subsidy to the system is not maintained in some way, it should not necessarily be assumed that all forms of energy subsidy have been eliminated.

Evidence on the ecological soundness of changes in farming practices remains contradictory and may be based on insufficient ecological information.

## Preparation of land for rainfed crops

### Clearing the land

During the process of clearing forest or well developed bush, such as still exists in eastern parts of Cross River and to a lesser extent in the more land rich parts of Imo and Anambra States, the plot to be cultivated is cleared of all but the largest trees and those of economic value which provide fruit, rope, or building materials, for example, are conserved. Tree felling is normally the task of the men. After the trees are felled, branches are cut off and heaped on the plot, a task which may also involve women and children. If labour resources are available, the tree trunks and largest branches are cut up and removed for firewood but if labour is in short supply, the trees are simply left on the plot and the nutrients are returned gradually to the ecosystem by decomposers, or more rapidly when the plot is burnt. Rarely are stumps or roots of the cut vegetation removed as few farmers have the equipment necessary for this, or the time. Leaving roots in the soil can help to stabilize the substrate, limit erosion, and increase soil organic matter. This approach is particularly successful in savanna environments where many species, such as certain *Acacias*, have shallow but wide spreading rooting systems.

The process of clearing the bush for upland rice, the main crop in much of Sierra Leone, is very similar. The farmer selects an area that has been in fallow for at least four years (Gwynne-Jones *et al.*, 1978) and preferably no more than 11 years, as after this bush regeneration is so dense that the labour required to clear it is too great (Richards, 1986). The length of the fallow varies according to pressure of population. The brushing of the land starts in December or January, following a break after the harvest. Carried out entirely by hand, this is an enormous task which has to be completed swiftly so that the vegetation has time to dry out and can be burnt before the next rains. This is usually achieved by the skilful organization of labour (Richards, 1986).

Organizing labour for clearing the fields is driven by social, economic, and political factors. This requires cooperation between those whose plots are near together and may involve the use of numerous types of work groups. The simplest is where farmers form informal groups, working together on the fields of each member, according to a rota. No compensation is involved. As full-time, hired labour is in short supply, more formal work groups are formed in response to labour shortage. These may consist of farmers together with young men from the locality and possibly from nearby urban areas. These groups can be hired for certain tasks, for example, brushing or hoeing (locally called ploughing in Sierra Leone). Demand for such work groups is considerable and not all farmers have the resources needed to hire these groups. Work groups are paid in wages and food, the latter being of considerable importance (Richards, 1986). In contrast to south-eastern Nigeria, the problem in areas of low population density in Sierra Leone is not with the physical side of the ecological equation, but with the human side. Farmers who can either participate in one of the work groups or hire a group to get a particular job done have better prospects of preparing and seeding their land before the rains than a farmer who does not have access to additional labour. Large families may have greater access to labour, smaller families may have to pay or to become involved in work groups and where farmers have to make use of work groups who are paid for their labour, political factors such as a family's importance may determine who gets access to the resource at the prime time and who is left until last.

Such methods of land preparation are relatively rare in those parts of Imo and Anambra States where population densities are at their highest, though even here shortage of agricultural labour can be a problem in spite of high densities of population. Where land is permanently cultivated, preparation along the lines described above no longer occurs. Instead, the soil is prepared for planting every time a crop is harvested. In spite of the density of population, labour is still a scarce resource and again, the autecological approach of the farmer is much influenced by socio-economic and political conditions at a local level.

## Burning

In spite of high humidity and rain in most months of the year in south-eastern Nigeria, the vegetation dries out well in the high temperatures and a few weeks after being cut is ready for burning. Burning has both advantages and disadvantages: it does result in the loss of almost all the organic carbon, nitrogen, and sulphur in the vegetation and litter, though the mineral nutrients are returned to the surface soil in the ash. Burning does not normally result in an appreciable loss of humified organic matter in the surface soil and on account of the ash, top soils are normally richer in phosphates and exchangeable calcium, magnesium, and potassium. The quantity and type of nutrients in the ash are a direct product of the quantity and type of the vegetation cover that was burnt. For example, in soils with a low base status, the nutrient value of ash tends to be lower in calcium, magnesium, and potassium (Pieri, 1992; Sanchez, 1976; Vine, 1968; Webster and Wilson, 1966). pH increases after burning and decreases as bases are leached out. In addition to cycling nutrients and giving the crops a good start, burning is a highly efficient method of clearing the plot of plant diseases and of vermin, making it a safer and easier place to work. It also discourages weed growth in the early stages of cultivation, a factor which soon makes significant demands on limited labour supplies (Moody, 1974). One of the main disadvantages is when fires get out of control and cause extensive damage. In those parts of Imo and Anambra States where population densities are very high and the land is under permanent cultivation, burning is much more limited in its use and is frequently reduced to little more than a garden bonfire. The nutrients derived from the ash are supplemented in such intensive systems by animal manure, by night soil, by other sources of composted organic matter (Goldman, 1992; Okafor, 1992) and by inorganic fertilizers, where they are accessible.

In Sierra Leone, where land is plentiful and labour is in short supply, burning is a tool of critical importance and is used very effectively to clear the land ready for planting. Because of the scarcity of labour, farmers frequently work together to cut the vegetation prior to burning. As it is critical that the burn should be good, and as there is advantage in burning large areas at one time because the fire can become well established, fields are burnt in blocks. Such a corporate strategy also reduces the danger of fires getting out of control. It is critical to complete the burn before the rains become established. Burning therefore takes place towards the end of March (Richards, 1986). According to Swindell and Hewapathirane (1969), this is a very stressful time for farmers. Within any block, whose field is cleared first and whose last is of considerable importance because the one burnt last will have the least drying time before burning.

If the vegetation is not adequately dry before being burnt, the fire may not reach a sufficiently high temperature, weeds may not be suppressed adequately and through the life of the rice crop keeping weeds at bay is a larger and more time and labour consuming task than it may have been, had the burn been efficient. In fields where the fire has not burnt at a sufficiently high temperature, farmers have to make bonfires to clear the unburnt debris, as unless this is properly cleared, the progress of the rice seedlings will be impeded (Richards, 1986). Socio-economic and political factors at a local level thus influence the order of work when farmers cooperate to prepare their rice farms for seeding. While there is undoubtedly competition for a better place in the queue, there is nevertheless a desire to complete a successful burn for everyone involved before the onset of the rains, because in small communities it is to no one's advantage if harvests are poor.

As the savannas become drier, the rules on fire use become more stringent as the following example from the Bassila region of Benin shows. In an attempt to keep errors to a minimum, local rules on burning are very precise and are laid down by village elders. Bassila receives approximately 1,300 mm of rain a year in a single rainy season lasting from March to October, approximately half the amount of the average rainfall in Calabar. Although Bassila's place in a discussion on the humid zone is questionable, the example is useful. Most of the rain falls in July and August and much of the year is dry. Burning in Bassila is very much a dry season activity which is highly regulated (Schreckenberg, pers. comm. 1995). Fires in the Bassila area are set for similar reasons to those in Cross River: to clean the area around houses and villages, to rid an area—field or compound—of snakes and vermin, or to create a firebreak to protect against late fires. Firebreaks are extremely important, being burnt around plantations often of cashew, to protect the trees. Firebreaks are burnt around future yam fields to ensure that the vegetation develops so that when the field is cleared and burnt, a substantial amount of ash will be returned to the soil. Fields to be cultivated are burnt to replenish nutrient levels in the soil. In order to control the amount of fuel on land to be cultivated, fields are burnt early. However, weed growth is so rapid that a second burn is usually necessary just before the onset of the rains. Peulh herders also burn the vegetation to promote the growth of fresh grass for their cattle.

Fires are lit from October and are legal up to 31 December. After this, the heart of the dry season, fires burn at a higher temperature and the risk of fire spreading is greater. Fires are usually set early in the mornings or late in the evenings when they can be more easily controlled because winds have dropped and temperatures are lower. As a rule, people inform their neighbours a few days before they start burning, giving them time to burn firebreaks of their own. When the burn commences people tend to work together, one setting fire while other family members or neighbours keep it under control with green branches. In spite of attempts to control the use of fire, invariably fire damage occurs as the dry season progresses with frequent heated conflict between cultivators and pastoralists as to who is to blame (Schreckenberg, pers. comm. 1995).

Beyond 31 December, fires are set by hunters, usually without warning. As soon as an area of bush is fired, animals and birds rush out to escape and are shot as they do so. The second reason for lighting fires nearer the onset of the rains and prior to planting, is to rid the fields of the late growth of weeds. Once the rains arrive, burning virtually ceases and cultivation commences. This lengthy discussion of burning demonstrates that it is of major ecological importance, that local people are fully aware of the importance of burning, and that its successful implementation as an effective tool in farming is ensured by an understanding of the physical environment and by socio-political decisions at village and household level.

## Managing vulnerable soils

The soil is at its most vulnerable after the vegetation has been cleared and burnt. Nutrient losses through burning and leaching are greatest in the phase immediately after a plot has been cleared (Ahn, 1970; Lal, 1979; Sanchez, 1976). Unlike soils in temperate areas, soils of the humid tropics are not major stores of nutrients and are prone to leaching by the heavy rains (Ahn, 1970; Longman and Jeník, 1987; Whitmore, 1990). Soils cleared from forest undergo major changes in physical and chemical properties. Humus decomposition is accelerated by the alternating humification and dessication which characterizes much of the humid tropics and the decline in the colloidal humus content is linked to a major decline in nutrient levels. Of particular significance to cultivators is the decline in nitrogen and phosphates. A reduction in the organic matter content can also leave the soil vulnerable to physical damage such as splash erosion. The soil thermal regime may be altered markedly by clearing. Diurnal fluctuations in temperature following clearing can be of a magnitude of 20–30 °C which affect the soil flora and fauna. High soil temperatures, a decline in organic matter, and a loss of the colloidal organic fraction can reduce the water holding capacity of the soil. Changes in soil thermal and moisture regimes can cause compacting of surface and sub-soil. Pans induced by human activity are evident in commercial farms in West Africa and can render soil unfit for cultivation (Lal, 1974). It is notable that indigenous methods of cultivation work simultaneously to meet the needs of the plant and to minimize soil degradation.

In south-eastern Nigeria the cleared and burnt plot is sown with a series of crops the first of which are usually fast-growing species with a low, creeping habit which quickly provide ground cover, keeping nutrient loss to a minimum. Soon after the burn and before the heavy rains begin the seed is scattered or sown directly into the ash layer and the soil is hoed lightly to cover the seed. It is in the farmer's interest to re-establish vegetation cover as soon as possible. Tillage is minimal for many crops, though not for yams which may be planted on mounds, a method which increases the depth of top soil and which involves covering the products of weeding with soil to compost it, a process which increases the nutrients available to the crop. Yam mounds are prone to erosion and in south-eastern Nigeria may be covered with leaves to minimize soil loss (Dunn, fieldwork, 1993; Morgan and Pugh, 1969). Mounding is labour intensive and where labour is in short supply, it is avoided or alternatively, yams are replaced with other crops, in particular cassava. For field crops such as maize, beans, pumpkins, and okra, limited disturbance of the soil is linked to low levels of erosion although some disturbance of the soil may be necessary if the soil has a tendency to form indurated layers (Lal, 1974; Mouttapa, 1974). Minimum tillage is a valuable technique which can raise topsoil organic matter content and improve moisture retention capacity. It can also reduce excessive heating of the soil and aid soil stability. The disadvantage of minimum tillage is that the density of the soil may increase and total porosity decrease which has implications for the capacity of the soil to absorb rainfall, or to be eroded by run-off (Ellis and Mellor, 1995). On upland rice farms in Sierra Leone, minimum tillage limits soil erosion and the intermix of crops sown adds to the protection of the soil. The leaching of nutrients remains high in the early weeks of cultivation but as plant cover builds up, so the loss of nutrients by this route is slowed (Ahn, 1970).

Inputs into indigenous farming systems such as those in Cross River and Sierra Leone are generally limited though there are marked differences among the systems under discussion. The main inputs during the cultivation phase are the ash-fertilizer from the burn, roots of the cut vegetation, and of harvested cereals which contribute to soil organic matter, and droppings of itinerant animals. In the case of Cross River, where population is beginning to increase, Dunn and Agom (1993) note the use of organic mulches on the fields. This does not usually occur in Sierra Leone. Young (1986) observes that the maintenance of fertility through soil biological processes requires skilful management to control the quantity, quality, and timing of decomposition of plant residues. If the timing is not correct, plant disease may spread.

Inputs in the form of nutrient subsidies into the dryland fields during the cultivation phase are few both in Cross River and in more densely populated areas where rainfed fields continue to exist. The ash-fertilizer from the burn gives seedlings and cuttings a good start though a high proportion of the nutrients are leached away (Ahn, 1970). If root crops have been left in the soil over the drier phase, weeds may be hoed and left to compost before a second planting of groundnuts or maize. Such composted material undoubtedly enriches the soil and according to Dunn and Agom (1993) is used in areas where demands on land are increasing and where access to fallow land is becoming ever more scarce. Although such composting techniques may improve both the physical quality of the soil as well as its nutrient status, much added benefit could also be derived from the appropriate use of inorganic fertilizers. However, the removal of subsidies is proving a disincentive to further use (Igbokwe, 1996). In the past, poor distribution systems have also hampered uptake of chemical fertilizer (Dunn and Agom, 1993). In the longer term, where land is still available, the fallow period remains the most effective means by which the nutrient status of the soil is restored, though its length is critical. Fallow systems are discussed further, below.

In central Sierra Leone energy subsidies are very similar to those used in south-eastern Nigeria though additional inputs such as compost are less often used. Application of inorganic fertilizers is also limited, but here, because of the availability of land, fertility is not the problem it is in south-eastern Nigeria as this is restored during comparatively lengthy fallow periods. It is access to labour which is the problem.

The situation is very different for the compound gardens which, apart from depending on the skilful spatial and temporal arrangement of crops, depend for their success on very high levels of inputs, mainly organic matter. Heavy mulches are applied to soils in the form of small branches, twigs, and leaves from trees and shrubs. In compound gardens near Maku, Enugu State, Nigeria, palm leaves and leaves from *Acioa Barteri* fallows are collected and spread on cocoyam plots. The palm leaves are rich in potassium which is essential for the formation of tubers (Igbokwe, 1996). Compost made in pits from a combination of household refuse such as ash, sweepings, leaves, other waste material, and grasses collected from fallow areas is also added to the soil, and animal manure, mainly dung from goats, sheep, and poultry is applied regularly. Where they can afford it, there is evidence that farmers use inorganic fertilizer though this is still very limited (Goldman, 1992; Lagemann, 1977; Okafor, 1992). In this way, farmers tailor resources to the specific needs of crops. Igbokwe (1996) notes that the use of inorganic fertilizer can be counter-productive because it promotes the growth of weeds which increases labour demand. Since this district is within the tsetse zone, cattle are comparatively few in number and with the acute land shortage there is little room to pasture animals though some animals are kept, particularly small stock. Organic residues represent a substantial nutrient subsidy to the agroecosystem, maintaining soil fertility, and ensuring that the productivity of compound gardens measured in output per hectare is higher than from dryland fields further away. In all these systems, in south-eastern Nigeria and in Sierra Leone, farming has been sustained by indigenous expertise. None of the systems import significant levels of inputs from outside the system except, perhaps, labour in peak periods.

## The role of fallow and bush land in dryland farming

After the dryland fields in the eastern Cross River region of south-east Nigeria have been cultivated for a period of perhaps three or four years, they are left to fallow. According to Nye and Greenland (1960), on land under mature, secondary forest cleared for cultivation in southern Ghana, yields fell by 50 per cent after six years of cultivation without the use of fertilizer. The fallow is thus of critical importance in low input systems as it is the main means of replenishing nutrients and improving soil structure. During the fallow period the land is no longer sown with crops as natural vegetation is allowed to regenerate. Although the nutrient status of the soil is poorer by the end of the cultivation period, weed growth has already become well established and the extent of weed growth is one of the determining factors on whether a plot should be left to fallow (Moody, 1974). With the cessation of harvesting as the fallow period begins, positive feedback forces operative during the cultivation phase turn to negative feedback as the system begins to return towards a more natural vegetative state. The net outflow of nutrients from the system is greatly reduced and as vegetation dies and decays, nutrients are returned to the soil by the decomposers and the ecosystem begins to accumulate. Whether the original plant cover can ever be re-established is difficult to demonstrate in practical terms, but the system usually recovers remarkably well, given time: the biomass which is not harvested builds up again and in tropical ecosystems this represents a growing store of nutrients in the system. Increased vegetative cover which is not harvested by humans also allows a build-up of the soil organic fraction which in turn increases the capacity of the soil to retain both moisture and nutrients (Ahn, 1970; Fitzpatrick, 1974; Jordan, 1989; Pitty, 1979; Pieri, 1992; Sanchez, 1976).

The early years of fallow growth are the most beneficial in terms of nutrient replenishment because leaf growth which approaches a maximum in this period has a far higher nutrient status than the woody parts of plants that develop later (Vine, 1968). This accords with the view of Nye and Greenland (1960), that after 5 years many ecosystems have recovered sufficiently for cultivation to begin again, though according to Lagemann (1977), a fallow period of 6 years or more is needed to replenish the fertility of soils in south-eastern Nigeria. According to Mouttapa (1974) soil nitrogen levels recover after the fallow but evidence suggests a pronounced decline in available phosphate following cultivation, an artificial subsidy usually being necessary if levels are to be restored. For the most part, however, indigenous, low input systems of cultivation are productive and not necessarily environmentally destructive. Although species composition of the vegetation following cultivation may be different, the chances are that forest will re-establish given time, though the constituent species are difficult to predict with any accuracy on account of the non-equilibrial nature of the environment (Chapter 1).

## Declining fallow periods

There is mounting evidence that throughout West Africa the length of fallow periods has declined, particularly in areas where population densities have increased (World Bank, 1981, 1989; Amanor, 1992, 1994; Goldman, 1992; Okafor, 1992). In consequence it is frequently stated in the literature that the environment is becoming degraded (Glaeser, 1995; World Bank, 1981, 1989). While declining fertility is inevitable if fallows are shortened and energy subsidies are not increased, the evidence on what is happening is far from clear cut. There is evidence to suggest that farmers are modifying their cultivation systems to compensate for a reduction in the fallow period. According to Dunn and Agom (1993), farmers in Cross River are using composted weeds on their fields to compensate for shorter fallows. Although no reliable empirical evidence is available to chart changes in soil fertility levels, Dunn and Agom's field information also suggests that the land in Cross River which is fallowed less and cultivated more frequently, in spite of positive measures to maintain its fertility, is declining in its productive capacity. This could be because the quantity of organic compost is insufficient or that its quality is inadequate. Compost making is time consuming and hard work, and easier access to inorganic fertilizers is perceived by farmers as a solution to their problems.

In those parts of Imo and Anambra where there is still land for more than one plot per family to be cultivated, Goldman (1992) and Okafor (1992) have identified declining soil fertility as a problem. With pressure on land there is an increasing tendency for the outfields or dryland fields to be reduced. Where the extent of the dryland fields away from the village or compound is limited by population pressure, only part of the field may be interplanted with the major staples while the remainder is left to fallow. This represents a significant modification of the traditional fallow pattern, as the area under fallow is just a part of a plot rather than an entire plot, and thus is a vestige of what it was in the past. This trend of reduction in the fallow period is taken to further extremes in parts of Imo and Anambra where land is so scarce that farmers are limited to the cultivation of a single plot. In the compound farm, cultivation of this site is permanent and intensive with high levels of inputs largely compensating for the virtual loss of the fallow period (see above). Part of the plot is reserved for compost heaps and bonfires. This is possibly a relic of the rotational bush fallow system as after a few years this area is incorporated in the cultivation system, another patch in the garden being set aside for compost heaps and bonfires.

But it must not be assumed that fallows everywhere have contracted, nor have they contracted uniformly. Goldman (1992) has shown that in Imo State, the extent to which fallow periods have contracted is far from uniform. He has shown that a wider range of fallow periods—1–9 years—existed about 20 years ago with a mean of about 4.5 years. This range has now narrowed from zero to 6 years with a mean of 2.7 years, which may be insufficient to allow sustainability of the system (Lagemann, 1977; Nye and Greenland, 1960). The rate of change in the fallow period has not been uniform but Goldman (1992) suggests a probable link with population increase. Though this seems plausible, it cannot be verified because of the paucity of available data on both the number and distribution of people in southern Nigeria.

The situation is very different in Sierra Leone. Here, where land is abundant and labour in short supply, fields are left to fallow for a minimum of four years and a maximum of 11 years. Beyond this the human resources to clear the land are not available. It is because of this that organization of labour discussed above is so critical.

## Changes in the use of fallow vegetation

Reduction in the fallow period has brought about changes in the use of fallow vegetation. Besides its ecological function in regenerating soil productivity, the fallow provides a variety of resources important to local people (Falconer, 1990; Goldman, 1992; Osemeobo, 1987*a, b*). Dunn and Agom (1993) in Cross River State note that the use of trees is very much taken for granted and that in areas where forest is still plentiful, forest trees have a wide range of uses, food being an important one. For example, leaves of the trees *Pterocarpus* spp, *Myrianthus arboreus*, and *Ceiba pentandra* are highly valued because they flush at the end of the dry season, providing a vegetable in this 'hungry period'. Similarly, the fruits of *Chrysophyllum albidum* and *Dacryodes edulis* are popular because they mature with the early rains when the crops are being planted (Falconer, 1990). Other uses include timber, the provision of shade, decoration, pillow filling, seeds for games and counting, chewing sticks, fish poison, and many more besides. Dunn and Agom (1993) observe that where forest is sparse, people tend to rely on local markets for goods once obtained 'free' from the forest but even in such villages, so strong is the tradition of using trees that many are planted in and around the villages to supply at least some local needs. Hunting for grass cutter, monkey, porcupine, and bush rat also supplements the diet which otherwise may be low in protein. Fish, another dietary supplement, are caught in the waterways.

Goldman's conclusions that reduction in both the area and duration of the fallow period have resulted in a decline in products, such as those listed above, are confirmed by Osemeobo (1987*a*) in Bendel State, Nigeria. In turn, this has led to the commoditization of fallow resources and restriction of access rights to products such as yam stakes and fuelwood on land under fallow. Access restrictions have become graded from full privatization of fallow crops such as *Acioa barteri* prized for yam stakes, to special restrictions applied only to preferred species or to people from outside the village or community, to virtually no restrictions. Sanchez (1976) quoting Nye and Stephens (1962) describes how *Acioa barteri* has been used as an artificial fallow in southern Nigeria and Richards (1985) also describes the planting of *Acioa barteri* on fallow land. *Acioa* seedlings were planted in rows after clearing and grew while food crops were produced for 1–2 years. The *Acioa* accumulated more calcium and magnesium but only about half the phosphorus and potassium that would be developed under secondary forest. Phosphorus and potassium were the limiting nutrients in the Alfisols of southern Nigeria and the artificial fallow was shown to be no better than the natural fallow.

In ecological terms removal of certain species from fallow plots will inevitably change species composition as will strategies of fallow enrichment. However, light harvesting of selected fallow products seems not to hinder the regeneration of a range of bush species which over time contribute organic matter to the soil and replenish both its physical structure and its nutrient status. Thus by taking a light harvest from fields in fallow, indigenous farmers are extracting a little more from the system without seriously compromising its future.

## Conclusions

Indigenous farming systems have survived because of their ecological soundness. Figs. 3.3, 3.4, and 3.5 which trace the major flows of energy through a generalized bush fallow farming system and through a compound garden suggest some of the reasons why these farming systems are, in principle, sustainable. The bush fallow system consists of two distinct phases: the cultivation phase (Fig. 3.3) and the fallow (Fig. 3.4). During the cultivation phase inputs into the system are minimal and those within the system are efficiently tapped to supply a range of cultivated crops. However, the consistent removal of nutrients, mainly through harvesting and the flow of energy out of the agroecosystem results, to some extent, in soil degradation, though it is often increased labour demand to keep pernicious weeds at bay which is the main reason for plot abandonment. Positive feedback forces come into operation and moving the system ever further from its natural state (one cannot with any justification call it an equilibrium), results in instability in the agroecosystem. The fallow period redresses the balance. Negative feedback forces come into play as nutrients are recyled, the ecosystem accumulates and recovers its nutrient status.

The compound farming system (Fig. 3.5) is different in that inputs of energy into the soil are high as it is this that enables regular harvests to be taken. Mechanisms for holding nutrients in the system would appear to be more sophisticated than in the cultivation phase of the bush fallow system as the mix of crops, permanent and annual, establishes a dense root network which, it seems likely, in similar fashion to tropical forest systems, generates an efficient nutrient cycling system. Efficiency of nutrient recycling is similarly well developed in the fallow stage of bush fallow systems.

But if indigenous farming systems are ecologically sound and hence sustainable, why is the development of domestic agriculture so slow? As discussed more fully in Chapter 2, the development of indigenous farming has been impeded by a range of factors most of which are well beyond the spheres of influence of West African farmers. This is evident from Fig. 3.6 which shows how decisions taken at an international and a national level can influence the decisions farmers make at field level. Moreover, many higher level decisions have done little to promote smallholder agriculture.

At the international level, social factors such as tastes for certain commodities can translate into economic factors and thus influence African agriculture. The establishment of trade blocs such as the EU can influence what farmers cultivate. For example, Europe's ability to provide most of its own vegetable

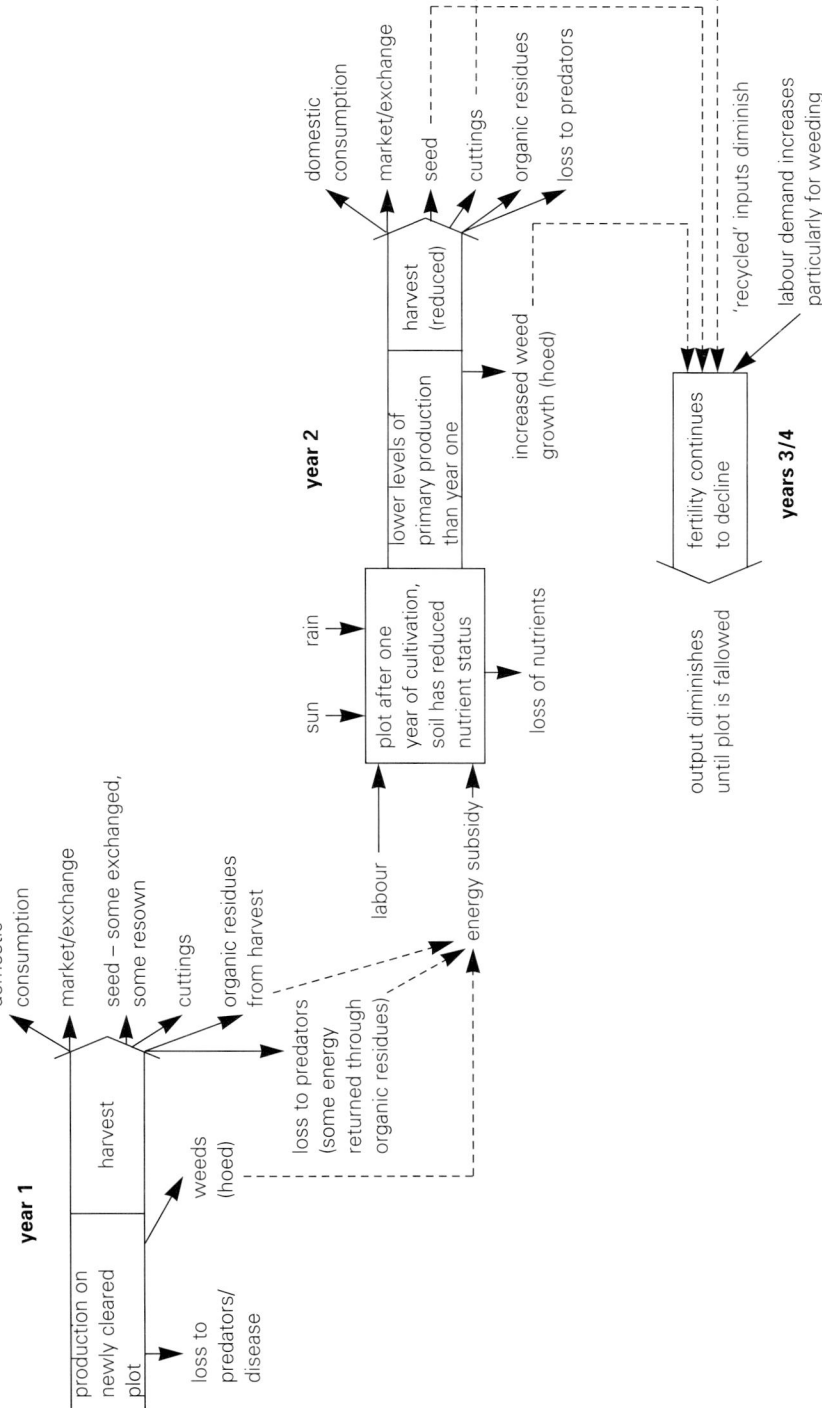

**Fig. 3.3.** Bush-fallow cultivation phase one—the progressive decline in energy during the cultivation phase

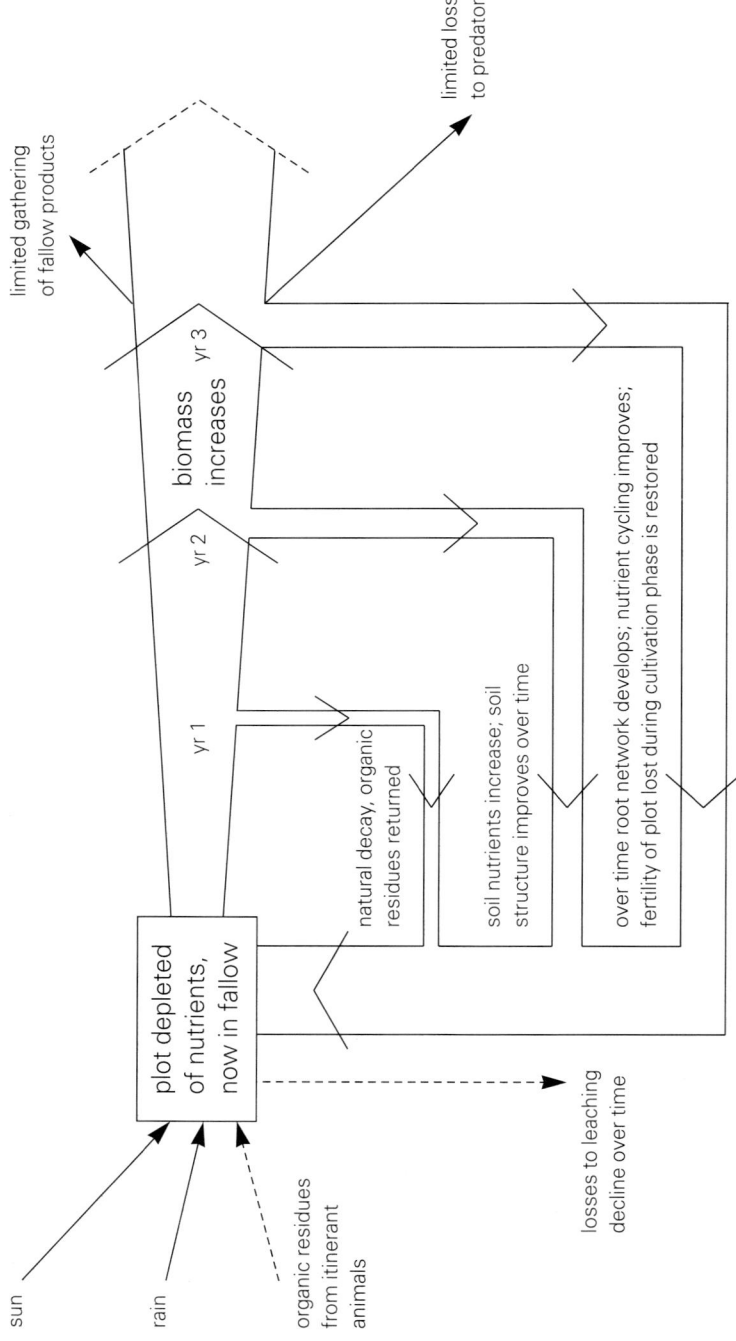

**Fig. 3.4.** Bush-fallow cultivation phase two—the accumulation of energy during the fallow period

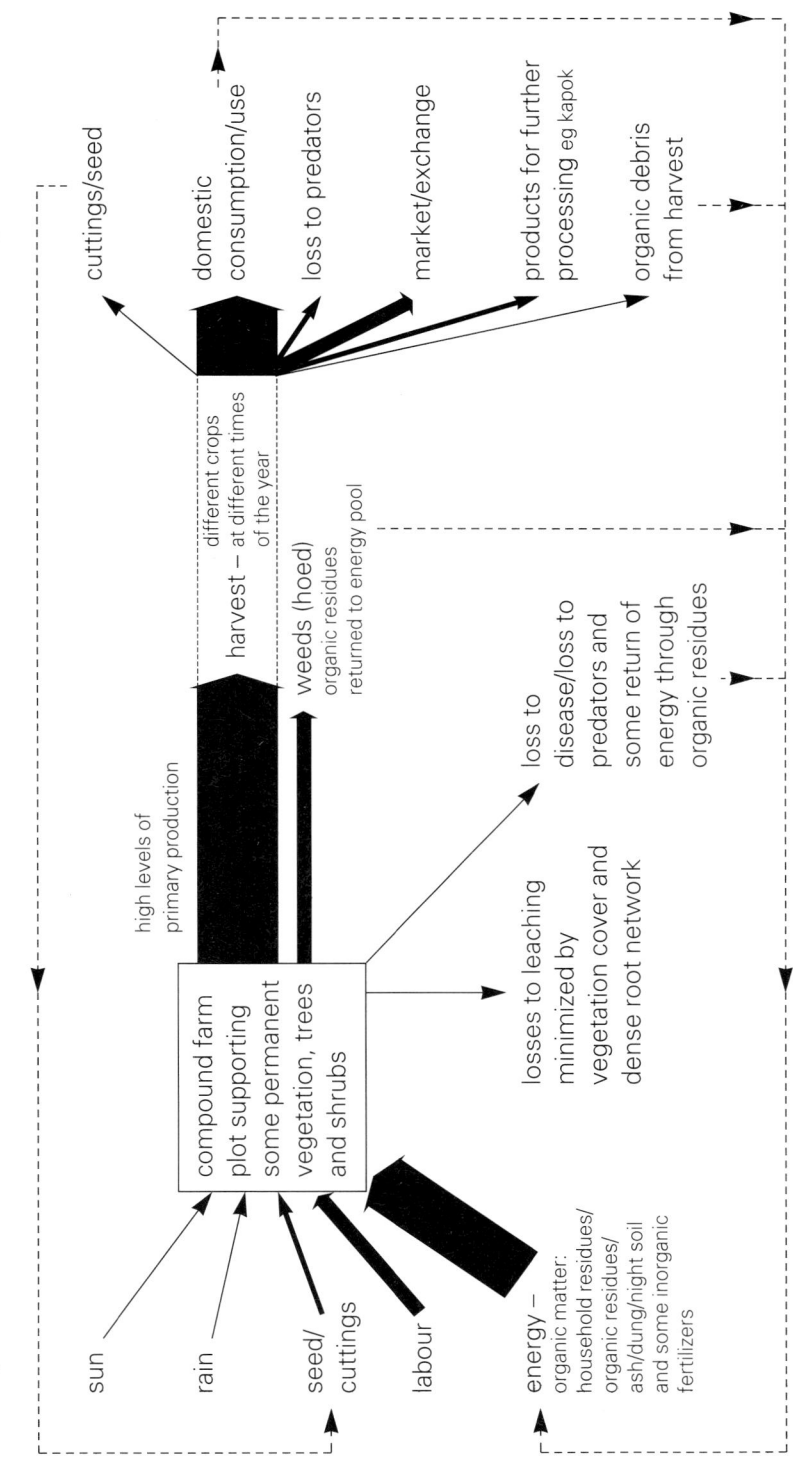

**Fig. 3.5.** The flow of energy through a compound farm/garden

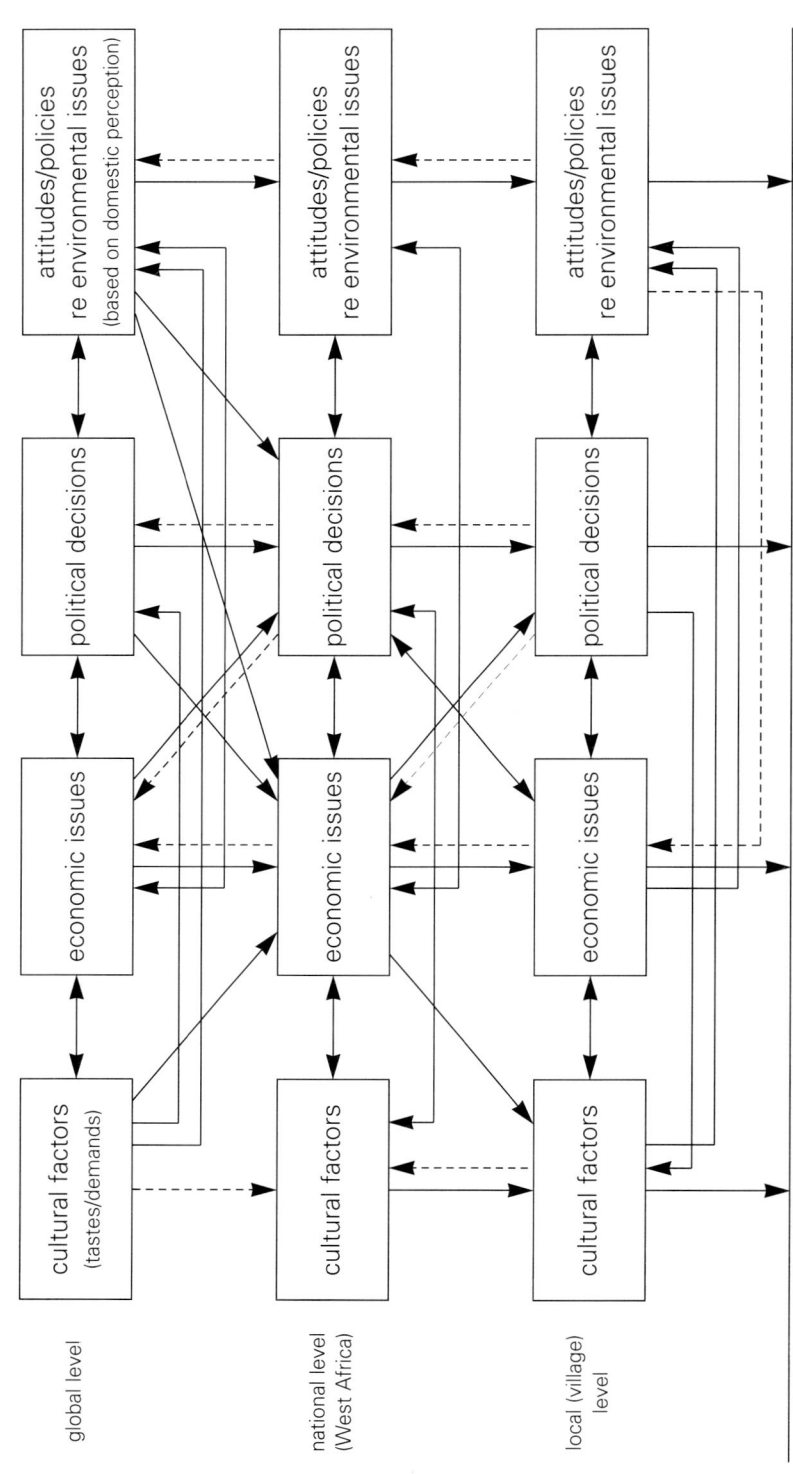

Fig. 3.6. Hierarchy of influences on decision-making at field level

oils and fats has undoubtedly worked to the disadvantage of African producers of vegetable oils and oil seeds. Political and economic decisions such as the setting of prices for cocoa and coffee inevitably have an impact at farm level and international decisions on aid and lending can also influence what is grown. Cash crops, for example, are one means of generating foreign exchange and incentives to produce cash crops have in some areas acted as a disincentive to the production of food crops. The promotion of cotton production in Mali is one such example. In recent years pressures exerted by SAPs have, in some areas, stimulated local food production (Chapter 2). These are just some indicators of how farm decision-making is influenced by factors which are taken with little thought for their impact on the African farmer. What is particularly unfortunate is that so many international decisions have proved disadvantageous to the West African farmer.

At a national level, the disregard by West African governments for smallholder agriculture is a relic which has persisted since colonial times and there have been few policies in the region designed to stimulate smallholder agriculture which have been implemented with lasting success. Much of the investment in agriculture has been directed at changing the nature of farming. Success in achieving this has been minimal not least because proposed changes have largely been based on externally constructed perceptions of develoment needs and on inappropriate assumptions about tropical ecology. Furthermore, policies such as those on markets and prices, together with government manipulation of national currencies, have been detrimental to the success of agriculture. As privatization becomes increasingly important, the land tenure issue is bound to loom large with those with less than secure tenure suffering. Growth in the number of landless people is likely to increase in the future in areas of high population density. Even the new focus on agriculture which is being articulated through Structural Adjustment Programmes has decided the course that policy on agricultural development should take, with, it would seem, insufficient reference to the smallholder cultivator.

At a more local level, there are pressures from environmental unpredictability, from limited technology and from increasing population pressure. The impact of years of low producer prices and frequently high inflation have contributed to lower incomes in the agricultural sector. Farm incomes have been supplemented in a variety of ways, in many cases by remittances from urban migrants. However, change in environmental and human circumstances is frequent and farming communities juggle resources as best they are able. Farming has always provided only part of the needs of a family and many people have more than one job in order to meet their basic needs. This trend towards an increasing portion of income from non-farm sources seems set to continue and is aptly demonstrated by Bryceson and Vali (1997) in East Africa. However, measures implemented under Structural Adjustment Programmes have seen a reduction in public sector employment, a loss of urban jobs and the increase of urban–rural migration. In some parts, rural areas are seeing reduced remittances from urban areas and are now having to house urban migrants whose rural skills are less well developed than in the past. There was some evidence of this occurring in the Western Division of The Gambia.

The ultimate success of smallholder farming depends on the farmers' skill and their ability to grow a range of crops successfully in the face of an array of oscillating physical and human-related variables, most of which are beyond their control. At times, pressures on farmers are so great that they have been forced to adopt seemingly unsound techniques such as a reduction in the fallow period, which could lead to land degradation and compromise farming in the future. While this might well be true, there are many assumptions about indigenous farming which have never been adequately tested. For instance, a reduction in the fallow period is not necessarily synonymous with reduced productivity, and if compound gardens are considered, then reduction of the fallow need not have the dire consequences predicted by the 'ecodoomsters'. There is evidence that farmers from other ecological zones moving into the humid zone do use techniques which may cause land degradation, techniques such as leaving the land exposed to the elements after it has been cleared and failing to generate plant cover rapidly. This does not alter the basic argument in this book that development strategies, where they involve outsiders, ought to look carefully at success stories in indigenous farming and utilize indigenous expertise and ideas in the development of more robust and appropriate schemes for increasing agricultural output.

This chapter has shown that indigenous farming, although far from perfect and with much room for improvement, has succeeded because indigenous cultivators have developed farming systems based around the ecology of the plants they cultivate. Although farming decisions are greatly influenced by a range of social, economic, and political factors operating at different levels (Fig. 3.6) and much of the ecology of farming systems is deemed to be 'political', it is a knowledge and understanding of the ecological requirements of plants and the characteristics of the physical environment which is responsible for the success they have achieved with the minimum of inputs. There is thus a very great need for increasing land productivity in Africa. The key to success, it would seem, is dependent on developing ecologically sound farming techniques for the production of economically and socially desirable species. Many of these already exist, but where additions to farming systems are being made, indigenous cultivators are best placed to evaluate the sustainability of potential changes. Once changes at field level become established, the next stage is to establish appropriate institutions to promote successes at field level and for this, cooperative decision-making at different levels is required.

*References*

AGBOOLA, S. A. (1979), *An Agricultural Atlas of Nigeria* (Oxford: Oxford University Press).

AHN, PETER M. (1970), *West African Soils* (London: Oxford University Press).
AKORODA, M. O., OYINLOLA, A. E., and GEBREMESKEL, T. (1987), 'Plantable stem supply system for IITA cassava varieties in Oyo State of Nigeria', *Agricultural Systems*, 24: 305–17.
AMANOR, KOJO SEBASTIAN (1992), 'Ecological knowledge and the regional economy: Environmental management in the Asesewa District of Ghana' (Paper presented at a Conference of the United Nations Research Institute for Social Development (UNRISD) and The Foundation for International Studies (FIS) on the social dimensions of environment and sustainable development, Valletta, Malta, 22–25 April).
—— (1994), *The New Frontier: Farmers' Response to Land Degradation, A West African Study* (London: Zed Books).
BAKER, KATHLEEN M. (1985), 'The Chinese agricultural model in West Africa', *Pacific Viewpoint*, vol. 2: 401–14.
BARROW, C. J. (1991), *Land Degradation: Development and Breakdown of Terrestrial Environments* (Cambridge: Cambridge University Press).
BARROWS, RICHARD and ROTH, MICHAEL (1990), 'Land tenure and investment in African agriculture: Theory and evidence', *The Journal of Modern African Studies*, 28 (2): 265–97.
BEETS, WILLEM C. (1990), *Raising and Sustaining Productivity of Smallholder Farming Systems in the Tropics. A handbook of sustainable agricultural development* (Alkmaar, Holland: AgBe Publishing).
BLAIKIE, PIERS (1985), *The Political Economy of Soil Erosion in Developing Countries* (London: Longman).
BOERS, T. M. (1990), 'Controlling erosion in south-eastern Nigeria', *The Courier*, 119: 38–40.
BOSERUP, E. (1965), *The Conditions of Agricultural Growth* (Chicago: Aldine).
BRÄUTIGAM, DEBORAH (1992), 'Land rights and agricultural development in West Africa: A case study of two Chinese projects', *The Journal of Developing Areas*, vol. 27: 21–32.
BRYCESON, DEBORAH and JAMAL, VALI (eds.) (1997), *Farewell to Farms: De-agrarianisation and Employment in Africa* (Aldershot: Ashgate).
CORNIA, GIOVANNI ANDREA (1994), 'Neglected issues in the decline of Africa's agriculture: Land tenure, land distribution and R&D constraints', in G. A. Cornia and G. K. Helleiner (eds.), *From Adjustment to Development in Africa* (London: Macmillan).
DUNN, JUSTINE (1996), 'The role of indigenous woody species in "farmer-led" agricultural change in south east Nigeria, West Africa' (unpubl. Ph.D. thesis, University of London).
—— and AGOM, DANIEL (1993), Fieldwork in Cross River State, Nigeria, unpublished.
ELLIS, S. and MELLOR, A. (1995), *Soils and Environment* (London: Routledge).
FALCONER, J. (1990), 'Hungry season food from the forests', *Unasylva*, 160 (1): 14–19.
FAO (1995), *Fertilizer Yearbook* (Rome: FAO).
FITZPATRICK, E. A. (1974), *An Introduction to Soil Science* (Edinburgh: Oliver and Boyd).
FLOYD, B. N. (1965), 'Soil erosion and deterioration in Eastern Nigeria: A geographical appraisal', *Nigerian Geographical Journal*, 8: 33–44.
FORDE, C. D. (1937), 'Land and labour in a Cross River village, Southern Nigeria', *Geographical Journal*, 90: 24–51.

GLAESER, BERNHARD (1995), *Environment, Development, Agriculture: Integrated Policy Through Human Ecology* (London: UCL Press).
GOLDMAN, ABE (1992), 'Population growth and agricultural change in Imo State, Southeastern Nigeria', ch. 8 in B. L. Turner II, Goran Hyden, and Robert W. Kates (eds.), *Population Growth and Agricultural Change in Africa* (Gainesville: University Press of Florida), 250–301.
GROVE, A. T. (1951), 'Soil erosion and population problems in south-eastern Nigeria', *Geographical Journal*, 117: 291–306.
——(1989), *The Changing Geography of Africa* (Oxford: Oxford University Press).
GWYNNE-JONES, D. R. G., MITCHELL, P. K., HARVEY, M. E., and SWINDELL, K. (1978), *A New Geography of Sierra Leone* (London: Longman).
HARRISON CHURCH, R. J. (1980 edn.), *West Africa* (London: Longman).
IDACHABA, F. S. (1985), *Rural Infrastructures in Nigeria* (published for Federal Department of Rural Development, Ibadan: Ibadan University Press).
IGBOKWE, E. M. (1996), 'A soil and water conservation system under threat: a visit to Maku, Nigeria', ch. 27 in Chris Reij, Ian Scoones, and Camilla Toulmin (eds.), *Sustaining the Soil* (London: Earthscan), 219–27.
IRVINE, F. R. (1934), *A Text Book of West African Agriculture: Soils and Crops* (London: Oxford University Press).
JALABI, S. A. (1978), *Agriculture in Sierra Leone* (New York: Vantage).
JORDAN, C. F. (ed.) (1989), *An Amazonian Rain Forest: The Structure and Function of a Nutrient Stressed Ecosystem and the Impact of Slash-and-Burn Agriculture* (Paris: UNESCO).
LAGEMANN, J. (1977), *Traditional African Farming Systems in Eastern Nigeria: An Analysis of Reaction to Increasing Population Pressure* (Munich: Weltforum Verlag).
LAL, R. (1974), 'No tillage effects on soil properties and maize production in Western Nigeria', *Plant and Soil*, 40: 21–31.
——(1979), 'Role of physical properties in maintaining productivity of soils in the tropics', in R. Lal and D. J. Greenland (eds.), *Soil Physical Properties and Crop Production in the Tropics* (New York: Wiley), 3–5.
——(1993), 'Soil erosion and conservation in West Africa', in D. Pimentel (ed.), *World Soil Erosion and Conservation* (Cambridge: Cambridge University Press).
LONGMAN, K. A. and JENÍK, J. (1987) (2nd edn.), *Tropical Forest and its Environments* (Harlow: Longman Scientific and Technical).
LOWE, R. G. (1986), *Agricultural Revolution in Africa* (London and Basingstoke: Macmillan).
MACGREGOR, JENNY (1990), 'The crisis in African agriculture', *Africa Insight*, 20 (1): 4–16.
MARTIN, SUSAN (1992), 'From agricultural growth to stagnation: the case of the Ngwa, Nigeria, 1900–1980' ch. 9 in B. L. Turner II, Goran Hyden, and Robert W. Kates (eds.), *Population Growth and Agricultural Change in Africa* (Gainesville: University Press of Florida), 302–25.
MOODY, K. (1974), 'Weed control in shifting cultivation', in *Shifting Cultivation and Soil Conservation in Africa* (Papers presented at the FAO/SIDA/ARCN regional seminar on shifting cultivation and soil conservation, Ibadan, Nigeria, 2–21 July 1973, Rome: UNFAO, 155–66).

MORGAN, W. B. (1955), 'Farming practice, settlement pattern and population density in south-eastern Nigeria', *Geographical Journal*, 121: 48–64.
—— and PUGH, J. C. (1969), *West Africa* (London: Methuen).
MOSS, R. P. (ed.) (1968), *The Soil Resources of Tropical Africa* (Cambridge: Cambridge University Press).
MOUTTAPA, F. (1974), 'Soil aspects in the practice of shifting cultivation in Africa and the need for a common approach to soil and land resources evaluation', in *Shifting Cultivation and Soil Conservation in Africa* (Papers presented at the FAO/SIDA/ARCN regional seminar on shifting cultivation and soil conservation, Ibadan, Nigeria, 2–21 July 1973, Rome: UNFAO, 37–47).
NETTING, ROBERT MCC. (1993), *Smallholders, Householders: Farm Families and the Ecology of Intensive, Sustainable Agriculture* (Stanford, California: Stanford University Press).
NIÑEZ, V. (1987), 'Household gardens: Theoretical and policy considerations', *Agricultural Systems*, 23: 167–86.
NORMAN, M. J. T., PEARSON, C. J., and SEARLE, P. G. E. (1984), *The Ecology of Tropical Food Crops* (Cambridge: Cambridge University Press).
NYE, P. H. and GREENLAND, D. (1960), *The Soil Under Shifting Cultivation* (Farnham Royal: UK Agricultural Commonwealth Bureau).
—— and STEPHENS, D. (1962), 'Soil Fertility' in J. B. Wills (ed.) *Agriculture and Land Use in Ghana* (London: Oxford University Press).
ODEMERHO, F. O. and AVWUNUDIOGBA, A. (1993), 'The effects of changing cassava management practices on soil loss: A Nigerian example', *Geographical Journal*, 159 (1): 63–9.
OFOMATA, G. E. K. (ed.) (1975), *Nigeria in Maps: Eastern States* (Benin City: Ethiope Publishing House).
OKAFOR, FRANCIS, C. (1992), 'Agricultural stagnation and economic diversification: Awka-Nnewi region, Nigeria, 1930–1980', ch. 10 in B. L. Turner II, Goran Hyden, and Robert W. Kates (eds.), *Population Growth and Agricultural Change in Africa* (Gainesville: University Press of Florida), 324–57.
OKIGBO, B. N. (1981), 'Alternatives to shifting cultivation', *Ceres*, 15 (6): 41–5.
OLALOKU, F. A. with ADEJUGBE, A. (1979), *Structure of the Nigerian Economy* (Lagos: Macmillan Press).
OLUWASANMI, H. A. (ed.) (1966), *Uboma: A Socio-economic and Nutritional Survey of a Rural Community in Eastern Nigeria*. World Land-use Survey, Occasional Papers, No. 6 (Bude, Cornwall: Geographical Publications Limited).
OSEMEOBO, G. J. (1987*a*), 'Smallholder farmers and forestry development: A study of rural landuse in Bendel, Nigeria', *Agricultural Systems*, 24: 31–51.
——(1987*b*), 'The migrant Igbira farmers and food crop production in Bendel State, Nigeria', *Agricultural Systems*, 25: 105–24.
PETER, GREGOR and RUNGE-METZGER, ARTUR (1994), 'Monocropping, intercropping or crop rotation? An economic case study from the West African Guinea Savanna with special reference to risk', *Agricultural Systems*, 45: 123–43.
PIERI, J. M. G. (1992), *Fertility of Soils: A Future for Farming in the West African Savannah* (Berlin: Springer-Verlag).
PITTY, A. F. (1979), *Geography and Soil Properties* (London: Methuen).
POULSON, RUDULPH, A. and SPENCER, DUNSTAN, S. C. (1991), 'The technology

adoption process in subsistence agriculture: The case of cassava in southwestern Nigeria', *Agricultural Systems*, 36: 65–78.

REIJ, CHRIS, SCOONES, IAN, and TOULMIN, CAMILLA (1996), *Sustaining the Soil: Indigenous Soil and Water Conservation in Africa* (London: Earthscan).

RICHARDS, P. (1985), *Indigenous Agricultural Revolution* (London: Hutchinson).

—— (1986), *Coping with Hunger: Hazard and Experiment in an African Rice-Farming System* (London: Allen & Unwin).

RUTHENBERG, H. (1980) (3rd edn.), *Farming Systems in the Tropics* (Oxford: Clarendon Press).

SANCHEZ, PEDRO A. (1976), *Properties and Management of Soils in the Tropics* (New York and London: Wiley).

SCHRECKENBERG, K. (1996), 'Forests, fields and markets: A study of indigenous tree products in the woody savannas of the Bassila region, Benin' (unpubl. Ph.D. thesis, University of London).

SHIMADA, SHUHEI (1993), 'Migration and change in agricultural production systems in rural Nigeria: A case study', *Science Reports of the Tohoku University*, 7th series (Geography, 43 (2) December: 63–90).

SPENCER, DUNSTAN S. C. (1975), 'The economics of rice production in Sierra Leone, Upland Rice' (Department of Agricultural Economics and Extension, Njala University College, University of Sierra Leone, Bulletin no. 1, March).

STOCKING, M. (1996), 'Soil Erosion', ch. 18 in W. M. Adams, Andrew S. Goudie, and Antony R. Orme (eds.), *The Physical Geography of Africa* (Oxford: Oxford University Press) 326–41.

SWINDELL, KEN (1985), *Farm Labour* (Cambridge: Cambridge University Press).

—— and HEWAPATHIRANE, D. W. (1969), 'Rice', ch. 34 in J. I. Clarke (ed.), *Sierra Leone in Maps* (London: University of London Press).

TIFFEN, M., MORTIMORE, M., and GICHUKI, F. (1994), *More People, Less Erosion. Environmental Recovery in Kenya* (Chichester: Wiley).

UDO, REUBEN (1982), *The Geography of Tropical Africa* (London, Ibadan: Heinemann).

UPTON, MARTIN (1967), *Agriculture in South-Western Nigeria, a Study of the Relationship Between Production and Social Characteristics in Selected Villages* (Development Studies no. 3, University of Reading, Department of Agricultural Economics).

UZOZIE, L. C. (1971), 'Patterns of crop combination in the three eastern states of Nigeria', *Journal of Tropical Geography*, 33: 62–72.

VINE, H. (1968), 'Develoments in the study of soils and shifting agriculture in tropical Africa', ch. 5 in R. P. Moss (ed.), *The Soil Resources of Tropical Africa* (London: Cambridge University Press), 89–117.

WATTS, M. (1983), *Silent Violence: Food, Famine and Peasantry in Northern Nigeria* (Berkeley, CA: University of California Press).

WEBSTER, C. C. and WILSON, P. N. (1966), *Agriculture in the Tropics* (London: Longmans, Green).

WHITMORE, T. C. (1990), *An Introduction to Tropical Forests* (Oxford: Clarendon Press).

WIERSUM, K. F., ANSPACH, P. C. L., BOERBOOM, J. H. A., DE ROUW A., and VEER, C. P. (1985*a*), 'Integrated rural development project, eastern province, Sierra Leone', in FAO Forestry Department, *Changes in Shifting Cultivation in Africa: Seven Case Studies* (FAO Forestry paper 50/1, Rome: UNFAO), 27–54.

——(1985b), 'Smallholder plantation agriculture of immigrant Baoulé farmers in south west Ivory Coast', in FAO Forestry Department, *Changes in Shifting Cultivation in Africa: Seven Case Studies* (FAO Forestry Paper 50/1, Rome: UNFAO), 83–108.

WILLIAMS, DONALD C. (1992), 'Measuring the impact of land reform policy in Nigeria', *The Journal of Modern African Studies*, 30 (4): 587–608.

World Bank (1981), *Accelerated Development in Sub-Saharan Africa: an Agenda for Action* (Washington: World Bank).

——(1989), *Sub-Saharan Africa: From Crisis to Sustainable Growth* (Washington: World Bank).

YOUNG, ANTHONY (1986), 'The potential of agroforestry as a practical means of sustaining soil fertility', ch. 5, part 2 in R. T. Prinsley and M. J. Swift (eds.), *Amelioration of Soil by Trees* (London: Commonwealth Science Council), 121–44.

# 4

# Non-Equilibrium and the Cocoa Sector

Cocoa was virtually unknown in West Africa in the mid-nineteenth century when it was imported from South America, but by the last decade of the century it was being exported to Europe from what is now Ghana and Côte d'Ivoire. Declining prices for palm oil and for rubber stimulated the uptake of cocoa but it was the skill of indigenous cultivators who made it a success (Berry, 1975; Chauveau and Léonard, 1996; Hill, 1963; Mikell, 1989). Few farmers had any prior knowledge of cocoa but by combining their innate knowledge of the environment with advice from colonial agricultural officers and by trial and error, they largely succeeded in meeting the ecological demands of cocoa and added it to indigenous cultivation systems. According to Hill (1962), 'It is no exaggeration to say that this dramatic and sustained response to the commercial possibilities of an unfamiliar tree is one of the great events of the recent economic history of Africa . . .' (Hill, 1962: 281). How farmers adopted cocoa and adapted their farming systems around it cannot be omitted from a book concerned with agriculture in West Africa.

Environmental non-equilibrium is a characteristic of the humid zone in West Africa. As evidence of this, Chapter 2 has shown that rainfall is unpredictable and that in consequence, so are other environmental variables. In the production of cocoa, indigenous cultivators have coped almost completely unaided with environmental uncertainty. Their solutions to environmental problems such as the existence of deep, hard pans, of patches of soil of low fertility and of disease may not have been perfect. Nevertheless, indigenous farmers have managed to produce over two-thirds of the world's cocoa and have made West Africa the world's leading producer of the crop (Table 4.1). In spite of the widespread adoption of cocoa, the sector has long been plagued by instability, due partly to the physical environment, but much more to the effects of human action. The nature of uncertainty included the volatility of cocoa prices on international markets, uncertainty about returns on cocoa and the levels of taxation that unstable governments, particularly in Ghana and Nigeria in the 1960s and 1970s, were likely to impose, and uncertainty about the accessibility of essential inputs and machinery: in particular, crop sprayers. There has also been a great deal of uncertainty and caution about accepting the advice of colonial experts whose knowledge of cocoa may have been greater

**Table 4.1(a).** Changes in world cocoa production

|  | 1969/70 ('000 tonnes) | 1996 ('000 tonnes) | change % (rounded to nearest figure) |
|---|---|---|---|
| World Production | 1,421.7 | 2,954 | 108 |
| **Africa** | **996.8** | **1,909** | **92** |
| N.C. America | 90.8 | 138 | 52 |
| S. America | 295.2 | 455 | 54 |
| S. Asia | 3.1 | 11 | 255 |
| S.E. Asia of which | 7.5 | 406 | 5,313 |
| Indonesia | 1.1 | 274 | 24,810 |
| Malaysia | 2.0 | 125 | 6,150 |
| Oceania | 28.3 | 35 | 24 |

Source: Derived from data in *FAO Production Yearbook* (1971), vol. 25; (1996), vol. 50.

**Table 4.1(b).** Cocoa production in West Africa

|  | 1996 ('000 tonnes) |
|---|---|
| Cameroon | 126 |
| Côte d'Ivoire | 1,254 |
| Ghana | 340 |
| Guinea | 4 |
| Nigeria | 145 |
| Sierra Leone | 10 |
| Togo | 3 |
| Total West Africa | 1,882 |
| Total Africa | 1,909 |
| World | 2,954 |

Source: *FAO Production Yearbook* (1996), vol. 50.

than that of indigenous African farmers, but whose knowledge of the African environment was undoubtedly inferior to that of indigenous farmers. These are the problems that have really harassed cocoa producers; they have coped with the vagaries of the physical environment.

To return to the opening pages of this book: it was argued that for rural development to succeed, the initial focus needed to be on the plant or animal in its ecological context. Once it was ensured that this basic element could be produced and incorporated into the farm system, it could then be integrated into the wider agricultural system. This required higher level political involvement and interdisciplinary responses with governments facilitating production for markets, and where necessary, facilitating storage, processing or whatever was required if the crop or animal was to be a socio-economic asset from local to national level. Returning to cocoa, production increased dramatically in the first half of the twentieth century in West Africa but the increase has not been smooth. Very frequently incentives have not been sufficient to stimulate planting, not least because of fluctuations in world market prices. In an attempt to increase and stabilize production by increasing the confidence of the producer, colonial governments in anglophone countries introduced marketing boards in the post-war period, and institutions with parallel functions in the francophone area. In essence, the boards fixed producer prices and guaranteed a market for all cocoa produced. A chain was thus established where high level political decisions facilitated the production of cocoa by enabling indigenous cultivators to meet the physical demands of the crop. Initially, this system worked fairly well. World market prices at the time were buoyant, demand for cocoa was high and so was farmer confidence. From the 1960s the relentless removal of funds from marketing boards, volatile and often low world market prices for cocoa and the failure of boards to support producer prices when they fell saw the destruction of the chain or framework which facilitated cocoa production. The result was the development of a pernicious crisis in the cocoa sector.

Since the mid-1980s efforts have been made, largely through Structural Adjustment Programmes, to reverse the downward trend in the cocoa sector. Whether the creation of producer-friendly policies is sufficient to revive the cocoa sector is uncertain, as the physical environment at the production end is now very different from what it was when cocoa cultivation was first adopted. Land is much more scarce. No longer are there vast areas of forest to be converted into cocoa farms. Replanting former cocoa farms with cocoa raises serious technical problems, many of which remain unresolved. Whether the interest of farmers in cocoa can be sustained is yet another question and where farmers are interested they will have to adapt to new methods of cultivation as former extensive methods are forced to give way to more capital intensive methods of production. Furthermore, whether seemingly producer-friendly decisions are sufficient to stimulate the revival of cocoa where environmental, social and economic conditions at field level have changed significantly, remains to be seen.

This chapter is divided into three parts: the first focuses on how farmers have used an autecological approach to the production of cocoa. It examines some of the cultivation practices used by farmers to show the care that has been taken to include cocoa in the agricultural system. Cultivation methods used may be judged far from perfect but they are well adapted to getting the most out of a difficult and unpredictable physical environment, while using the minimum of material inputs. The second part of the chapter focuses on Ghana, Nigeria, and Côte d'Ivoire, the three main cocoa producers in West Africa. It shows how decison-making at a higher level has undermined achievements by farmers at field level and has created instability in the economic lives of cocoa producers. The third part of the chapter looks to the future of cocoa and considers what may lie in store for West Africa. It also considers whether political decisions may be sufficient to bring about change at field level, where new techniques are being sought in response to new ecological problems.

## Meeting the ecological needs of cocoa

The adoption of cocoa represented a major change in indigenous farming, which formerly focused predominantly on field crops. Tree planting involved long term use of land and in order to make it viable, subsistence crops were cultivated with cocoa to create a farming system which was different from the traditional one. Thus at farm level the agricultural system was modified to accommodate cocoa, though the crop was not integrated into it (Morgan and Pugh, 1969). In spite of information being available on how and where to grow cocoa, many wrong decisions were made about where the crop should be grown. Much of West Africa's cocoa is still to be found in far from optimal conditions (Ruf, 1991; Ekanade, 1991). Cocoa needs rainfall of at least 1270 mm a year, accompanied by constantly high humidity and as important as the total rainfall is its distribution through the year (McKelvie, 1962). The more seasonal the rainfall becomes towards the savannas in West Africa, the more precarious is cocoa cultivation but as long as rainfall in the dry season, November to March, exceeds 250 mm, cocoa can be grown (McKelvie, 1962). At the other extreme, yields are depressed in areas where rainfall exceeds 2,500 mm a year. In such areas, soils tend to be more severely leached of their nutrients and are more acidic (Ahn, 1970). Thus cocoa tends to be confined to the zone of semi-deciduous forest in West Africa.

Cocoa thrives on soils with good moisture retention capacity, good drainage, and good aeration. The most suitable soils for cocoa are loams with clay, sand, and silt in the approximate proportions 30–40 : 50 : 10–20. pH is also critical. While cocoa grows well on alkaline soils where the pH is 6 and over, the more acidic soils with a pH of 5 or less, supported little cocoa (Ahn, 1970; Adams, 1962). As Ahn (1970) observes, in Ghana the more acid soils receive the higher rainfall which correlates with lower yields in cocoa, so it is not possible to say whether cocoa production is limited by either soil or rainfall, as the two variables are so interrelated. Even in areas where rainfall levels are suitable for cocoa, the tree cannot thrive if a hard pan exists below the surface and where such pans are 30 cm or more below the surface, they may easily be missed by farmers.

Until recently, most of the cocoa originally grown in West Africa took 7–9 years to come to fruition. The chief varieties cultivated were the Central American Criollo, the Amazonian Forastero, and the hybrid Trinitario. However, most West African cocoa was of the Amelonado variety of Forastero, which yielded rather low quality beans but which was considered by some to be less susceptible to insect attack and fungoid disease (Morgan and Pugh,

1969). In recent years there has been a great increase in the planting of hybrid cocoa, largely derived from wild stock from the Amazon region. Much of this matures in 3 years though many cocoa experts believe that its quality is poorer than the slower maturing varieties. It is common now to see plots planted to several types of cocoa where farmers have attempted to replace trees that have died (Ekanade *et al.*, 1990).

## From bush to cocoa farm

Preparing a plot for cocoa normally involves the clearance of brush wood under the forest canopy, followed by removal of the largest forest trees and the selective thinning of small and medium trees. The largest trees are cut at this stage, though rarely are tree stumps dug out, cocoa and other crops being planted around them. Some forest trees are left in the more moist areas, usually those with an economic value, and the tree cover that remains must provide shade but at the same time must allow filtered light through to the cocoa. Shade is critical. It is essential when plants are young but becomes less needed as trees mature (Hammond, 1962). On the northern, drier margins of cocoa production, all forest trees are usually removed, as in the few months without rain cocoa is unable to compete successfully with these for limited supplies of moisture. Here shade is provided entirely by food crops such as cocoyam or cassava. Cocoyam is a good indicator of soil suitable for cocoa, for where the cocoyam flowers, cocoa usually thrives (Hammond, 1962).

Once the forest has been thinned, multi-purpose nurse shade species are established. These provide protection for the young cocoa during the dry season and if possible, should produce an additional harvest for the farmer. They must also be easy to remove as the cocoa grows up (Kolade, 1991). In much of the cocoa belt where the moisture deficit period is at a minimum and drought is not a problem, plantain (*Musa* spp.), banana and cocoyam (*Xanthosoma* and *Colcasia*) are the most common forms of nurse shade, others of note being pawpaw (*Carica papaya*) and tree cassava (*Manihot* spp.). Where food production has priority over cocoa, a crop of maize may be sown on freshly cleared forest land prior to the establishment of nurse shade.

Once the nurse shade is established, cocoa seedlings, increasingly of hybrid varieties germinated in a nursery, are transplanted into the cleared forest (Thresh *et al.*, 1988; Oduwole, 1995). The density of cocoa seedlings planted has been highly variable, generally ranging between 600 and 1000 per hectare. Smallholders normally planted at a much higher density than was recommended by Agricultural Officers, to offset the destructive effects of rodents, the effects of impenetrable soils due to the unpredictable presence of hard pans and also soils of low fertility. Having been heavily criticized for dense planting, it is now recognized that farmers' actions were more ecologically appropriate than the methods being proposed by the Colonial Agricultural Officers.

According to Lass and Wood (1985), wide spacing of trees is not conducive to high yields in the case of Amelonado cocoa, the dominant variety grown in West Africa prior to 1950 (Austin, 1996). Nevertheless, there are disadvantages of spacing trees too closely. The reduced air flow among the trees, for instance, can increase their susceptibility to black pod disease (discussed below). Still, from the farmers' viewpoint it would seem that the potential for higher yields more than offsets the risk of black pod disease.

The nurse shade is removed from the Amelonado when it is about 5 years old and after only 2 years for the F3 Amazon hybrids. Once all available land on the farm was planted to cocoa, a new farm would be established on a new area of bush (Amanor, 1994; Schaff and Manshard, 1989). Labour requirement for cocoa is high in the early stages of the life of the tree, as capital formation is achieved by planting and nurturing of trees until they are productive (Szereszewski, 1965, cited in Austin, 1996). Once mature, labour input decreases. Hammond (1962) estimated that 0.4 ha of mature cocoa absorbed little more than 35 man/woman days per year. Any reduction in labour input may have been detrimental to the crop, as maintenance, which includes regular harvesting, pruning and removal of damaged or diseased parts of the tree, is critically important for sustained production (Eshett et al., 1991).

In Western Nigeria nursery preparation begins in October when the rains are receding and when insect pests and diseases are least virulent. In the following July the seedlings are planted into the main field. At much the same time as the nurse shade and the cocoa are planted, farmers start intercropping with food crops. In south-eastern Nigeria cocoa is the responsibility of the men, food crop cultivation being the preserve of women in collaboration with their children (Eshett et al., 1991). In the spaces between the cocoa and the nurse shade, yam mounds may be made. White yams are the first crop planted in November/December. Water yam is planted soon after and while the yams are maturing a range of quick crops is planted. These include okra, pepper, and other vegetables. Cassava, maize, and possibly cowpeas are planted towards the end of the yam growing season. As the yams are harvested, cocoyams are planted on remoulded mounds and a second maize crop may be sown on the mound. While there is adequate light space, quick crops of vegetables including okra and peppers are cultivated. After the second maize crop has been harvested, cassava is the only remaining crop and this might be left unharvested until it is needed. The plot now covered in nurse shade and cocoa is left to mature. By this time it is the third year since the cocoa was planted out. The F3 Amazon hybrids which are much more vigorous cultivars than the ordinary Amelonado are no longer in need of the nurse shade which is progressively removed to give the cocoa adequate room in which to develop. There is little light space left for food crops and the farmer's main return is from cocoa (Eshett et al., 1991). Although there may be some variation in the food crops planted, the conversion of bush to cocoa normally followed this pattern throughout the cocoa belt, the autecological emphasis being apparent.

It is evident from the above description why, after the initial adoption of cocoa in Ghana, food production declined as cocoa production increased. It is also evident why farmers needed to move to new land, both to keep producing food and to extend the area under cocoa. This explains the pattern of land acquisition described by Hill (1962) in Ghana, when Akwapim cocoa growers migrated to northern and western Akwapim, southern Akim Abuakwa, and later to Brong in the north west of Ghana. Thus a farmer might own three farms in different locations, which had major implications for the supply of labour. It also had implications for population distribution, as areas which had been very sparsely populated prior to the commercial production of cocoa were settled rapidly and the Accra–Kumasi railway which was built in stages between 1908 and 1923, facilitated the export of the crop (Harrison Church, 1980).

A similar pattern of land 'ownership' to that in Ghana developed in the former Ivory Coast, where farmers would bring one cocoa farm to fruition and would then open another, somewhere between 10 km and 300 km away from the first (Ruf, 1991). Expansion of cocoa depended on the availability of labour and the task of management was delegated increasingly to women and to hired labour as men went further afield to establish new farms. One of the reasons that smallholders rarely became major plantation owners in the British colonies was the problem of managing farms and labour in different locations. Plantations did exist in both the British and the French territories but in the former they could not compete successfully with smallholder producing units, because the cost of labour was too high. Austin (1996) argues that it was not the mere reliance of plantations on wage labour but that the agricultural practices they employed were excessively labour intensive and, to a lesser degree, capital intensive. Thus it was not the mode of production, but the mode of cultivation, that was their weakness. The strength of Ghanaian and other African smallholders, on the other hand, resulted from autecological approach and from relatively beneficial economic conditions (Austin, 1996). From the 1930s, women became more deeply involved in the running of cocoa farms. Many women set up in business on their own, a trend that has continued and, increasingly, women involved in divorce settlements began to lay claim to cocoa farms they had managed. The spread of cocoa thus resulted in a significant change in the role of those involved, but its production remained firmly within the smallholder domain.

## Pests and Diseases: ecological constraints

The auspicious start that cocoa had in West Africa did not continue unhampered as little known pests and diseases took their toll. In 1936 swollen shoot was first reported, when patches of dead and dying trees were found in the Eastern Region of Ghana centred on Nankese, where cocoa farms were almost continuous over several hundred square miles (Dale, 1962). By 1938 a virus was identified as its cause at the West African Cocoa Research Institute (WACRI) in the then Gold Coast, and from then on many viruses were identified in cocoa, not only in Ghana, but throughout the West African cocoa belt. Their virulence differed: the New Juaben strain for example was one of the worst and led to the death of the tree in as little as 3 years. With other viruses trees could survive for 12 years, though they were debilitated. By the time the swollen shoots were in evidence, it was too late to save the tree.

The vectors for the swollen shoot virus are mealy bugs (*Pseudococcidae*), which feed on the fluids in the cocoa tree. The mealy bugs themselves are delicate, but they are ministered to by certain species of ant which feed on the sweet, waxy secretion from the bug and protect the latter from fungal growth. Some species of ant go further in their protection of the mealy bug. They build 'tents' over mealy bug colonies, using vegetable debris or particles of earth cemented together by oral secretions (Dale, 1962). This protects the mealy bug from predators and from the elements. It enables populations to build up and move from tree to tree through the canopy, which is more or less continuous.

No real solution to the problem of swollen shoot has yet been reached. The most common prescription by agricultural officers was cutting out and burning diseased trees along with their neighbours but this did not find favour with African cocoa farmers. By 1986 almost 169 million infected cocoa trees had been removed in the Eastern Region and 18 million cocoa trees elsewhere in Ghana (Thresh *et al.*, 1988). This had not stopped reinfection and bears out the view of Lass and Wood (1985) that cutting out was not a solution to the problem. Chemicals did work, but sprays and sprayers were in short supply and those supplies that did exist were not reliable. Attempts have been made to breed strains of cocoa more resistant to disease than the highly susceptible Amelonado. Over the past 30 years there has been a shift to new strains, largely upper Amazon derivatives otherwise known as F3 hybrids but resistance of these to swollen shoot has been variable and most are still susceptible. Thresh *et al.* (1988) review the difficulties involved in breeding disease resistant strains of cocoa.

As soon as a gap occurs in the canopy, the cocoa may be invaded by capsids, plant-feeding insects with needle-like mouth parts which pierce the plant tissue and suck out the sap or cell contents. Once damaged by capsid attack the trees are prone to deep-seated fungal infection and while neither capsid attack nor fungal infection normally results in severe damage to the tree, in combination, their effect can be devastating (Johnson, 1962). Capsids have probably taken a far greater toll from West African cocoa than has swollen shoot because the former are more widespread. It has been particularly serious in parts of Côte d'Ivoire and along the savanna borders of the cocoa belt, where conditions are relatively open. In the early decades of Ghanaian cocoa farming, capsid attack was perceived by the farmers as the major threat to production (Austin, 1996).

Over the past 40 years there have been advances with spraying to control capsids but all too often spraying simply destroys part of the insect population which increases significantly later.

Black pod disease (*Phytophthora palmivora*) is a third major problem for cocoa producers. In contrast to swollen shoot, which is a virus and capsids, which are insects, black pod is one of the commonest pathogenic fungi in the humid tropics and its effects are worst where humidity is highest (Wharton, 1962). The first sign of black pod is a brown, necrotic spot on the husk which spreads rapidly. Fungicides can limit the spread of black pod but in the absence of these, crop losses may be reduced by cutting back shade above cocoa to lower the air humidity to a level unfavourable for development of the pathogen. Whether cutting out diseased trees, or parts of the tree, is the best method of disease control remains a contentious issue. Faulkner and Mackie (1933) claimed that such measures were of little value in practice and Nyanteng (1995) observes that farmers never accepted this as a solution. Many were hostile to Colonial Agricultural Officers who suggested that they should adopt this course of action (Dale, pers. comm., 1998). It was unpopular partly because it depleted the cocoa farm and also because it was labour intensive, and labour was a commodity constantly in short supply. Instead, farmers preferred to allow the cocoa farm to revert to bush, a means that is claimed to have achieved greater success than cutting out trees (Austin, 1996). Understandably there has been a reluctance by Colonial Agricultural Officers to acknowledge the comparative success of indigenous methods of disease control (Faulkner and Mackie, 1933). Trees with swollen shoot virus would have been unlikely to survive if the farm was left to revert to bush, but the build-up of weeds, shrubs, and climbers would, in theory, have impeded the movement of the vectors of the virus and hence would have limited the spread of swollen shoot. Attack by insects and fungus usually abated after the farm had been left unweeded for a minimum of 3 years.

According to Eshett *et al.* (1991), careful maintenance can help limit the spread of disease but this is often far from adequate, largely because labour is in short supply and access to sprays is both difficult and frequently beyond the purse of the smallholder. This is particularly so now that subsidies on inputs have been removed or at least reduced, with the result that in Ghana, Nigeria, and Côte d'Ivoire there has been a reduction in the use of sprays (Nyanteng, 1995; Konan, 1995; Chauveau, 1995). Thinning and pruning of mature cocoa are often neglected although on projects promoting the use of hybrid varieties such practices are encouraged.

## An alternative to artificial inputs

Where access to inputs such as fertilizers and sprays are limited, recent research has shown the benefits of tree-crop agroforestry involving cocoa (Ekanade and

Egbe, 1990). As yet there is remarkably little research data on the relative merits of different crop combinations (Egbe and Adenikinju, 1990) but results from initial trials are proving encouraging. Research conducted at the Cocoa Research Insititute of Nigeria (CRIN) has shown that interplanting cocoa and kola (*Cola nitida*) to a specified tree pattern could help to maintain soil fertility and could be adopted by farmers as it involves neither mulches nor chemical fertilizers (Ekanade and Egbe, 1990). However, the potential benefits of combining cocoa with kola does not necessarily hold true if the trees are placed too closely. Under these circumstances there may be competition between the surface feeding roots of the two crops and cocoa yields may also be depressed by overhead shade cast by kola (Egbe and Adenikinju, 1990). Interplanting can help reduce the spread of disease though Dale (pers. comm., 1998) observes that this does not always hold true as kola is susceptible to certain cocoa viruses.

Research at CRIN found that when cocoa is combined with oil palm in a particular pattern and spaced at particular intervals, yields of cocoa rose by as much as 67 per cent over monocropped plots. There is apparently little competition between the restricted, fibrous roots of the oil palm and the extensive superficial feeding roots of cocoa. The relatively high and open canopy of the oil palm does not cast as much shade over the cocoa as does kola, and these could be reasons for the relative success of the cocoa/oil palm combination. Egbe and Adenikinju (1990) note that contrary to the advice given by agricultural research stations, farmers in South Western Nigeria have long been intercropping cocoa with other tree crops such as oil palm, kola, rubber, and cashew. However, the benefits farmers obtained from intercropping cocoa and oil palm have been less than those realized by CRIN. This is attributed to the haphazard spatial arrangement of the intercropped species and frequently inadequate maintenance of the crop (Egbe and Adenikinju, 1990). Possibilities do exist for involving cocoa in more ecologically sound cropping systems which utilize accessible resources and which do not necessarily require capital inputs beyond the reach of most smallholders. Intensive extension work would be necessary to promote the adoption of such methods but there is scope for combining resources of aid donors with expertise of development practitioners and indigenous cultivators.

## Mixed fortunes of cocoa

In spite of ecological problems and far from perfect field techniques for the production of cocoa, smallholders have demonstrated considerable skill and initiative in establishing West Africa as the world's leading producer of cocoa. The second part of this chapter moves to a larger scale. It demonstrates that rather than enhancing the achievements at field level and assisting farmers with their problems, events at a national scale, together with ill-conceived political decisions, adversely affected cocoa production and seriously damaged a once thriving sector. Since the mid-1980s attempts have been made to revive cocoa production throughout the region. The situation in each of West Africa's main cocoa-producing countries, Côte d'Ivoire, Ghana, and Nigeria is examined separately as there are differences between them. Although Côte d'Ivoire is now the world's largest producer, a chronological approach will be adopted with Ghana being examined first, followed by Nigeria and finally Côte d'Ivoire.

## Ghana

Cocoa has played a major role in Ghana's economy, although its contribution has declined since the mid-1960s. Cocoa is still Ghana's main cash crop and accounted for just over 31 per cent of export earnings in 1997. However, it no longer retains its status as Ghana's outstanding, principal export. Its position is being challenged by gold (EIU 1998*a*, Country Profile). After the introduction of cocoa to Ghana in the late nineteenth century, uptake of the crop by smallholders was very rapid (Gordon, 1974) and from 1910 to 1920 the cocoa industry grew rapidly as the price paid for cocoa on the international market continued to rise. The large number of new cocoa trees that male farmers had planted prior to 1929 when prices were rising continued to bear into the 1930s, but with the marked fall in cocoa prices during the worldwide depression of 1929–30s farmers either could not afford to replant, or decided against replanting to increase future cocoa production on account of low prices. By 1936/7, the impact of low prices on production was apparent (Mikell, 1989). The decline was further exacerbated by the cocoa boycott of 1936/7 which was organized by the big cocoa farmers and the chiefs and involved the withholding of cocoa from the world market in an attempt to push world market prices up (Mikell, 1989). This did not succeed.

The producer price of cocoa was determined by competitive market forces until 1939, but the combination of fluctuating world prices, the cocoa boycott, and probably the desire by the colonial government for assured supplies of cocoa, led to the establishment of the Cocoa Control Board of West Africa. This was converted into the Gold Coast Cocoa Marketing Board in 1947, a state monopsony that purchased all cocoa produced in the country. Most of the cocoa merchants of the time were licensed as buying agents (LBAs) for the marketing board and purchased cocoa on behalf of the Board throughout the country. This system remained in operation until 1961 (Gyimah-Brempong and Apraku, 1987). The decline of colonial rule and the relaxation of regulations on access to forest land in much of West Africa were a stimulus to the increased production of cocoa (Clarence-Smith and Ruf, 1996). Higher prices for cocoa in the early 1950s encouraged the planting of cocoa and subsequently production increased.

The year 1961 saw vital changes in policy that adversely affected cocoa producers. The first change concerned the LBAs. Before 1961 the marketing board, by now the Ghana Cocoa Marketing Board (GCMB), authorized a number of commercial houses as LBAs to purchase cocoa from the farmers for a commission. To compete for farmers' cocoa the LBAs went to the farms where they paid cash at the time of purchase for the cocoa they bought. LBAs could not compete on price, but they could compete on non-price services such as the sale of inputs, the provision of consumer goods and particularly of loans, especially at times in the year when cash flow was a problem. In 1961 the licences of all LBAs were revoked and and one of them, the Cocoa Purchasing Company, was placed under the United Ghana Farmers' Council (UGFC), supposedly a farmer's cooperative, and made the only cocoa buying agent in Ghana. The UGFC was then made a wing of the ruling Convention People's Party (CPP).

This structural change in the marketing system had many effects: first, farmers no longer had access to the range of services offered by the LBAs. Secondly, the government began to tax farmers heavily, deducting 6*d*. per load national development tax on cocoa (Mikell, 1989). This marked the beginning of the transfer of funds from cocoa farmers to the industrial sector in Ghana. It was also announced by the government that cocoa farmers were making 'voluntary contributions' amounting to 10 per cent of the producer price of cocoa to the second development plan. In 1961 the UGFC announced that each farmer would be responsible for a 'compulsory savings' scheme equal to 15 per cent of the producer price of cocoa bought by the Council. Later in 1961 the UGFC executives announced that the compulsory savings had been given to the government as a gift from the farmers. The UGFC was abolished after the 1966 coup that overthrew Nkrumah but purchasing power remained with the monopolistic Cocoa Purchasing Company, later renamed the Produce Buying Agency (Gyimah-Brempong and Apraku, 1987).

A further disadvantage of the monopoly buying of cocoa was the introduction of the 'chit system'. Chits were IOUs given by the purchasing agent to the farmer at the time of sale, to be redeemed at a later date. Waiting periods proved to be long, between 6 months and a year, during which farmers suffered severe cash flow problems. Furthermore, farmers had to travel to cash their chits which could be a major inconvenience to small-scale farmers (Nugent, 1990). And if the chit was lost, not an uncommon happening, they received nothing at all. Gyimah-Brempong and Apraku (1987) note that these changes seriously disadvantaged the Ghanaian cocoa farmer. First, they were not familiar with IOUs and cheques and farmers were cheated after the introduction of the chit system.

In consequence, smaller cocoa farmers could not afford to acquire new land for cocoa while larger farmers expanded into maize production while maintaining their cocoa farms (Mikell, 1989). Furthermore, the real producer price of cocoa production fell in relation to food crops such as maize between 1958 and 1962. By the mid-1960s it was becoming economically more worthwhile for the farmer to grow and sell maize rather than to plant more cocoa (Kusi, 1991). Cocoa producers had come to depend on hired labour, much of it from the north, but low prices for cocoa put hired labour beyond the reach of many cocoa producers and instead, placed greater reliance on hard pressed family labour (Mikell, 1989). The situation was made worse by the overvaluation of the *cedi* since the mid-1970s. Some of the most immediate effects of an overvalued exchange rate were that export producers who used official channels were effectively taxed while importers benefited. This seriously distorted the comparative advantage formerly enjoyed by productive sectors such as cocoa (Kusi, 1991). One major outlet for cocoa farmers was through smuggling cocoa into neighbouring francophone states where prices were higher and where the CFA franc was convertible. Nugent (1990) argues that smuggling was a response not only to higher prices in Franc Zone states but because essential commodities were virtually unavailable in rural Ghana in the early 1980s and were more accessible in neighbouring states. Furthermore, Ghanaian farmers (in Ghana) were issued with a cheque for their cocoa which could only be redeemed at a bank. Rural banks were relatively inaccessible and after travelling considerable distances the farmer was by no means guaranteed that the bank could clear the cheque. This encouraged cross-border trade and many in the Volta region, for example, chose to sell their cocoa in Togo (Nugent, 1990). Such unofficial trade was seriously disadvantageous to Ghana's development.

Under the exchange controls of 1961, the allocation of foreign exchange was towards government and industry, with the result that essential imports such as insecticide were in very short supply (Gyimah-Brempong and Apraku, 1987). In the early 1960s the government also reduced its investment in cocoa, and research and extension services were cut back. Disease spread through the cocoa forests, substantially reducing production. The government was to provide free seedlings, extension services, and loans to farmers, but producer prices were not increased and turning former cocoa farms over to food appears to have become a popular move. By 1965 it was reported that virtually all new planting of cocoa had ceased (Gyimah-Brempong and Apraku, 1987).

In an attempt to reverse the decline of cocoa, the government of Ghana launched a programme to rehabilitate the cocoa sector under the first phase of its Economic Recovery Programme (ERP I) which commenced in 1983. Funded partly by financial assistance from the World Bank, farmers were given cash incentives to plant cocoa; producer prices were increased significantly and three million hybrid cocoa pods were distributed to private cocoa farms to ensure the planting of seed of good quality. Attempts were made to improve the supply and availability of fertilizer, insecticide, and pesticide and facilities for transporting the crop were improved. In spite of this many farmers chose to plant food crops. Droughts in the 1970s and in 1982–3, the effects of bush fires, and the return of over one million Ghanaian migrants from Nigeria resulted in food shortages in 1983 and confirmed in the minds of many that the cultivation of food crops was a safer option than growing cocoa (van Buren, 1999). Food shortages saw prices escalate and in 1983 cocoa production fell to an all-time low of 160,000 tonnes, a decline of over 65 per cent on 1971 levels (derived from FAO statistics, 1996). Despite the ERP, the problems of the cocoa sector persisted: ageing trees, shortage of inputs, the high cost of labour and services, and comparatively low returns saw the focus on food crops maintained. De Frece (pers. comm., 1998), working in the Nkwanta District of the Volta region of Ghana in 1997/8, confirmed that many farmers were still of the view that planting food crops was preferable to cocoa, their experience of the latter having been so bitter. Attempts to revitalize cocoa production continued under ERP II (1987–90) and a target of 350,000 tonnes per year has come close to being realized not least because rainfall has been favourable. Between 1991 and 1996, production averaged 300,000 tonnes a year, the 1995 crop reaching 404,000 tonnes (EIU 1998*a*).

Commitment to higher producer prices has been difficult to maintain because of a decline in world market prices, though the Ghana government is due to raise the domestic price of cocoa to 60 per cent of the world market price by 2000 (EIU 1998*a*). Currently it resides at 56 per cent. Output of cocoa has increased but abolition of subsidies on fertilizer has seen a decline in their use since the beginning of the 1980s (EIU, 1998*a*; FAO, 1995*a*). In addition, high interest rates have discouraged farmers from seeking credit from commercial banks (van Buren, 1997).

The mid-1980s also saw the reorganization of the Cocoa Marketing Board as COCOBOD. In 1993, under a decade later, the heavily overmanned and inefficient COCOBOD was deprived of its monopoly over the internal marketing of cocoa and its labour force was substantially reduced in line with the demands of Structural Adjustment. Although COCOBOD has retained its control over Ghana's cocoa exports, three trading companies have now been licensed to purchase cocoa directly from the farmers and private buyers now account for 30 per cent of purchases (EIU, 1998*a*). History seems to be repeating itself with regard to marketing but whether the production trends of earlier years can be replicated remains to be seen. Recent trends are promising. In 1995 cocoa achieved the highest rate of growth of any agricultural subsector (11.1%). In 1996, export earnings from cocoa reached $508.6 m, almost 20% above the target (van Buren, 1999). Whether it is reasonable for the Ghana government to invest so heavily in the rehabilitation of cocoa when supply currently exceeds demand on the world market is also questionable (Fig. 4.1). There will be much scope for increased production once the cocoa processing industry in less developed countries takes off but as yet, this stage has not been reached.

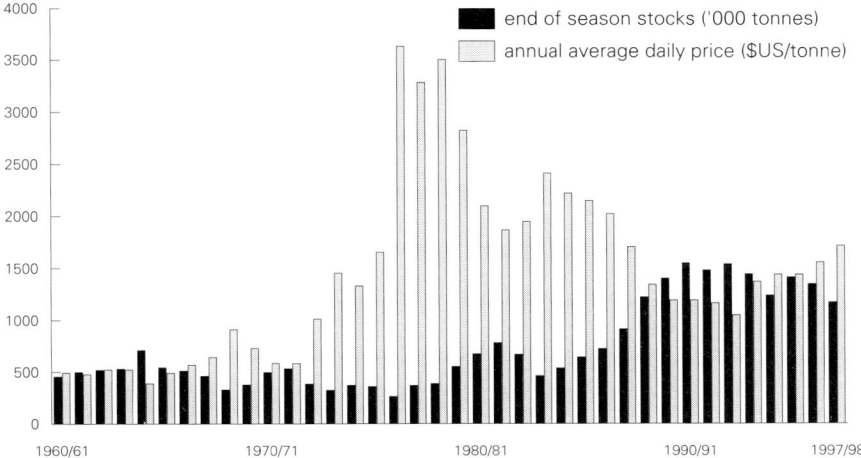

**Fig. 4.1.** Cocoa—annual average daily price and end of season stocks

*Source*: Derived from ICCO (1997/98) World Cocoa Bean Position, summary data 1960/61–1997/98, *Quarterly Bulletin of Cocoa Statistics* 24 (3), Table 1.

## Nigeria

The pattern of decline in cocoa in Nigeria bears similarities to Ghana and in Nigeria can be divided into two main phases, the 1960s and the period from 1970 to 1985. Post-1985 there have been attempts to rehabilitate the cocoa growing industry.

After independence in 1960 funds available to Nigeria's new regional governments for their development programmes were smaller than had been anticipated and in the Western Region, the country's main cocoa growing region, the considerable resources of the cocoa marketing board represented a potential source of wealth to the newly independent country. The regional government increased its access to these funds through the 1960s with the result that the marketing board was no longer able to support prices and farmers had to bear the brunt of price fluctuations such as a fall in price of £50/ton in 1961–2 (Bates, 1981). In spite of this, by the end of 1970 conditions were not as bad as they were to become for cocoa farmers. Producer prices were not high but real wages were still low and a range of subsidized inputs were available to producers. Estimates suggest that returns to cocoa farmers were around 12.9 per cent, which for many farmers meant that cocoa was still a viable proposition (Dorosh and Akanji, 1988).

From 1971 conditions in the cocoa sector deteriorated and production continued to decline as the oil boom progressed. Ecological problems were the ultimate cause of declining production and according to a study by Dorosh and Akanji (1988) in ten villages in Oyo state, more than 25 per cent of cocoa trees were over 40 years old and well past their peak. Unproductive trees were not being replaced (Dorosh and Akanji, 1988). Added to this was the spread of disease in Nigeria, in particular, black pod. Two factors contributed to this: the use of chemical sprays against black pod had declined as inputs became increasingly difficult to obtain, and the labour force needed to do the spraying and to cut out infected wood had also declined as urban areas beckoned them away from rural life. Migrant labour available for hire in most villages was used in peak periods of demand but the cost of this reduced returns from cocoa. The more aged farmers remained in the villages and, unable to cope with the physical demands of cocoa, left many fields to return to bush. In this way the cocoa was 'disinfected'. Many such 'wild' areas have continued to be harvested though they are not as productive as well-maintained cocoa.

But the question is why, when cocoa had been so successful, did the quality of management decline? In answer to this it is essential to look more closely at certain macro-economic, trade, and price policies which combined to discourage cocoa production in Nigeria.

Major increases in dollar oil revenues in the 1970s and 1980s led to a sharp increase in inflation, as these revenues were spent in the domestic economy. With continued access to foreign exchange, there was no pressure on the Nigerian government to devalue the *naira*. High domestic inflation, together with constant and sometimes appreciating nominal exchange rates, resulted in an appreciation of the real exchange rate, especially after the second major rise in world oil prices in 1979 and 1980 (Dorosh and Akanji, 1988). The appreciation in the real exchange rate of the *naira* had the effect of reducing prices of tradeable goods internally, relative to non-tradeable goods. Whereas the price of tradeables is largely determined by the exchange rate and world market prices, the price of non-tradeables rises at roughly the rate of overall domestic inflation. The result is stagnation in the tradeable goods sectors in the midst of an apparently booming economy, a situation known as 'Dutch disease'. We now look at the effect of overvaluation of the *naira* on the real price of cocoa and the viability of the Nigerian Cocoa Board. Much of the detailed information below is drawn from the work of Dorosh and Akanji (1988).

In the early 1970s the Nigerian Cocoa Board was able to operate profitably because there was a sufficient margin between the producer price and the world market price of cocoa. Between 1970 and 1974, the world market price was nearly 50 per cent higher than the producer price in Nigeria. World cocoa prices rose sharply in 1976 so that the appreciation in the real value of the *naira* by about 40 per cent between 1975 and 1977 did not lead to a fall in real producer prices. Instead, nominal producer prices were increased at the level of inflation and so real producer prices fell only slightly between 1976 and 1979. Most of the direct benefits of higher world prices were secured by the government, as the marketing margin between producer prices and world market prices more than doubled to an average of N3,244/tonne (at 1985 levels) (Dorosh and Akanji, 1988). When cocoa prices fell in the 1980s to levels equivalent in real terms to those of the early 1970s, the margin between the buying price of the Cocoa Board and its selling price fell to N290 in 1980 and was actually negative in 1981 and 1982 (Fig. 4.2). In these years the Nigerian Cocoa Board was buying cocoa for more than it could get for it in London or New York. The Board's position improved slightly with the small increase in the world market price of cocoa in 1984 and 1985.

With nominal producer price increases limited by the world market price, converted to *naira* at an increasingly overvalued exchange rate, the real price of cocoa paid to producers fell by nearly 50 per cent from 1979 to 1985. Had the currency not been allowed to appreciate to such an extent and had it been held at 1970 levels, the US dollar value would have translated into considerably more *naira* and the Nigerian Cocoa Board could have maintained a profit, even in the years when the world market price fell, and nominal producer prices could have been maintained. By the end of 1985, low real producer prices for cocoa and high real wage rates did much to discourage futher planting of cocoa or other forms of investment in the crop. Even improved technology did not increase returns significantly (Dorosh and Akanji, 1988).

Many farmers avoided some of the worst effects of the overvalued *naira* by smuggling their cocoa into the neighbouring Franc Zone country of Benin,

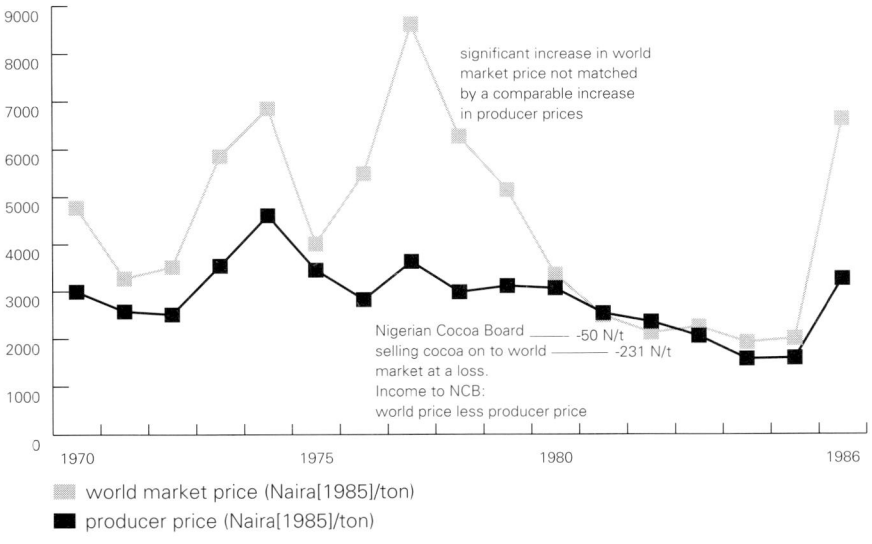

**Fig. 4.2.** World market price of cocoa and producer price in Nigeria, 1970–86
*Source*: After Dorosh and Akanji (1988), 30.

where prices they received were considerably higher than in Nigeria. Attracted by the oil wealth, many migrated to urban areas leaving the villages short of labour. Unlike Ghana, there appears to have been no significant move out of cocoa and into food crops (Dorosh and Akanji, 1988). One possible reason was that the overvalued *naira* enabled the import of foodstuffs very cheaply, hardly making it worth the while of many farmers to grow food for market in Nigeria where the cost of both hired labour and of transport was very high, having risen in line with inflation. What was evident in Oyo was that in the cocoa fields that were being maintained farmers did not replant trees when they died; instead, food crops were grown in the gaps. Food crop production probably did increase, but it stayed 'at home', avoiding the worst effects of rampant inflation in the economy. The consumer price index, set at 100 in 1970, reached a level of 992 by 1985 (IMF, cited in Dorosh and Akanji, 1988).

In 1986 two major changes resulted in a doubling of the producer price. First, the Nigerian Cocoa Board was abolished, though only formally in 1987. Secondly, a devaluation of the *naira* resulted in a fourfold increase in the *naira* price of cocoa. These changes were in line with recommendations by the World Bank in their Structural Adjustment Programme which Nigeria had not yet adopted but which it was following closely. According to Dorosh and Akanji (1988), had the Nigerian Cocoa Board not been abolished, devaluation of the *naira* would not necessarily have led to an increase in the producer price of cocoa because the government had a monopoly on exports and set the producer prices itself. Equally, abolishing the Cocoa Marketing Board without devaluing the *naira* would not have provided profitable trading opportunities at 1985 prices. Both were therefore necessary to raise the producer price of cocoa.

Since 1986 a virtual doubling of domestic cocoa prices has given some incentive to producers (Synge, 1997), but reviving cocoa production is not as simple as all that. Nigeria has done little to promote smallholder agriculture during much of the independence period and cocoa remains the only agricultural export of significance in the country. It now represents less than 1 per cent of the total value of exports (EIU, 1998*b*). Resurrecting cocoa is proving far from easy as farmers have too many bitter memories of the past. Farmers need to be confident of receiving adequate prices in the future to ensure that their investment is worthwhile. Even with price guarantees there are problems as land is not as easy to come by as it was in the past. The effects of ageing trees, widespread disease, the eternal problem of accessibility to essential inputs and now their increased cost with the removal of subsidies is likely to further impede the revival of cocoa (Synge, 1997). Unlike Ghana where farmers have replaced cocoa with food crops on a significant scale, in Nigeria the replacement of cocoa with food crops has been far less pronounced.

Major technical problems with replanting are something which affects all cocoa-producing countries. During the life of a cocoa plantation little is done to maintain the quality of the soil and usually, cocoa does not thrive on land formerly planted with cocoa (Jarriage and Ruf, 1990; Ruf, 1991). If replanting is to be encouraged, much effort needs to be invested in replenishing the decline in the physical and chemical properties of the soil resulting from some 30 years of cocoa cultivation before the new crop is planted. Also it is evident that many of the areas initially planted with cocoa were far from the best areas for the crop as soil conditions were not appropriate. Such areas have not yielded well even at a relatively early age and hence should not be replanted. There is, nevertheless, much land that could be planted with cocoa and with the technical expertise available, it should be possible.

## Côte d'Ivoire

Côte d'Ivoire is the world's leading producer of cocoa having taken over from Ghana in 1976/7. Currently, cocoa accounts for almost 45 per cent of the country's export earnings (EIU, 1998c). Although cocoa cultivation had a false start at the end of the nineteenth century in the south-west of Côte d'Ivoire, with the assistance of the French Colonial government, the south-east of the country became the focus in the early twentieth century and from there cultivation moved westwards. Increased planting in the forests of the centre and south-west since independence has been encouraged by the government and has been responsible for the vast increase in production (Chauveau and Léonard, 1996).

The cocoa sector which grew so rapidly after independence has been severely affected by the collapse in commodity prices since 1980. With the exception of price rises in 1986 cocoa prices have fallen from an annual average price of $US 3,632 per tonne in 1977, to $US 1,051 per tonne during the 1992–93 season (International Cocoa Organization 1997/8). (The average daily price has since risen to $US 1,556 per tonne in 1996/97 (International Cocoa Organisation 1997/8).) In addition to the effects of collapsing prices, ill-fated decisions taken by the Ivorian government have adversely affected cocoa producers. There is much discussion about the crisis in the cocoa sector and there is grave concern that the future for cocoa in the next 20 to 30 years in Côte d'Ivoire is bleak (Chauveau, 1995; Chauveau and Léonard, 1996; Konan, 1995; Léonard and Oswald, 1995; Ruf, 1995). The reasons for the problems in the sector are reviewed below. At the same time, however, cocoa production is relatively buoyant at the moment and the question remains: is this a response to good rains in recent years, or has cocoa production increased because of increased investment in cocoa due to a lack of economic alternatives?

Crucial to the story of cocoa in Côte d'Ivoire is *Caistab* or the *Caisse de stabilisation et de soutien des prix des produits agricoles*. The *Caisse* was an extremely powerful body. At a national level it worked closely with the Presidency and the Ministries. It appointed 30–50 big companies, some Ivorian, some Lebanese, but most French, to export the cocoa crop each year.

At a local level the *Caisse* also played an important role. Together with the *prefets* from the *Administration Préfectorale* who were responsible for policing the cocoa marketing system with officials from the Ministry of Agriculture and with members of the farmers' cooperatives, the *Caisse* appointed traders and bush buyers, each of whom was allocated a *carte d'acheteur* in order to operate in specific prefectures. In addition, every year the *Caisse*, in consultation with the Presidential office, would fix producer prices for cocoa throughout the country and produce an officially determined export reference price system for both cocoa and coffee. Exporters were obliged to pay the *Caisse* if they got more than the official export reference price. Equally, the *Caisse* was duty bound to reimburse exporters if prices fell below the reference level. By setting the producer price and controlling the activities of the export houses, the *Caisse* had sufficient influence to direct the marketing of cocoa (and similarly coffee) without ever handling the crop (Crook, 1990; Curtis *et al.*, 1987: 64). This represents a marked difference from the anglophone states where marketing boards were normally the sole buyers of the crop.

The Ivorian export houses were expected to export only cocoa of good quality, though standards have fallen short of those in Ghana where a great deal of attention has been paid to quality control. It should be noted that Ghana too has had problems with quality control since the restructuring of the marketing sector. As long as the *Caisse* continued to meet its obligations, there was no way that traders could lose. In addition to supporting producers and traders at all levels, government support was guaranteed to Ivorian transport enterprises which conveyed crops from village to port. The system worked well in the years when world cocoa prices were relatively high, largely because the state system ensured that it did. In boom years the surpluses from cocoa sales accumulated in the government coffers but as in Ghana and Nigeria, although 60 per cent of profits from cocoa sales should have been added to the stabilization fund, this rule was regularly ignored. This undoubtedly increased the vulnerability of cocoa producers, because instead of the *Caisse* retaining surplus funds with which to stabilize producer prices in the event of a decline in world prices, they were spent on a variety of development schemes—many of them quite successful but unrelated to cocoa production (Crook, 1990).

The system operated successfully for many years and took Côte d'Ivoire to first place among the world's producers. However, conditions deteriorated from 1985/6 when it was decided to maintain producer prices at CFA400 $kg^{-1}$ for both cocoa and coffee right through the slump in world prices. Given the obligations of the *Caisse* to maintain traders' margins, it quickly went into deficit with world market prices so low. The *Caisse* became indebted to the export houses, a debt which filtered back to the traders who bought cocoa from the producers and sold it to the export houses. This led to traders paying producers less than the official price for their cocoa, which was illegal, and where they were stopped from doing this, they simply did not collect the cocoa at all. The situation was made worse by Houphouët Boigny, the then Ivorian

President, who attempted to push world market prices up by withholding Ivorian cocoa from the international market. This was unsuccessful and the result was depression in the cocoa market as the *Caisse* refused to allow cocoa to be moved and whatever was moved was piled up at depots near the ports. Finally, a French company, Sucres et Denrées, bought Côte d'Ivoire's cocoa for an undisclosed sum, allegedly subsidized by the French government. By this stage the farmers could no longer be protected and the *Caisse* was forced to reduce the producer price for cocoa to CFA250 kg$^{-1}$ in 1989. Even so the total cost of a kilo of cocoa to the *Caisse* which included costs beyond the producer price, meant that it was still selling at a loss and prices were reduced again to CFA200 kg$^{-1}$ at the onset of the 1989–90 main season (Crook, 1990).

In reality, these changes were not as precipitate as the the prices and dates above suggest, as the situation in the field had been deteriorating for some time. Chauveau and Léonard (1996) show that between 1988 and 1992 the farm gate price of cocoa fell by nearly one-third of former levels owing to organizational problems. Many farmers could not shift their cocoa crop in 1988 or 1993 owing to disruption at every level of the marketing chain, and by the mid-1990s, they faced a reduction of 60 to 80 per cent in their monetary income. Although the *Administration Préfectorale* was intended to police operations in the cocoa market, they were powerless at the time when the *Caisse* became indebted. Farmers and traders resorted to accepting lower than official prices, promissory notes, and anything they could to earn a return on their cocoa. The insolvency problem of the *Caisse* was overcome by loans from the World Bank and from France which enabled the *Caisse* to survive for another year. However, as the World Bank increased its control over the operations of the *Caisse*, so it reduced its power in accordance with the free market principles which pervade Structural Adjustment Programmes throughout Africa.

In the latter part of the 1980s disillusionment grew among those who had any connection with the the cocoa market and while this may be attributed to errors in high level decision-making, other factors contributed to the problems of the Ivorian cocoa sector. First, as in other parts of West Africa, cocoa trees are growing older and the productivity of many is declining. In addition to age, much cocoa has been planted in unsuitable soils, in itself a cause of premature ageing of trees and reduced production. According to Jarriage and Ruf (1990), this is already true of many plantations in the south-west of Côte d'Ivoire. Nevertheless, the area planted to cocoa has continued to expand in the far south-west of the country where land is still available (Hodgkinson, 1997; Chauveau and Léonard, 1996).

As returns to cocoa farmers have declined since the mid-1980s, so phytosanitary problems have worsened as resources to invest in plant protection have been limited. Furthermore, cocoa has been severely hit by labour scarcity particularly as the Burkinabé labour force in Côte d'Ivoire has declined. Without adequate labour to weed, to harvest the crop at the correct times, and to cut out infected wood, the health of the trees has declined and with it, yields.

According to Jarriage and Ruf (1990), out of 1.5mn ha of cocoa in Côte d'Ivoire, around 50 per cent are dependent on non-family labour, formerly, much of it Burkinabé. Economic problems in the cocoa sector have seen the heightening of ethnic tensions the country. In the past, the Burkinabé supplied labour to the Baoulé, the pioneers of cocoa production in Côte d'Ivoire, and in consequence, the Burkinabé were second to them as producers. In view of the increased cost of labour, the Burkinabé now have an advantage as they control a significant portion of the labour supply. Furthermore, the Baoulé method of expanding production was by extending the area under the crop. With forest land diminishing, the Burkinabé show signs of developing more intensive methods of cocoa cultivation and are thus strengthening their position. Tension between the two groups is inevitable (Chauveau and Léonard, 1996).

Visions of the future for the cocoa sector in Côte d'Ivoire conflict. On the one hand the internal problems of a shortage of new land for the expansion of cocoa cultivation, the ageing of trees, ageing of the farmers, the shortage of inputs, and the problems of replanting suggest a bleak future for the cocoa industry. On the other hand, production has risen since the 1980s. Production of around one million tonnes a year leaves Côte d'Ivoire the current undisputed world leader. Increased production since the mid-1980s has been due to the opening up of new forest but also the conversion of coffee to cocoa (Hodgkinson, 1997). There is also evidence of replanting in fallow land and the reuse of old plantations (Chauveau, 1995; Chauveau and Léonard, 1996). With more intensive methods of land management and the increased use of hybrid cocoa, production, in theory at least, could be sustained. However, as Berry (1975) has argued with regard to Ghana, investment in cocoa when market conditions are not favourable, may merely reflect the lack of economic alternatives. A further problem in the future concerns the supply and demand for cocoa at a global level.

## Future prospects for cocoa

### Balancing supply and demand

A major advantage of cocoa producers is that cocoa is a product of the humid tropics and that it cannot be produced in the temperate zone, where demand for it is greatest. A second advantage is that a viable alternative to cocoa has never been found, so it has not suffered the same fate as some of the tropical oil seeds. Carob, for instance shows no signs of achieving the success of cocoa. Over the past 20 years, consumption of cocoa beans has grown at an average rate of 1.88 per cent per year, though the trend has been far from smooth. According to Curtis *et al.* (1987), consumption cannot outstrip production for long, unless large stocks exist. For producers, it is desirable that consumption increases but at the same time, a structural surplus helps to hold prices down. Cocoa prices have been volatile historically (see Fig. 4.1). Periods of oversupply have been followed by periods of under-supply, each phase lasting three to four years. The comparatively low prices of the 1990s have been much related to structural oversupply, though other factors contribute to price instability as well. For example, there has been uncertainty about the effects of El Niño on the 1997/8 cocoa crops in Ecuador and Malaysia. During the first part of 1998 the world market has also been affected by uncertainty about the current economic and political instability in Indonesia and, more locally, over the sales strategy of farmers in Côte d'Ivoire which will follow the privatization of cocoa marketing in October 1999. Unexpected demand for chocolate and cocoa products earlier in the year, followed by rumours of a late start to the 1998/9 main crop in West Africa, saw the market rise. However, improved predictions for the 1998/9 main cocoa crop in West Africa and the prospect of supply continuing to exceed demand, saw the market fall again. Several Member States of the European Union have agreed to the EU proposal to allow the use of 5 per cent non-cocoa fat in chocolate, thereby reducing the amount of cocoa required. Although this is being resisted by many EU members, in a single market, where goods are allowed to circulate freely, such legislation could influence demand patterns for cocoa and cocoa products (EIU, 1998*c*; International Cocoa Organization, 1998; Horner, 1994). Reports suggesting that the Netherlands Cocoa Association had now dropped its objection to the EU proposal have recently seen the market continue its bearish trend (EIU, 1998*c*; International Cocoa Organization, 1998).

Volatility of cocoa prices has done little to help the economic situation in producer countries and in 1973, the International Cocoa Organization (ICCO)[1] was established to stabilize the market by limiting prices to a range agreed between the major producing and consuming countries. There have now been five International Cocoa Agreeements (ICAs). Under the first two agreements, producing countries were allocated export quotas which matched as closely as possible the import requirements of the consuming countries. Prices for these quotas were to be contained within an agreed price range though there has been little success in agreeing a minimum price which is universally acceptable. At one extreme is the USA which is the world's largest consumer of cocoa and is anxious to keep prices as low as possible, while at the other extreme is Côte d'Ivoire, the world's largest producer whose objectives are exactly the opposite. In additon to quotas, the ICCO, the body implementing the Agreement, was to establish a buffer stock of 250,000 tonnes to absorb any production that exceeded the quotas. The Agreement also required that its members should not trade cocoa with non-signatories to the Agreement. In contrast to its predecessors, the fifth ICA aimed to stabilize prices not through the operation of a buffer stock and export quotas, but by regulating and

---

[1] ICCO is the acronym adopted by the International Cocoa Organization. It is thus distinguished from its near neighbour the International Coffee Organization (ICO).

promoting consumption. As the conclusion of the fifth ICA draws near, the market remains oversupplied with cocoa (Brown, 1997).

Cocoa Agreements have been hampered in their attempts to stabilize prices by a number of factors, not least the opposing attitudes of producers and consumers (Brown, 1997). Conflict has raged around the size of quotas, the management of buffer stocks and the payment of levies by members. Crop and stock surpluses have continued to depress the market in the 1990s (Fig. 4.1), the ICA being powerless in its attempts to balance production and consumption because not all producers and consumers are members. For example, the USA, the world's largest consumer, is not a member, nor is Indonesia which continues to increase its production and supply on to the world market (Brown, 1997). But why, if prospects for cocoa are so poor, are farmers producing more? Price clearly has an influence, though as Berry (1975) observes, there have been periods in Western Nigerian history when investment in cocoa farms proceeded at a high rate despite a steady downward trend in the real price of cocoa. On the other hand, Clarence-Smith and Ruf (1996) argue that increased prices have always seen cocoa prooducers respond positively. If cocoa production is increasing, it may be, as Berry suggests, that few alternatives are open to investors.

## Perceptions of a bleak future for West African cocoa

The long-term future of West African cocoa does not look particularly hopeful. The world market is oversupplied with cocoa and whether West Africa will be able to maintain its market share seems doubtful as South-East Asian producers come much more into prominence (Table 4.1) (Newsletter, Tropical Agriculture Association, 1992; Ruf, 1991, 1993, 1995, 1996). Investment in cocoa has increased in West Africa since the mid-1980s, when producing countries adopted Structural Adjustment Programmes, but whether economic measures are sufficient to enable Côte d'Ivoire to remain the world's leading producer and other major producers in West Africa to increase their output is questionable, when some of the most pressing problems in the cocoa sector are ecological and social.

Ruf (1991, 1993, 1995, 1996) and Jarriage and Ruf (1990) are extremely pessimistic about the future of West African cocoa and largely for ecological and social reasons. Ruf's model shows the movement of centres of production first within individual countries, and then from one country to another and one region to another. Cocoa originated in South America where Brazil and Colombia have remained major producers, but after being introduced to West Africa, this became the next major centre of production, first with Ghana dominating the world's producers, and now Côte d'Ivoire. According to Ruf (1991, 1993, 1995) the 'golden age' of cocoa in West Africa is virtually over and the next world leader is likely to be Indonesia—as long as disease can be controlled and the deterioration of cocoa beans can be halted in the very humid conditions. Although this replacement of West Africa by South-East Asia as the world's dominant producing region is not imminent, Ruf notes the decline in production from a peak in all the major producing countries in West Africa except the Côte d'Ivoire. It is assumed by Ruf that Côte d'Ivoire will follow the same pattern as its neighbours, shortly after the beginning of the twenty-first century.

According to Ruf, three major factors determine the success of cocoa at any time: first, there is what Ruf calls the 'forest rent', or the benefits of the forest to cocoa. Secondly, is the value of the tree as a capital resource, and thirdly, is the life cycle of the growers, together with their resources. First, forest rent. When cocoa is planted in recently cleared forest, it benefits from the forest environment (Ruf, 1991). According to Ruf, much of the physical protection afforded by the forest has been lost owing to extensive forest clearance, not only for cocoa but for other crops and for logging. Whether forest clearance has led to increased desiccation, or whether this is due to a trend of reduced rainfall over much of the region is not clear, but along the northern fringes of the cocoa belt, the dry season has been even more marked in certain years and the survival of cocoa has been difficult partly because of moisture stress and partly because of damage from fire.

Secondly, the tree may be regarded as a capital resource. Like all living things the cocoa tree has a finite life and its value as a resource is only as great as its capacity to produce cocoa. Ruf (1993) argues that trees planted in optimal conditions such as those in Bahia, North East Brazil, can remain productive for as long as 50 years. In West Africa, the productive life of a tree is frequently less. In the haste to increase the area under cocoa, seedlings have frequently been planted in suboptimal conditions with the result that the trees are of poor quality and are likely to suffer premature ageing and die even before reaching 25 years. In south-west Côte d'Ivoire, for example, cocoa planted in gravelly soils where there is no forest protection is dying after only 15 years (Ruf, 1993). Declining yields can to some extent be reversed by the use of fertilizers and insecticides and pesticides, though the use of these is still limited. Assuming that a plot is planted to cocoa at one time, then all the trees, which are the capital resources of the farmer, are likely to cease production or go into a phase of lower production at much the same time. Replanting of former cocoa fields is fraught with technical problems and so areas of production tend to move from one place to another. As forest land for conversion to cocoa becomes increasingly limited, the future for cocoa in West Africa is perceived to be bleak.

Thirdly, the age and resources of the planter. Farmers who planted cocoa when they were around 25 years old may be disinclined to replant when they are 45 to 50 years old. First, their age is a disincentive, secondly, a substantial part of their profits may have been invested in the education of their children who have left farming. Thus the age of cocoa farmers is on average quite high and this is a further reason why smallholder production of cocoa will continue to decline. Ruf argues that there are relatively few 'young farmers' in cocoa in comparison with 30 years ago. Higher producer prices could be an incentive to younger people, but Ruf (1991) notes that scarcity of suitable land may well restrict expansion.

But West Africa's position as the world's major cocoa producing region is not lost yet. Since the adoption of Structural Adjustment Programmes, incentives for farmers to plant cocoa have been greater than for many years and there is evidence that the area under cocoa is expanding and that production is increasing. The prospects of making cocoa a more sustainable form of landuse are being investigated and in Nigeria (Kolade, 1991) the value of interplanting cocoa with crops such as coconut, kola, and oil palm are being studied and have been outlined above. Farmers in Côte d'Ivoire are beginning to replant cocoa in fallow land, often cocoa fields abandoned for at least 10 years (Chauveau and Léonard, 1996; Ruf, 1991). Although freshly cleared forest land is preferable, these attempts represent an important move by smallholders to transform cocoa cultivation into a sustainable system. Because the problems of sustainability are largely ecological, smallholders would be greatly helped by improved access to credit and to essential inputs. These changes are, however, very limited and according to Ruf (1993) are unlikely to avert what seems an almost inevitable decline in cocoa production in West Africa in the future. Whether Ruf's historical model is applicable to West Africa is debatable. In times past there has been little concerted effort to regenerate cocoa, but there is now. Ghana may have declined from its peak of production in the mid-1960s but there is evidence that production of cocoa has been climbing. Whether the attempt to revive cocoa succeeds will depend very much on whether profits to the farmer are adequate and assured. Under such circumstances it is possible that by combining the skill of the farmer with technological achievements, the problems of replanting cocoa farms with cocoa may be overcome. Were this to be the case, Ruf's historical model may be unduly pessimistic. However, perceptions of a bleak future for cocoa producers are also shared by some Economists. Koester *et al.* (1989) argue that World Bank policy of increasing exports of crops such as cocoa may be self-defeating, leading to large, price depressing effects, particularly where a commodity is of prime importance such as cocoa. While this has been dismissed by the World Bank, it has been shown by Koester *et al.* (1989) to have some validity. Finally, it should not be forgotten that a potentially enormous market for cocoa and cocoa products exists as and when consumption increases in the developing world.

## Conclusions

It was shown in Chapter 1 that environmental conditions in the humid zone were unpredictable, at least with regard to rainfall. Means were virtually meaningless, deviations about the mean often were considerable and highly varied, giving weight to the argument that ecological equilibrium does not exist. Uncertainty exists with regard to numerous environmental variables. The presence or absence of hard pans in the soil is one example of an abiotic variable which has greatly influenced the success of cocoa cultivation. Able to cope with hard pans when they are near the soil surface, farmers, owing to their limited technology, have often been unable to detect the presence of hard pans a metre or so from the surface. The presence of such pans can impede the growth of the cocoa tree. So, while it has been argued that farmers cope well with a difficult physical environment, skilfully making the best use possible of scarce resources, it might also be argued that the success of cocoa cultivation is directly proportional to available resources of land, labour, and capital.

Cocoa is an excellent example of how farmers who had the incentive to succeed adapted their autecological methods to a difficult physical environment and to limited resources, and managed to grow the crop sufficiently well to earn a living. This chapter shows how ill-fated political decisions have wrought havoc in an arguably thriving cocoa sector in West Africa and have seen its decline since independence. Essentially, the physical ecology and the political ecology of cocoa have not been integrated to any degree over much of the past 40 years.

The current wisdom, urged initially by the World Bank, is trying to reverse earlier political decision-making in order to stimulate a revival in the cocoa sector. Raising producer prices has been perceived as one method of reviving cocoa—along with other crops. While political attempts to control the development of resources in the cocoa sector may now appear more producer friendly, they may not be sufficient to generate new growth. Farmers in Ghana, Nigeria, and Côte d'Ivoire have bitter memories of the changes wrought in the cocoa sector. Furthermore, circumstances have changed at field level: many of the farmers who developed the cocoa sector are too old to start again. Their children, many of whom were educated out of cocoa farming, are reluctant to return to cultivation of the crop, not least because of its history. Fifty years ago cocoa was a new crop to the humid zone, but now that the trees need to be replaced, regeneration of cocoa is proving difficult. It is not beyond the wit of indigenous farmers to overcome problems of physical ecology and now that aid to assist with such problems is forthcoming, a solution seems highly probable. As Hasan (1995) argues, cocoa replanting will only begin to reap benefits from technical progress when deforestation is no longer an option in cocoa cultivation. The environment in which the cocoa industry developed over the past century cannot necessarily be re-created by the collection of political changes now being instigated. Nevertheless, there is evidence that cocoa production has increased since the mid-1980s when many such reforms were introduced. These claim success but whether farmers are responding to a more favourable economic and political environment or whether increased cocoa production is the result of higher rainfall in the region in the 1990s, is still difficult to discern.

For a sustained revival, it is of critical importance not only that technical problems of cocoa cultivation are resolved at field level but also that political decisions are of such a nature that farmers are confident that their investment will receive adequate reward. In short, the physical and political determinants of cocoa cultivation, operating at different scales, should be integrated, or at least linked, to ensure that cocoa can be grown successfully and can be sold on for profit. If farmers can be motivated to increase cocoa production, Ruf's pessimism might be misplaced. If not, it seems highly likely that predictions of West Africa being superseded by South-East Asia as the world's largest producer of cocoa in the next 15 to 20 years, will be borne out.

## References

ADAMS, C. D. and BAKER, H. G. (1962), 'Weeds of cultivation and grazing lands', ch. 22 in Brian J. Wills (ed.), *Agriculture and Land Use in Ghana* (London: Oxford University Press), 402–15.

AHN, PETER (1970), *West African Soils* (London: Oxford University Press).

AMANOR, KOJO (1994), 'Ecological knowledge and the regional economy: Environmental management in the Asesewa District of Ghana', in Dharam Ghai (ed.), *Development and Environment: Sustaining People and Nature* (Oxford: Blackwell).

AUSTIN, GARETH (1996), 'Mode of production or mode of cultivation: Explaining the failure of European cocoa planters in competition with African farmers in Colonial Ghana', ch. 9 in W. G. Clarence-Smith (ed.), *Cocoa Pioneer Fronts since 1800: The Role of Smallholders, Planters and Merchants* (Basingstoke and London: Macmillan), 154–75.

BATES, ROBERT H. (1981), *Markets and States in Tropical Africa: The Political Basis of Agricultural Policies* (Berkeley: University of California Press).

BERRY, SARA S. (1975), *Cocoa, Custom, and Socio-Economic Change in Rural Western Nigeria* (Oxford: Clarendon Press).

BOURKE, GERALD (1988), 'The cocoa conundrum', *Africa Report*, Sept.–Oct.: 29–31.

BROWN, RICHARD (1997), 'Major commodities of Africa', *Africa South of the Sahara 1997* (London: Europa), 27–59.

CHAUVEAU, JEAN-PIERRE (1995), 'Land pressure, farm household life cycle and economic crisis in a cocoa-farming village (Côte d'Ivoire)', ch. 5 in François Ruf and P. S. Siswoputranto (eds.), *Cocoa Cycles: The Economics of Cocoa Supply*. International cocoa conference on cocoa economies 1993, Bali (Cambridge: Woodhead Publishing), 107–24.

—— and LÉONARD, ERIC (1996), 'Côte d'Ivoire's Pioneer Fronts: Historical and political determinants of the spread of cocoa cultivation', ch. 10 in W. G. Clarence-Smith (ed.), *Cocoa Pioneer Fronts Since 1800: The Role of Smallholders, Planters and Merchants* (Basingstoke and London: Macmillan), 176–94.

CLARENCE-SMITH, WILLIAM GERVASE (ed.) (1996), *Cocoa Pioneer Fronts Since 1800: The Role of Smallholders, Planters and Merchants* (Basingstoke and London: Macmillan).

—— and RUF, FRANÇOIS (1996), 'Cocoa pioneer fronts: the historical determinants', ch. 1 in W. G. Clarence-Smith (ed.), *Cocoa Pioneer Fronts Since 1800: The Role of Smallholders, Planters and Merchants* (Basingstoke and London: Macmillan), 23–44.

CROOK, RICHARD (1990), 'Politics, the cocoa crisis, and administration in Côte d'Ivoire', *The Journal of Modern African Studies*, 28 (4): 649–69.
CURTIS, BRONWYN, N. (principal author), SCHEU, J. J., SIGG, OSCAR, DAND, ROBERT J., MAXWELL M. C., ROCKETT, C. J., and LASS, R. A. (collaborating authors) (1987), *Cocoa: A Trader's Guide* (Geneva: International Trade Centre UNCTAD/GATT).
DALE, W. T. (1962), 'Virus diseases', part of ch. 19, 'Diseases and pests of cocoa', in Brian J. Wills (ed.), *Agriculture and Land Use in Ghana* (London: Oxford University Press), 286–316.
DICKSON, KWAMINA V. and BENNEH, GEORGE (1970), *A New Geography of Ghana* (London: Longman).
DOROSH, PAUL and AKANJI, BOLA (1988), *Impacts of Exchange Rate Changes on the Cocoa-Food Crop Farming Systems of Southwest Nigeria* (Ibadan: International Institute of Tropical Agriculture).
DUNN, JUSTINE (July 1996), 'The role of indigenous woody species in "farmer-led" agricultural change in south east Nigeria, West Africa' (unpubl. Ph.D. thesis).
ECKERT, ANDREAS (1996), 'Cocoa Farming in Cameroon, *c*.1914–*c*.1960: Land and Labour', ch. 8 in W. G. Clarence-Smith (ed.), *Cocoa Pioneer Fronts Since 1800: The Role of Smallholders, Planters and Merchants* (Basingstoke and London: Macmillan), 137–53.
EGBE, N. E. and ADENIKINJU, S. A. (1990), 'Effect of intercropping on potential yield of cacao in south western Nigeria', *Café Cacao Thé*, vol. xxxiv, no. 4, Oct.–Dec., 281–4.
EIU (Economist Intelligence Unit) (1998*a*), *Ghana: Country Profile, 1998–99* (London: EIU).
EIU (1998*b*), *Nigeria: Country Profile, 1998–99* (London: EIU).
EIU (1998*c*), *Côte d'Ivoire; Mali: Country Profile, 1998–99* (London: EIU).
EKANADE, OLUSEGUN (1989), 'An assessment of the temporal variations of soil properties under cocoa (*Theobroma cacao* L.) and fallow towards increasing cocoa production in Nigeria', *Soil Technology*, vol. 2: 171–84.
——(1991), 'Degradation of the physical elements of the rural environment resulting from tree crops cultivation in the Nigerian cocoa belt', *Singapore Journal of Tropical Geography*, 12 (2): 82–94.
——and EGBE, N. E. (1990), 'An analytical assessment of agroforestry practices resulting from interplanting cocoa and kola on soil properties in south-western Nigeria', *Agriculture Ecosystems and Environment*, 30: 337–46.
ESHETT, EBONG T., AY, P., OMUETI, A. I., and JUO, A. S. R. (1991), 'A cocoa-based cropping system on basaltic soils (typic tropohumult) in southeastern Nigeria', *Beiträge zur Tropischen Landwirtschaft und Veterinärmedizin*, 29 (1): 13–24.
FAO (1970, 1972, 1975, 1978, 1981, 1984, 1987, 1990, 1993, 1996), *Production Yearbook*, vols. 24, 26, 29, 32, 35, 38, 41, 45, 47, 50 (Rome: FAO).
——(1995), *Quarterly Bulletin of Statistics*, 8 (1/2) (Rome: FAO).
——(1995), *Fertilizer Yearbook 1995* (Rome: FAO).
FAULKNER, JULIA (1990), *The Major Importance of 'Minor' Forest Products: The Local Use and Value of Forests in the West African Humid Zone*, prepared by Julia Faulkner, ed. Carla R. S. Koppell (Rome: UNFAO).
FAULKNER, O. T. and MACKIE, J. R. (1933), *West African Agriculture* (Cambridge: Cambridge University Press).
GORDON, SARA L. (1974), 'The role of cocoa in Ghanaian development', ch. 3 in Scott

R. Pearson and John Cownie (eds.), *Commodity Exports and African Economic Development* (Lexington, Mass. and London: D. C. Heath and Company).

GYIMAH-BREMPONG, KWABENA and APRAKU, KOFI KONADU (1987), 'Structural change in supply response of Ghanaian cocoa production: 1933–1983, *Journal of Development Areas*, vol. 22, Oct.: 59–69.

HÄGERSTRAND, TORSTEN (1968), *Innovation Diffusion as a Spatial Process* (trans. from Swedish) (Chicago: Chicago University Press).

HAMMOND, P. S. (1962), 'Agronomy', part of ch. 18, 'Cocoa', in Brian J. Wills (ed.), *Agriculture and Land Use in Ghana* (London: Oxford University Press), 252–6.

HARRISON CHURCH, R. (1980 edn.), *West Africa* (London: Longman).

HASAN, I. (1995), 'Preface', in François Ruf and P. S. Siswoputranto (eds.), *Cocoa Cycles: The Economics of Cocoa Supply*, International cocoa conference on cocoa economies 1993, Bali (Cambridge: Woodhead Publishing), vii–xi.

HILL, POLLY (1962), 'Social factors in cocoa farming', part of ch. 18, 'Cocoa', in Brian J. Wills (ed.), *Agriculture and Land Use in Ghana* (London: Oxford University Press), 278–85.

——(1963), *The Migrant Cocoa Farmers of Southern Ghana: A Study in Rural Capitalism* (Cambridge: Cambridge University Press).

HODGKINSON, EDITH (1997), 'Economy of Côte d'Ivoire', *Africa South of the Sahara 1997* (London: Europa), 336–9.

HOPKINS, A. G. (1973), *An Economic History of West Africa* (Harlow: Longman).

HORNER, SIMON (1994), 'Country report—Ghana', *Courier*, 144 (March–April): 20–4.

HUBAND, MARK (1989), 'Ivorian cocoa, the cash crop nobody wants to buy', *Financial Times*, Wednesday, 22 Nov.

International Cocoa Organization (1993), *Quarterly Bulletin of Cocoa Statistics* (London: ICCO).

——(1997/98), 'Review of Developments in the World Cocoa Situation', *Quarterly Bulletin of Cocoa Statistics*, vol. 24 (3), (London: ICCO): v–ix and 5–8.

JARRIAGE, F. and RUF, F. (1990), 'Understanding the cocoa crisis', *Café Cacao Thé*, vol. xxxiv, no. 3, July–Sept.: 223–9.

JOHNSON, C. G. (1962), 'Capsids: a review of current knowledge', part of ch. 19, 'Diseases and pests in cocoa', in Brian J. Wills (ed.), *Agriculture and Land Use in Ghana* (London: Oxford University Press), 316–31.

KILLICK, TONY (1992), *Explaining Africa's Post-Independence Development Experiences*, Working Paper 60 (London: Overseas Development Institute).

KOESTER, ULRICH, SCHAFER, HARTWIG, and VALDÉS, ALBERTO (1989), 'External demand constraints for agricultureal exports', *Food Policy*, 14 (3): 243–54.

KOLADE, J. A. (1991), 'A review of agronomic practices in cocoa cultivation: with reference to Nigeria', *Agriculture International*, Oct., vol. 43 (10): 276–82.

KONAN, G. K. (1995), 'The present cocoa crisis and its impact on protection of cocoa crops against pests: The case of central-western Côte d'Ivoire', ch. 7 in François Ruf and P. S. Siswoputranto (eds.), *Cocoa Cycles: The Economics of Cocoa Supply*, International cocoa conference on cocoa economies 1993, Bali (Cambridge: Woodhead Publishing), 151–60.

KUSI, NEWMAN K. (1991), 'Ghana: can the adjustment reforms be sustained?', *Africa Development*, vol. xvi (3/4), 181–206.

LANE, D. A. (1962), 'The forest vegetation', ch. 10 in Brian J. Wills (ed.), *Agriculture and Land Use in Ghana* (London: Oxford University Press), 160–9.
LASS, R. A. and WOOD, G. A. R. (eds.) (1985), 4th edn., *Cocoa* (Harlow: Longman).
LÉONARD, ERIC and OSWALD, M. (1995), 'Cocoa smallholders facing a double structural adjustment in Côte d'Ivoire: responses to a predicted crisis', ch. 6 in François Ruf and P. S. Siswoputranto (eds.), *Cocoa Cycles: The Economics of Cocoa Supply*, International cocoa conference on cocoa economies 1993, Bali (Cambridge: Woodhead Publishing), 125–50.
MCKELVIE, A. D. (1962), 'Physiology', part of ch. 18, 'Cocoa', in Brian J. Wills (ed.), *Agriculture and Land Use in Ghana* (London: Oxford University Press), 256–60.
MIKELL, GWENDOLYN (1989), *Cocoa and Chaos in Ghana* (New York: Paragon House).
MORGAN, W. B. and PUGH, J. C. (1969), *West Africa* (London: Methuen).
NUGENT, PAUL (1990), 'Much ado about smuggling: The political economy of cocoa in the Volta Region of Ghana, *c*.1970–1986', Paper for the African Studies Association UK Conference (Birmingham, Sept.).
NYANTENG, V. K. (1995), 'Prospects for Ghana's Cocoa Industry in the 21st Century', ch. 9 in François Ruf and P. S. Siswoputranto (eds.), *Cocoa Cycles: The Economics of Cocoa Supply*, International cocoa conference on cocoa economies 1993, Bali (Cambridge: Woodhead Publishing), 179–208.
NYE, P. H. and STEPHENS, D. (1962), 'Soil fertility', ch. 7 in Brian J. Wills (ed.), *Agriculture and Land Use in Ghana* (London: Oxford University Press), 127–43.
ODUWOLE, O. O. (1995), 'The shifting of cocoa production areas in Nigeria: past, present and future trends', ch. 10 in François Ruf and P. S. Siswoputranto (eds.), *Cocoa Cycles: The Economics of Cocoa Supply*, International cocoa conference on cocoa economies 1993, Bali (Cambridge: Woodhead Publishing), 209–18.
RUF, FRANÇOIS (1991), 'Les crises cacaoyères. La malédiction des âges d'or?' *Cahiers d'Études africaines*, 121–2, xxxi, 1–2, 83–134.
——(1993), 'Will Côte d'Ivoire give up its position of world leading cocoa producer to Indonesia?', *Café Cacao Thé*, vol. xxxvii, no. 3, July–Sept.: 243–9.
——and SISWOPUTRANTO, P. S. (eds.) (1995), *Cocoa Cycles: The Economics of Cocoa Supply*, International cocoa conference on cocoa economies 1993, Bali (Cambridge: Woodhead Publishing).
SCHAFF, THOMAS and MANSHARD, WALTER (1989), 'The growth of spontaneous agricultural colonization in the border area of Ghana and the Ivory Coast', *Applied Geography and Development*, vol. 34, 7–22 (Tübingen: Institute for Scientific Co-operation).
SCHRECKENBERG, K. (1996), 'Forests, fields and markets: A study of indigenous tree products in the woody savannas of the Bassila region, Benin' (unpubl. Ph.D. thesis, March).
SYNGE, RICHARD (1997), 'Economy of Nigeria' (rev. Linda van Buren), *Africa South of the Sahara 1997* (London: Europa), 737–45.
SZERESZEWSKI, R. (1965), *Structural Changes in the Economy of Ghana, 1891–1911* (cited in Austin, 1996) (London: Weidenfeld & Nicolson).
SZOLNOKI, T. W. (1985), *Food and Fruit Trees of The Gambia*, edited by Stiftung Walderhaltung in Afrika und Bundesforschungsanstalt für Forst- und Holzwirtschaft (Hamburg: Bundesforschung).
THRESH, J. M., OWUSU, G. K., BOAMAH, A., and LOCKWOOD, G. (1988), 'Ghanaian cocoa varieties and swollen shoot virus', *Crop Protection*, vol. 7, Aug.: 219–31.

Tropical Agriculture Association (1992), 'Prospects for the world cocoa market until the year 2005', *Newsletter*, June 12 (2): 30–1.
VAN BUREN, LINDA (1999), 'Economy of Ghana', *Africa South of the Sahara 1999* (London: Europa), 513–20.
WHARTON, A. L. (1962), 'Black pod and minor diseases', part of ch. 19, 'Pests and diseases in Cocoa', in Brian J. Wills (ed.), *Agriculture and Land Use in Ghana* (London: Oxford University Press), 333–42.
World Bank (1989), *Sub-Saharan Africa: From Crisis to Sustainable Growth* (Washington: World Bank).

# 5

# Unpredictable Savanna Environments

## Savanna, the most extensive and unpredictable biome in West Africa

In spite of being the largest ecological zone in West Africa, savannas have provoked less interest and been researched less than the humid forests. This underemphasis on savannas throughout the world is now being corrected as the abundance of modern literature on the subject reflects. Savannas are now known to be far more productive than was previously thought (Kinyamario and Imbamba, 1992; Jones *et al.*, 1992), and are second only to moist forests in terms of their contribution to terrestrial primary production (Scholes and Walker, 1993). The West African savannas are rich and varied in terms of their flora and fauna though far less so than the savannas of central and southern Africa or elsewhere in the world (Cole, 1986). They are also of major socioeconomic importance as the majority of the region's population live there and depend heavily on the land for their livelihood. In West Africa, a belt of savanna vegetation extends very approximately from 7°N to 14°N, with a marked southern projection which reaches the coast in southeastern Ghana, Togo and Benin (Fig. 5.1).

It is the co-dominance of trees and grass in varying and unpredictable proportions that has caused complications in the understanding and classification of savannas. Only in recent years has it been accepted that the form of savannas is determined by a range of variables, notably plant available moisture (PAM), plant available nutrients (PAN), fire, herbivory, and human actions. The impact of these variables in different combination inevitably results in a highly varied landscape. The acceptance that a range of factors, both abiotic and biotic, influence the form that savanna landscapes take, represents a significant development in ecological thought. It marks a move from an unsatisfactory equilibrial approach where the explanations of patterns of savanna vegetation are based on an expectation that savannas are tending towards domination by woody species, to a wider acceptance that the attainment of such an equilibrium is a virtual impossibility owing to the range of factors determining the nature of the environment. The literature on savannas thus represents a growing acceptance of non-

**Fig. 5.1.** Vegetation zones of West Africa

*Source:* After White (1983). *Vegetation map of Africa* (UNESCO/AEFAT/UNSO).

equilibrial ecology (Menaut *et al.*, 1990, 1995; Scholes and Walker, 1993; Stott, 1991, 1994).

After a brief discussion of the varied character of savannas in West Africa and their distinguishing characteristics, the major ecological determinants of vegetation patterns in the biome are then considered through the use of a functional model which focuses on plant available moisture (PAM), plant available nutrients (PAN), fire, and herbivory. Human impact may be of greater importance than these four but as this is considered in subsequent chapters the focus here is on the four more 'natural' determining factors. The literature on West African savannas is limited so this chapter draws heavily on research on savannas in other parts of Africa. Extrapolation of information necessitates caution but it can be the best form of explanation available in the absence of empirical evidence from the West African savannas.

## Characteristics of savannas

### Vegetation zonation in the West African savannas

In spite of being classified as a single unit the West African savannas are highly varied (Fig. 5.2). To the north of the moist forest (shown as the 'rain forest domain' on Fig. 5.2) is a belt dominated by shrubs and grass where the proportion of woody biomass is low. Opinion is divided as to whether this humid savanna zone is derived from repeated clearing of a former moist forest zone and is maintained by fire, or whether the moist forest receded during a former dry period and that fire prevents the forest from re-establishing itself (Menaut *et al.*, 1995). Swaine *et al.* (1976) review a range of factors influencing vegetation in the boundary zone between forest and savanna.

North of this derived savanna zone the proportion of woody vegetation increases and both dry forest and savanna woodland are common (Fig. 5.2). It is noteworthy that the moist forests are physically separated from the core area of woodlands and dry forests which are located in the zone with 800–1,100 mm of rain a year. This area of higher woody biomass is on the borders of the more moist guinea savannas and the drier sudan savannas, though some woodland does extend into more arid areas, particularly along water courses.

The distribution of dry forest and woodland within this zone is not satisfactorily explained. It is to some extent moisture related but there is no clear evidence linking the distribution of woody vegetation to edaphic conditions. It is believed that this area was more heavily wooded in the past but extensive forest clearance largely for agriculture has reduced forest coverage. Once cleared, dry forests and woodlands do not regenerate easily. Atmospheric and soil humidity declines and burning reduces the number of non-fire-resistant species (Menaut *et al.*, 1995). The environment is unpredictable both in the short and long term and the similar unpredictability of savanna vegetation could

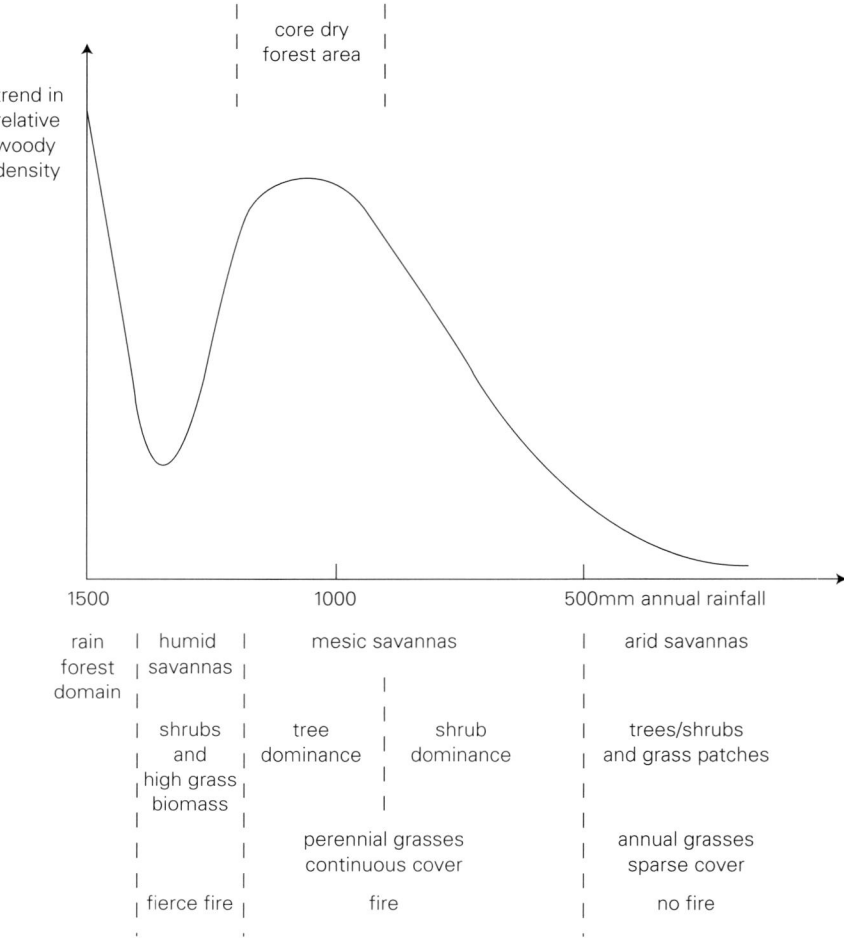

**Fig. 5.2.** Idealized scheme of vegetation types and woody plant density along a gradient of rainfall

*Source*: After Menaut *et al.* (1995).

be due to the prevalence of non-equilibrial conditions (see below and also Chapter 1).

As rainfall diminishes into the sudanic zone with a lengthening of the dry season, tree savanna merges with shrub savanna and tall trees are virtually absent. The vegetation is again lower and less dense than further south. Such landscapes are evident in northern Senegal, the very north of Ghana, and throughout much of northern Nigeria. Finally, in the sahel, shrub savanna gives way to a more sparse and even lower formation dominated by grass and thicket. While there is a direct relationship between vegetation and rainfall, the situation is far more complex. Local variation within these broad zones is considerable and reflects changes in physical factors such as geology, topography, soils, altitude, and of course, human influence. For instance, patches of pure grass savanna can be found as far south as 6°N, well south of the southern boundary of savanna (Hopkins, 1974), while forest may fringe water courses in The Gambia, Senegal, Mali, and Burkina. It must be stressed that the ecology of savannas is neither that of forest nor of grassland, the strong and complex interactions between woody and herbaceous plants giving the vegetation a character of its own (Scholes and Walker, 1993).

## Climate, soil, and vegetation

Rainfall in the savannas is seasonal and a dry spell is characteristic. Rainfall decreases from south to north in the West African savannas, and as it does, the length of the dry spell increases. The sahel savannas of the north are thus very different from the moist savannas of the south largely on account of moisture availability. The dry spell can vary from around two months in the southern savannas, as at Olokemeji in south western Nigeria, to over six months north of Dakar in Senegal. In the zone where the guinea savannas merge into the sudan savannas, mean annual rainfall is about 1,000 mm and is seasonally distributed so that during 4 or 5 months of the year there is a minimum of 100 mm and also no more than 5 or 6 months with less than 25 mm of rain. In the southern guinea savannas of Southern Guinea, for example, total annual rainfall may exceed that in parts of the forest zone but its distribution is concentrated as part of the year is virtually dry. Species composition of the forests of Guinea and Sierra Leone where there is a marked moisture deficit during the year thus differs from the moist forests of Côte d'Ivoire and southern Nigeria, or what remains of them. Mean monthly minimum and maximum temperatures are normally 13°–21 °C and 35°–41 °C respectively, a greater range than in the forest to the south (Hopkins, 1974).

The highly varied soils of the West African savannas are the product of inter-relationships between a combination of factors which include parent material, relief, climate, and vegetation, though according to Scholes and Walker (1993), vegetation itself does not have a profound effect on pedogenic processes. Nevertheless, a close relationship frequently exists between soil and vegetation. The major soil-forming processes in the savannas are thought to be clay illuviation and ion movement which result in the formation of soil horizons and catenary sequences. Some of the soil horizons, rich in iron, form hard pans under extensive areas of savanna. The organic matter content of savanna soils is frequently low due to the high rate of its decomposition in the savannas. Low organic matter levels are a characteristic of sandy soils and soils where the clays are predominantly kaolinite. Both these soil types are common in the savannas and neither has a high cation exchange capacity.

Trapnell *et al.* (1976), cited in Menaut *et al.* (1995), observed that in a

*Brachystegia* (miombo) woodland in Zambia, the mineral status of the top soil was inversely related to the biomass on a plot, which depended on the release of nutrients through burning. Unburnt plots had the highest biomass, but this retained a major part of the mineral nutrients of the system. The plant debris is consumed by termites and other decomposers, and mineral nutrients are concentrated in their bodies, so leaving the soil low in nutrients and protected from the worst effects of leaching. Once the vegetation is burnt the ash increases the calcium and magnesium content in the surface layers of the soil. Phosphorus increases too though to a lesser degree, while carbon, nitrogen, and sulphur are largely volatilized. Whatever the burning regime, the topsoil tends to be deficient in nutrients except in termite mounds (Menaut *et al.*, 1995). Although the structure and physical quality of savanna soils are frequently poor, water, not nutrients, tends to be the main limiting factor in primary production.

Savannas, like forests, conserve nutrients in their biomass and are efficient at cycling them though most biological activity is concentrated in the wet season (Menaut *et al.*, 1995; Scholes and Walker, 1993). This is markedly reduced during the dry season and the effect is intensified when the tree stratum is sparse. Problems of nutrient loss from the system begin to occur as soon as soils are disturbed, and vegetation is cleared.

The semi-evergreen forests of the south give way to deciduous forests which lose their leaves for part of the year, usually in the dry season. In Senegal, for example, entire forests may appear totally lifeless, dry, and perhaps charred if they have been burnt before the commencement of the rains. Even here there are a few evergreen species which provide a stark contrast to the bulk of the desiccated vegetation in the dry season. Grasses, the other major component of savanna vegetation, are also well adapted to the environment. Belonging predominantly to the C4 group of photosynthetic plants, these are more productive than their C3 counterparts, because the rate of photosynthesis in C4 plants increases with the intensity of photosynthetically active radiation. When moisture is in short supply, C4 plants can limit the amount of water vapour lost through transpiration by limiting the time their stomatal pores are open for gaseous exchange. C4 grasses are thus efficient operators in hot, bright, and fairly dry climates, while C3 grasses are better adapted to cool, moist, and shady environments (Deshmukh, 1986; Moore *et al.*, 1996; Stott, 1994). In biochemical terms C3 plants fix $CO_2$ as a three-carbon compound, phosphoglyceric acid, while C4 plants fix $CO_2$ as a four-carbon compound, oxaloacetic acid.

The morphology of savanna species differs from species of the forest zone. Trees and shrubs often have a gnarled appearance, distorted and stunted by fires and with bark commonly thicker than 1 cm, which partly explain their tolerance to fire. Most woody species have a tap root that goes deep into the soil and from which several laterals branch off horizontally 10–20 cm below the soil surface. Thus, in similar fashion to the forest zone, a dense network of roots from trees and shrubs may penetrate the top layers of savanna soils. Small woody plants often have extensive root systems. Although fairly small above ground, these plants may be very old, their growth having been hampered by frequent fires after which regeneration has occurred by suckers (Hopkins, 1974). This is common in the savannas and occurs in most woody plants up to 2 m high. Survival can be a problem where extreme environmental conditions prevail, and according to Raunkiaer's classification of plants, which is based on the vertical position of the resting bud during the unfavourable season—the dry season in West Africa—a high proportion of perennial species in the savannas are hemicryptophytes, chamaephytes, or geophytes, that is, their resting buds are at the soil surface, near the soil surface, or below the soil surface respectively (Raunkiaer, 1937). Most grasses are hemicryptophytes. They grow in tufts and during a fire it would appear that the outer parts of the plant protect the inner, perrenating buds. In the forests where fire is less of a problem and the environment is generally more constant, phanerophytes, the resting buds of which are well above ground level, tend to be the most commonly occurring life form among plants. Phanerophytes also occur in savannas but here they are normally deciduous.

## Seasonal contrasts

Vegetation and climate come together in the savannas in a landscape which is very different in the wet season from the dry. It is in the dry season that savanna fires occur—naturally, accidentally, and intentionally—and immediately after a fire the savanna may look devastated, black, and lifeless. Most of the grasses have been burnt and the soil surface is covered in ash and blackened by it. Within a week the picture changes, even more quickly if there has been rain. Grasses begin to shoot, larger trees come into bud and small trees, burnt down, throw up new suckers. The ash is washed away, the charred remains of the fire are disguised by new greenery, and primary production continues apace for the duration of the rainy season. After the rains the grasses which have flowered change from green to brown. As their seeds mature and are dispersed, the plants become dry and non-woody species gradually fall over producing a mass of vegetation, potential fuel for the next fire after which the cycle begins again. It is this marked seasonality which gives savannas their distinctiveness. If the savanna is not fired the cycle is similar, but new growth is slower and the dried remains of the previous year's growth decay more gradually, building up fuel for the next fire.

The floristic composition of the derived savannas on the margins of the forest is very similar to that of the southern guinea zone. Tree species are most varied in the derived and southern guinea savannas, species diversity decreases northwards and is parallelled by a declining trend in tree height from south to north (Menaut *et al.*, 1995). Many species are common to both northern and southern guinea zones though the southern guinea savannas are typified by a mixture of species while the northern guinea zone is characterized by almost pure communities of only one species of tree (Hopkins, 1974). *Isoberlinia doka* is the most common species occuring in pure stands though others are *Isoberlinia tomentosa*, *Monotes kerstingii*, and *Uapaca togoensis* (Hopkins, 1974; Menaut *et al.*, 1995; White, 1983).

Among the smaller trees and shrubs, species of *Gardenia*, *Piliostigma thonnongii*, and *Stereospermum kunthianum* are common to both zones while *Bridelia ferruginea*, *Maytenus senegalensis*, and *Psorospermum febrifugum* are more typical of the southern guinea zone. *Annona senegalensis*, *Combretum* spp. *Grewia mollis*, *Protea elliottii*, and *Zimenia americana* are more characteristic of the northern guinea zone (Hopkins, 1974).

In the drier sudan savannas (750–1,200 mm) which may have a dry season as long as seven months, *Isoberlinia* gradually disappears and dominant tree species include *Bombax*, the kapok or silk-cotton tree, *Adansonia digitata*, the baobab, *Hyphaene*, the dum palm, *Anogeissus*, *Sclerocarya*, *Balanites aegyptiaca*, and *Lannea*. *Combretum* and *Acacia* species produce a low scrub vegetation where the water-table is low while on the flood plains of the river valleys the fan palm, *Borassus* is common (Morgan and Pugh, 1969; Rattray, 1960).

Grass cover in the derived savanna and southern guinea savannas consists of the very tall *Pennisetum*, commonly known as elephant grass, *Andropogon* spp., *Beckeropsis uniseta*, *Monocymbium ceresiiforem*, and the genus *Hyparrenhia*. The same tall grasses, 1.5–3 m high in the moist southern savannas, continue into the drier northern guinea savannas though here they are not as tall. Further north into the sudan savannas the grass cover consists of yet shorter species, 90–120 cm tall, many of which are annuals. *Angropogon gayanus* and several other varieties of *Androprogon* are common as are *Digitaria perrottetii*, several varieties of *Panicum* and *Schizachyrium*. As the rainfall diminishes to below 750 mm in the sahel dominant grass species are *Cenchrus biflorus*, *Eragrostis tremula*, *Pennisetum pedicellatum*, *Ctenium elegans*, *Aristida longiflora*, and *Andropogon gayanus*. In this heavily grazed area most grass species are annuals though allegedly 6 years' protection from grazing and cultivation would see a reversion to perennial species such as *Angropogon gayanus* (Rattray, 1960; Crowder and Chheda, 1982). Whether this still remains true now after over 20 years of drought, desiccation, and dryland degradation remains to be seen. While the paragraphs above describe broad floristic zones in the savanna, the boundaries are far from precise and 'outliers' of one zone are frequently found in the next.

The focus here is on vegetation, but of equal importance is the fauna of the savannas which is as diverse as the flora. Populations of larger mammals are now much reduced in West Africa following centuries of land clearance and hunting but many lower species are still numerous and highly diverse. Rodents, for example, reflect considerable species diversity within the savannas as does the range of insect life. The importance of herbivorous animals in maintaining the savannas is discussed below.

## Fire

Until comparatively recently the importance of fire in savanna management has been underestimated in research work on savannas. However, the importance of fire in savanna management has long been recognized by indigenous land users. By non-African advisers, it has frequently been considered a constraint on development, a viewpoint which has resulted in errors of decision-making in management of vast areas of savanna. The role of fire is considered later in the chapter.

## The enigma of savannas

Savannas have proved difficult landscapes to understand. Most ecological models can cope with ecosystems dominated either by trees and shrubs or by grasses, but problems of explanation arise when these two types of vegetation are co-dominant (Stott, 1994). The literature on the savannas weighs heavily in favour of their being the product of human activity, though Schimper (1898, 1903), cited in White (1983: 50), suggested that grassland occurred in tropical Africa as a distinct zonal formation with a distinct type of climate. Schimper distinguished grassland with trees as savanna and grassland without trees as steppe. It is now recognized that much of the grassland believed by Schimper and others to represent a climatic climax was in reality maintained by fire, or was an edaphic climax where soil conditions were not conducive to the development of trees. It is also thought that P. W. Richards's view (1952) which doubts that any tropical grassland is a true climatic climax is too sweeping (White, 1983).

Historical evidence reviewed by Gritzner (1988) suggests that in the past the savannas were more densely forested in West Africa and that environmental degradation began with the cutting of trees for charcoal for the trans-Saharan camel caravans in the Middle Ages. Stebbing (1937) concluded that much of the present Guinea savanna would be forested were it not for clearance by farmers, a view reinforced by Aubréville (1949) who demonstrated that when savanna is protected from burning, it reverts to forest. Hopkins (1992, 1974) supports this view, stressing the importance of the human use of fire in the forest degradation process. Both these views suggest an allegiance to the Clementsian model of plant succession and climax vegetation. Menaut *et al.* (1995) conclude that experiments on protection from burning have not lasted long enough and are too few or have been conducted in such highly specific conditions as to make general conclusions impossible. Following observations on the impact of farming in the savanna forest boundary zone of Nigeria, Morgan and Moss (1965) argued that farming in forested areas does not necessarily lead to degradation of the vegetation cover. More recently it has been shown that patches of forest in this savanna zone are actually being created and enriched by farmers (Fairhead and Leach, 1996). While the role of human activity in savannas of West Africa has received considerable attention in the literature, Swaine *et al.* (1976) and Cole (1986) demonstrate that physical features such as geology and soil factors may also have an important influence on the characteristics of savanna. Explaining savannas has thus proved highly complex and explanations have been far from satisfactory. Deshmukh (1986) skilfully draws together the wide range of views on reasons for the existence of savannas. He states that it now seems indisputable that savannas were present before the intervention of humans but that their extent has changed in response to long-term climatic changes in the Quaternary period (Roberts, 1989; Nicholson, 1996), and that they have become much more widespread due to human influence.

As with all vegetation formations, savannas are dynamic systems but a major problem is their unpredictability. If an area of moist forest is cut, burnt, cultivated, and fallowed for a lengthy period the chances are that it will return to forest, although there may be changes in species represented. If, however, an area of tree/grass savanna is cut, burnt, cultivated, and fallowed, the pattern and ratio of trees to grassland that re-establishes during the fallow may be completely different from the savanna that existed prior to cultivation. While a multitude of questions about the dynamics of savanna landscape systems has always existed, it is not until relatively recently that the functional aspects of savannas have been the focus of research. In consequence, past literature tends to reflect conflicting, individual views on savanna ecology by experts often geographically and culturally isolated (Huntley and Walker, 1982). The literature on savannas, like forests, is also stifled by a wealth of imprecise terminology which is an additional hindrance to those trying to make sense of the literature. What is certainly evident from the literature is the relative paucity of material on West African savannas compared with savannas in other parts of the world (Cole, 1986). Owing to the socio-economic importance of the world's savannas, the need for greater understanding of their complex ecology has been acknowledged as an essential and pressing task (Stott, 1991) and hence the emphasis on functional analyses of savanna environments has much increased.

## Ecological determinants of savannas—a functional model

Most of West Africa's savannas, like its forests, are secondary. This is due largely to the effects of agriculture and pastoralism, but putting such human induced influences aside for the moment, four major factors have been identified as prime determinants of the nature of savannas. Two of these, water and nutrient availability, are of paramount importance but their effects may be much modified by two other variables, fire and herbivory. Savanna ecologists have recently developed a functional model of tropical savannas based on the two fundamental variables—plant available moisture (PAM) and plant available nutrients (PAN) (Belsky, 1990; Stott, 1991). The model integrates rainfall, water infiltration, evapotranspiration, soil texture, and hydrological variables into two measurements of moisture available to plants and nutrients available to plants during their period of growth. These aggregated measurements can be located within a plane defined by the axes of plant available moisture (PAM) and plant available nutrients (PAN) (Fig. 5.3). According to Stott (1991) this model allows a functional classification of savannas according to their abiotic characteristics, and thus permits comparison of savannas from different sites and even from different continents. They are not all equally important: PAM and PAN are seen to be the initial determinants of biomass whereas fire and herbivory are secondary modifiers of primary productivity in the savannas (Stott, 1994). PAM, PAN, fire, and herbivory may also exert different influences on savannas at different scales. Added to these four is a fifth—human activity which can exceed the effects of the others and frequently does. Human intervention is considered in the subsequent chapters while the following sections of this chapter are devoted to the first four ecological determinants of savannas incorporated in the model.

## Plant available moisture (PAM)

Moisture available to plants is greatly influenced by rainfall which, in West Africa, is tied closely to the movement of the Inter Tropical Convergence Zone (ITCZ) across the region from south to north from January to June and then south again in the latter part of the year. This is the junction zone between the warm dry continental air mass centred over the Sahara and the warm moist maritime air centred over the Gulf of Guinea. Because of their different properties the junction between these two air masses does not take place in a vertical plane. Instead, the relatively light, warm, dry continental air from the Sahara rises or is pushed up by the warm moist maritime air from the Gulf of Guinea which forms a wedge beneath it (Fig. 5.4). The ITCZ is thus the meeting point at ground level, more theoretical than real, of the two air masses but this is not the most interesting point in terms of 'weather' generated by the front. Rainfall occurs in a belt approximately 600–1,200 km behind the ITCZ. Here, the wedge of turbulent warm, moist air has usually reached a sufficient depth to produce rain. Rainfall is rarely continuous in the savannas and is more commonly confined to storms the frequency, intensity, and duration of which are greater further south than north. This is a cause of very uneven rainfall distribution over the season and over the region, as storms are highly localized. While some areas benefit, others close by may receive no rain at all, emphasizing the risks involved in land management, particularly in the more arid north of the region.

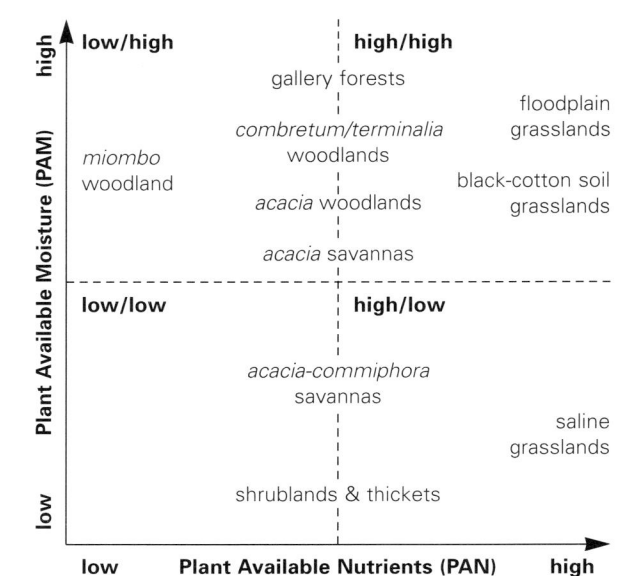

*Source*: After Frost *et al.* (1986); Belsky (1990), fig. 5; and Stott (1994) fig. 13.3.

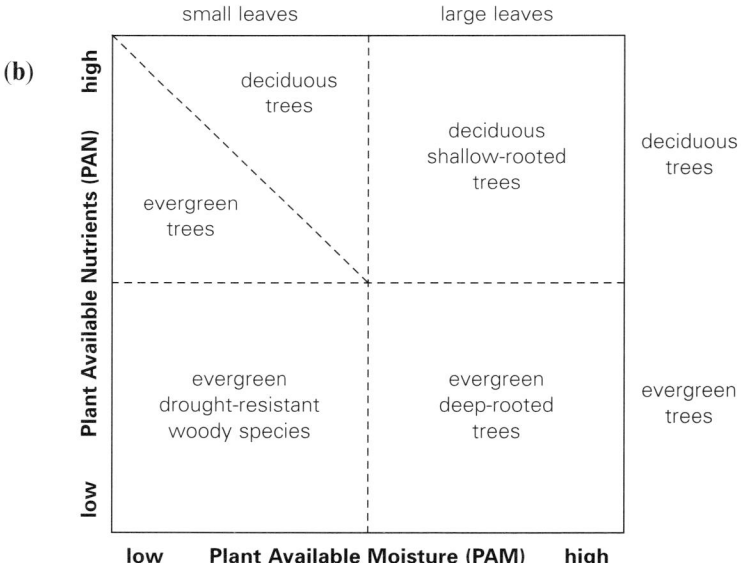

**Fig. 5.3.** (*a*) The disposition of East African savanna types within the PAM/PAN plane (*b*) The suggested phenological and physiognomic disposition of South American savanna types within the PAM/PAN plane

*Source*: After Walker and Menaut (1988), fig. 2.1; Stott (1991); and Stott (1994), fig. 13.4.

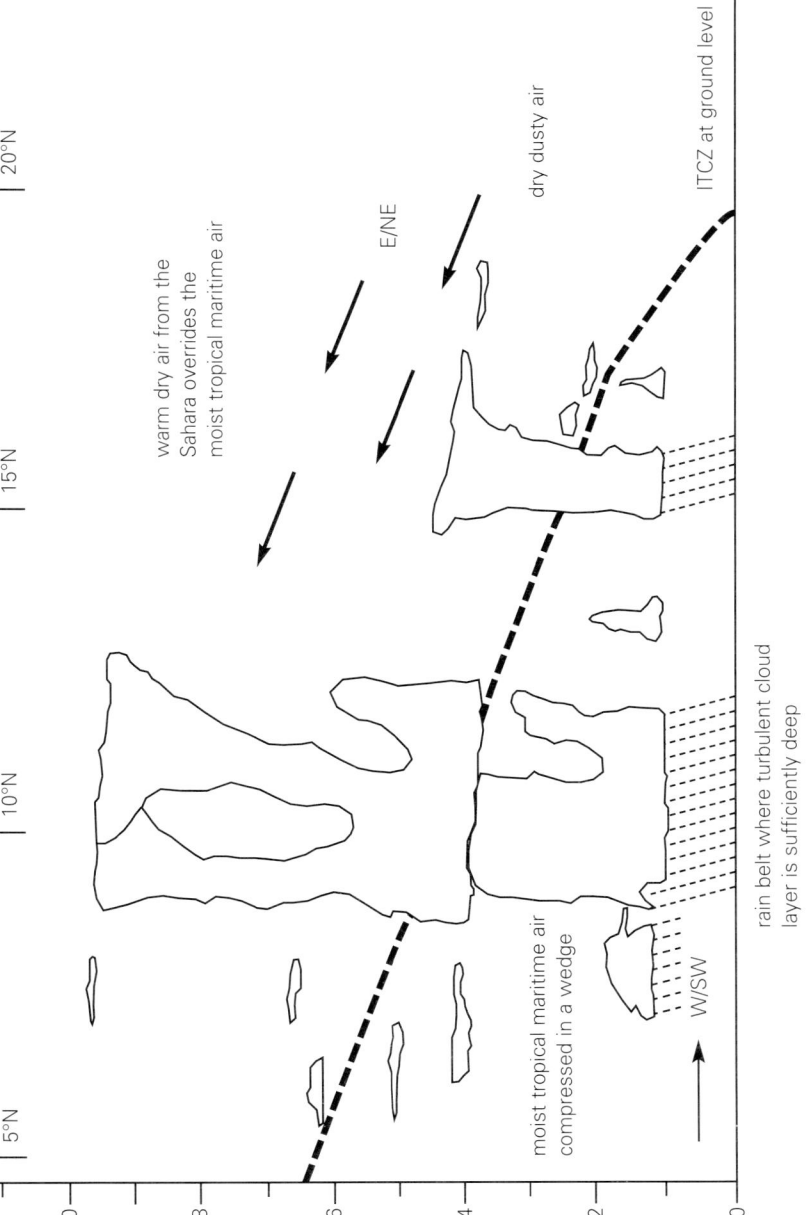

Fig. 5.4. Diagrammatic cross section of the intertropical convergence zone over West Africa

From mid-June to mid-December the ITCZ retraces its path and moves southwards. This explains why the more southerly savannas in West Africa are the more moist. They are crossed by the rain belt on both its northward journey and again on its return south while the more northerly regions receive only one, shorter rainy season. Also by the time the front has moved well into the region, even the compressed maritime air has lost much of its moisture, storms are less frequent and severe and the total volume of rainfall decreases. As the bulk of West Africa's rainfall is frontal rain, the movement of the ITCZ is of fundamental importance to understanding West Africa's rainfall regime and, without wishing to sound too deterministic, much of the geography of the region.

Many factors come together to influence the pattern of vegetation in the West African savannas. Bell (1982) notes a positive correlation over Africa between rainfall and plant biomass. Nowhere is the critical importance of moisture more evident than in the sahel where it is probably the major factor controlling the northern boundary of the savanna. Drought since the late 1960s has left parts of the sahel so short of rain that plant life has died as a result of prolonged desiccation. Particularly affected are trees normally resilient to drought. Compounded by overgrazing, parts of the sahel savannas have become degraded and sand dunes have moved southwards. Whether this is permanent is not yet certain.

But as important as the total rainfall is, its availability to plants is determined by the interaction of many factors: the seasonal distribution of the rains and the frequency, duration, and intensity of storms is of vital importance as is the proportion of rain lost to interception, runoff, or drainage. This is influenced by relief, plant cover, and soil type. Where soils are sandy and vegetation cover is poorly developed, particularly after the land has been cleared for cultivation, loss of moisture via the soil is greater than on land that has not been cleared for cultivation (Ahn, 1970; Scholes and Walker, 1993). Where either the clay fraction or the soil organic matter or both are high, water retention is greater. Factors such as the rooting depth and capacity of plants for making use of water are also of critical importance as is atmospheric evaporative demand. Savanna species are frequently highly conservationist, capable of coping with drought and also using moisture efficiently as soon as it becomes available.

Tinley's work (1982) is instructive in the importance of PAM. While acknowledging nutrient availability to be an important influence on vegetation in the southern African savannas, field level experiments have demonstrated why soil moisture is the fundamental control over primary production and how this may influence spatial patterns of vegetation (Tinley, 1982). Soil moisture is closely related to soil texture and structure and much of Tinley's work was focused on the nature of soil with an impervious layer or poor subsoil drainage. Where impervious layers or pans exist they may be of varying thickness according to their genesis. If covered by sandy shallow soils up to 50 cm in depth, the pan may act as a moisture barrier causing the overlying soil to become saturated in the wet season. This discourages the development of woody perennials but if they do survive the rainy season, their roots usually die back as the soils become desiccated in the dry season. The depth of the pan appears to be critical for the soil moisture balance. Where the pan is around 1 m below the surface the moisture balance of the overlying soil may be sufficient to permit the existence of woody species and vegetation other than grass. Impeded drainage, according to Tinley (1982) is thus a major factor explaining the presence of grassy savannas. Flat sheets of hard rock-like ironstone, or ferricrete, underlie extensive areas in the West African savannas and in the fringe forest areas (Ahn, 1970; Goudie, 1973; Grove, 1992).

Impermeable layers of calcic material, calcrete, and layers of siliceous material, silcrete, are also to be found in West Africa although they are more limited in their extent than ferricrete. It is thus conceivable—though not supported by field work—that arguments similar to those used by Tinley in southern Africa might help to explain the distribution of woodland and grass in the West African savannas. Referring to the savannas of north central Africa, Tinley (1982) did suggest that following the removal of surface soil through erosion, wooded savannas were progressively converted to grassy savannas, the argument being that the depth of remaining soil was insufficient to support woodland. Human impact does not seem to figure largely in Tinley's investigation.

Tinley also argued that if woody species do become established in shallow soils where drainage is impeded, and if dieback during the dry season is not complete, then it is possible that in consecutive years of high rainfall woody species may become established in shallow soils and their roots penetrate the pan horizon. The effect of this would be to improve drainage and encourage the invasion of former grassland by woody species. In an attempt to control for the effect of nutrients, Tinley demonstrated that sandy lithosols in central Mozambique with similar nutrient levels supported woodland vegetation where they were loamy and free draining, but only grassland where drainage was impeded. The scope of Tinley's work is greater than indicated in this chapter though the examples cited here serve to show the major importance of plant available moisture (PAM) in the distribution of woody and grass species in savanna lands (Tinley, 1982).

## Plant available nutrients (PAN)

Without water, plant life—or any other form of life—is impossible. But even with water and a poor nutrient supply the primary productivity of savanna may be much reduced as would be the higher trophic levels it supports (Bell, 1982). Plant available nutrients (PAN) are thus inextricably linked to plant available moisture (PAM) and also to the other major determinants of savannas, fire and herbivory.

The term 'plant available nutrients' is not to be confused with soil fertility.

Savanna soils are highly varied and as mentioned above, are frequently nutrient poor (Menaut *et al.*, 1995; Holt and Coventry, 1990). However, there are areas known to be nutrient rich as exemplified by the Nylsvley savannas (Scholes and Walker, 1993) (see below). In the drier savannas soils may be rich in basic cations such as $K^+$, $Na^+$, $Ca^{++}$, and $Mg^{++}$ but these are not accessible to plants throughout the year, moisture being a major limiting factor. Similarly, other nutrients of critical importance such as C, N, P, and S may all be present in savanna soils but again are not easily accessible to plants. The bulk of these are held in organic form and have to be altered to an inorganic form before they can be used by plants. For example, the nitrogen in urea has to be converted to nitrate before it is directly available to plants and the carbon in plant debris has to revert to $CO_2$ before it can be used in photosynthesis. Furthermore, the cycles of nutrients such as C, N, P, and S are interrelated and any factor which holds up the cycle of any of these and, for example, slows or temporarily halts the return of organic carbon to carbon dioxide, or organic nitrogen to nitrogen gas, may also cause delays in the cycling of the other nutrients. The basic cations $K^+$, $Na^+$, $Ca^{++}$, and $Mg^{++}$, also important in plant growth, are not held in the same manner as C, N, P, and S and can move more easily from organic to inorganic form. In consequence, the availability of nitrogen and phosphorus are frequently greater limiting factors on vegetation development in the drier savannas than the basic cations which are not strongly leached and hence are available to plants with the onset of the rains (Heady and Heady, 1982; McCown and Williams, 1990; Scholes and Walker, 1993).

In the savannas the mechanisms which make nutrients available to plants are thus as important as the presence of nutrients themselves. We now focus on selected aspects of the nitrogen cycle to demonstrate this. As the gas is converted from atmospheric nitrogen, the reservoir of the element, to inorganic nitrogen, $NH_4^+$ and $NO_3^-$, the latter being available to plants, processes such as ammonification, nitrification, and denitrification convert organic nitrogen compounds to inorganic compounds. But the process does not operate smoothly from the reservoir of nitrogen gas through the cycle and back again. The sequence of events is highly regulated by biological processes and the potential for feedback loops is considerable allowing shortfalls in any particular part of the system to be rectified. In the high temperatures of the tropics many of the processes which involve the conversion of nitrogen in organic or inorganic form to nitrates are stimulated by the presence of moisture and slowed in times of its absence thus emphasizing the link between PAM and PAN.

The bulk of soil nitrogen is held in organic plant residues such as leaf litter, twigs, and decaying roots and the release of nutrients—not only nitrogen—from their organic form in adequate quantities and at appropriate times is critical in the functioning of savanna ecosystems. In the case of nitrogen it involves both physical and biological processes. In the conversion of organic matter to nitrogen, carbon in the organic molecule is attacked by bacteria and this liberates ammonium ($NH_4^+$) compounds. Where C:N ratios are very high (>16) as would be the case where the soil is covered with fresh litter, the $NH_4^+$ ions are utilized immediately by the bacteria effecting the ammonification process and there may be no coversion of $NH_4^+$ to $NO_3^-$ for uptake by plants. This is known as 'microbial immobilization'. Moisture is necessary before most bacteria become active and this phase of breakdown usually coincides with the rains. So when organic matter is abundant in the soil it can lead to a period of nitrogen starvation for plants. In spite of the apparent disadvantages of this, the retention of nitrogen by bacteria can reduce nitrogen losses through leaching particularly at a time when plant roots are not well developed (Scholes and Walker, 1993). As deterioration of organic matter continues and the C:N ratio falls to between 10 and 12, nitrate is liberated and is available to plants. It is thus evident that the nitrogen cycle is very closely linked to the carbon cycle and to the availability of moisture. It is also controlled by herbivores in this case represented by organisms that process the organic matter.

De-nitrification involves the conversion of nitrate to gaseous nitrogen and oxides of nitrogen (NO and $NO_2$), mostly in anaerobic situations. Where soils are sandy and free draining, return of nitrogen compounds to the reservoir via this route is limited. However, de-nitrification does occur and in the savannas, fire is an important means of liberating oxides of nitrogen which then return to the more stable form of nitrogen gas, the major reservoir of the cycle. Fire mineralizes nutrients rapidly and speeds up the nutrient cycling process. Although the nutrients are returned to the soil in the ash and are thus accessible to plants, burning does cause losses to the system as carbon, nitrogen, and sulphur are lost to the atmosphere during the burn. They may also be lost through wind erosion when ash is blown away and through water erosion when nutrients are washed or leached away. However, losses through burning may not be as great as the dramatic change wrought by fire would lead one to expect. First, evidence from different tropical grasslands including those in Côte d'Ivoire have shown that tropical grasses are able to relocate nutrients from the above ground parts of the plants to the roots as the end of the growing season approaches. This reduces nutrient losses in dry season fires. Secondly, there is evidence that losses of nutrients through volatilization, particularly nitrogen and sulphur, are to some extent returned through rainfall. Evidence from Northern Australia has shown that concentration of nitrates, calcium, magnesium, and potassium compounds was normally highest at the beginning of the wet season when the atmosphere was more likely to be laden with debris from fires and dust storms. Ratios of nutrients in rainwater have been found to be similar to the ratios of nutrients in local soils and vegetation suggesting that the source of these nutrients was both terrestrial and local. It also adds to growing evidence that nitrogen compounds, translocated from soil and vegetation to the atmosphere are returned to the soil close to their source areas (Holt and Coventry, 1990). Hobbs *et al*. (1991), working in the North American prairies, argue that the effects of fire should not be studied independently of grazing as the two go hand in hand. They argue that grazing by herbivores results in conservation of nitrogen that would otherwise have been lost through burning. Furthermore, fire temperatures and energy release were reduced by grazing.

Clearly there is considerable capacity for loss of nitrogen through leaching in the wet season but as Scholes and Walker (1993) have shown, losses through leaching were minimal in undisturbed savanna in Nylsvley, reinforcing the view that savannas, like the moist forests, are geared to the conservation of essential resources. It is when these ecosystems are disturbed that losses occur.

Other forms of organic matter decomposition are not so closely tied to moisture availability. Termite activity, for example, continues through the year, consuming plant detritus and in the process converting organic compounds into inorganic. Although micro-organisms were undoubtedly the main mineralizing agents, none the less, studies by Holt and Coventry (1990) showed that in southern Africa termites could be reponsible for mineralizing up to 20 per cent of total carbon. Whether this is upheld in West Africa requires field investigation. The chances are that it is significant.

In any discussion of nutrient availability it should be noted that different nutrients become available at different depths. Savannas possess two major types of vegetation—grasses and woody species. Grasses are surface rooting and compete fiercely with woody species for available moisture and probably nutrients too. The survival of woody species is ensured by their roots invading a zone well below those of the grasses. It is quite likely that conditions of soil texture, moisture, pH, organic matter, cation exchange capacity, available nitrogen, phosphorus, and carbon vary considerably between the upper layers of the soil and those further down on which the woody species depend. For example, organic matter in the grass root zone is relatively high compared with further down and nitrogen content is always higher within the grass root zone than below (Medina and Silva, 1990).

## Fertility of savanna soils

The nutrient status of savannas has long been surrounded by controversy and according to Scholes (1990), classification of entire landscapes as either fertile or infertile according to their geology, as Bell (1982) has done, is far too sweeping when in reality savannas consist of a mosaic of fertile and infertile sites. Thus research on savannas, although still limited, has gone too far for us now to accept former broad generalizations that in Africa, wetter savannas are associated with infertile soils or dystrophic soils, classified as Ultisols and Oxisols by the USDA (Sanchez, 1976), and arid areas with their more fertile eutrophic Vertisols and Alfisols.

As evidence of this variation Scholes (1990), Blackmore *et al.* (1990) and Scholes and Walker (1993) cite the widespread occurrence of nutrient enriched patches in savanna lands which appear to be self-sustaining. Several factors are believed to contribute to these and are not necessarily interrelated: first, in areas of general low fertility and with homogeneous parent materials, geomorphological processes have somehow led to the creation of nutrient rich areas. A second explanation is that preference for certain types of vegetation by mammalian herbivores can lead to heavier grazing of these areas and the concentration of nutrients from animal dung further stimulates primary production. The area thus retains its nutrient enriched status. Thirdly, termite mounds can lead to the accumulation of nutrients in and around the mound and aerial investigations of 20-year-old maize fields in southern Africa have shown that the previous locations of termite mounds have had a persistent beneficial effect on crop growth. Fourth is the beneficial effect of trees the lateral roots of which spread widely, accumulating nutrients from an extensive area in the subcanopy habitat. And finally Blackmore *et al.* (1990) have shown that human activity in the form of precolonial settlements where cattle were corralled, where firewood was stored, charcoal burnt, and iron smelted led to significant nutrient increase.

For some reason organic matter is increased on these nutrient enriched patches or, in the case of termite mounds, the clay content is increased. This results in a greater ion exchange capacity of these areas which enables them to retain both moisture and nutrients. Some of the most important nutrients, nitrogen, and phosphorus are held in these areas in organic or some other non-leachable form and this retards the rate of nutrient loss from the area. Quite how these nutrient-enriched patches are maintained is not altogether clear, but it would seem to involve the presence of feedback loops in establishing the enriched patch as a new stable state. The evidence is largely for southern Africa, but Scholes observes that such patches can be created by humans and that once established they are extremely persistent. Perhaps the principles underlying their functioning, once better understood, could be incorporated in a role model for environmental improvement of poorer savanna lands in the future.

## Fire

Fire is an endemic force throughout the savannas (Stott, 1991) though its impact varies significantly in spatial terms. Major controlling influences are fuel, lightning strikes, and human activity and the first two of these are closely linked to rainfall. Thus the guinea savannas where moisture is abundant and where storms and lightning strikes are frequent and of considerable intensity are more prone to fires than are the semi-arid savannas further north. Vegetation is more dense in the guinea savannas than in the sudan or sahel savannas and the effects of fire are much more dramatic in the moist savannas where biomass is greater and fuel for fires is abundant. According to Swaine (1992) African guinea savannas owe their distinctive physiognomy more to frequent ground fires than to the effects of climate and grazing. Further north where there is less moisture and less fuel fires are not as important a determinant of the nature of savannas as they are in the guinea zone. With PAM of overriding importance and PAN in second place, fire might be the main modifier in the savannas with herbivory occupying this role in the semi-arid savannas. Fire is an important tool in savanna management but according to Scholes and Walker (1993), attitudes to it have varied from pyromanic to pyrophobic.

Fires may occur for a variety of reasons, four of which are important in West Africa: first, they can occur naturally as a result of lightning strikes during storms at the end of the dry season when fuel in the form of desiccated vegetation is at a maximum. Before humans became a significant force in the savannas it is likely that most fires occurred naturally at the end of the dry season. Secondly, fires may be lit deliberately when land is being cleared for cultivation. They may be lit at different times of the year to serve several purposes. Early in the dry season fires may be lit as a safety measure to reduce the fuel load which would burn at a higher temperature as the dry season progressed. Early fires may also be lit for fire breaks. Late season fires give the best burn because the vegetation is very dry but they are also the most dangerous and can get out of control easily, causing widespread destruction. Pastoralists are a third source of fires. They fire the guinea savannas in the dry season in the knowledge that this is followed by the growth of nutritious fresh green shoots. In the drier sudan savannas, grasses fired during the dry season do not produce new shoots until the rains arrive (Morgan and Pugh, 1969). Pastoralists also light fires as an aid to the control of parasites (Edwards, 1984). Finally, accidents are responsible for many fires. Land clearance fires get out of control, fires which are thought to have been extinguished continue to smoulder and come to life when the winds pick up, or people are simply careless lighting and extinguishing cigarettes. So, in terms of their distribution through the year naturally occurring fires which result from lightning strikes tend to be most frequent towards the end of the dry season.

Where vegetative matter is abundant such as in the guinea zone, farmers clearing land for cultivation often stagger their fires through the dry season, partly for safety reasons (Baker, fieldwork in Senegal and The Gambia, 1981 and 1991; Schreckenberg, pers. comm., 1996). The aim is to burn the bulk of the vegetation before it gets too dry. At this stage fires burn at lower temperatures and tend to be less destructive than fires later in the dry season when the vegetation is very dry and the heat of fires is much more intense. Patches burnt earlier can act as fire breaks for late season fires. Pastoralists in the guinea zone light fires throughout the dry season to stimulate the production of fresh grass though in the sudan savannas, fires are usually lit just before the rains arrive. The use of fire as part of a land management programme is thus of critical importance in the savannas and will be referred to again in subsequent chapters.

## Effects of fire

The effects of fire have not been as well documented as they might until recently and for the most part involve studies in savannas not in West Africa. Scientific evidence suggests that fire is not necessarily a destructive force as the biotic components of savannas are well adjusted to coping with fire. Essentially burning leads to the rapid mineralization of nutrients in plant residues but as Menaut and César (1982) note, at the time of burning, the plant matter may be poor in nutrients and, where this is the case, losses due to fire are low. Nitrogen in organic matter is rapidly mineralized through combustion from plant residues, from soil moisture and from the urine of animals. During this process ammonium compounds dissociate to liberate ammonia and oxides of nitrogen. The amount of nitrogen lost is a function of the fire temperature. In temperatures of over 600 °C, easily achieved in an intense savanna fire, there is a total loss of nitrogen. At fire temperatures of around 200 °C, the temperature possible where soil and litter meet, only 50 per cent of nitrogen is lost. In spite of such losses through volatilization, there is no evidence of the destruction of soil organic matter (Brookman-Amissah *et al.*, 1980; Scholes and Walker, 1993). Studies of the impact of fire at Nylsvley revealed no increase in the leaching of nutrients following a fire in comparison with unburned control areas, though at a depth of 35–45 cm below the surface soil moisture was notably lower than in similar locations under unburnt plots, a month after the burn. There was no significant difference in the soil moisture content of the surface layers as these had been replenished by rain. Evidence suggests that mineralization by the fire leads to an increase in nutrient availability for a short time after the fire. The productivity of the herbaceous layer increases and scientific evidence confirms knowledge long utilized by indigenous pastoralists, that there is an increase in the forage nutrient content in post-fire vegetation, new young leaves being higher in protein than old (Edwards, 1984; Scholes and Walker, 1993). This is all part of an autecological approach to herding by indigenous pastoralists (see Chapter 7).

Burning does lead to a reduction in vegetative cover. It does not normally kill plant species and savanna forests devastated by fire can usually recover (Swaine, 1992). It may, however, influence plant size. A reduction in leaf mass caused by a burn is normally offset by proliferation of new basal shoots which support proportionately more leaf than older, unburned shoots. For the most part savanna species—both trees and grass—are adapted to coping with fire.

Although many savanna species tolerate fire and some actually benefit from it, nevertheless fire can kill trees though not grasses. Research in Nylsvley (Scholes and Walker, 1993) revealed that trees died not as a direct result of fire but as a result of damage to the basal shoots through nibbling by porcupines. This left the plant vulnerable to fire attack and after erosion by subsequent fires, trees died. Human activity has also increased the vulnerability of trees to fire. Fieldwork in The Lower Gambia (Baker, 1990, 1991) confirmed that the use of fire year after year in savanna farming systems did lead to the death of trees (see Chapter 6) and similarly, severe and repeated fires in savanna forests can cause death in parent trees which can lead subsequently to the depletion of the soil seed bank (Swaine, 1992). While the ecology of savanna trees reflects their tolerance to fire under 'natural' conditions, it is possible that too much fire can be destructive of tree species. Equally, some herbaceous and woody species, normally robust in burnt areas, are weak in protected areas. Their growth may be impeded by an abundance of dead biomass (Menaut and César, 1982).

With regard to the animal population, most animals move out of an area as fire advances. The dramatic fall in the animal population following the burn is more a reflection of a lack of food than of the destruction of the population. In confirmation of this, regrowth of plant matter of high nutritive quality following the fire is accompanied by a massive rise in faunal communities. All this evidence confirms the view of Swaine *et al.* (1976) that savanna fires are more important for maintaining and intensifying vegetation patterns than for initiating them. In hierarchical terms they thus fall slightly below PAM and PAN in their importance in terms of savanna function.

## Herbivory

Last, though by no means least, herbivores are an important force in maintaining savannas at certain stages of development. Whilst mammalian herbivores and in particular cattle have received more attention than any other part of the grazing chain, the role of savanna insects in the grazing chain has been underestimated. It now appears that the invertebrate population may have a greater effect in maintaining savanna ecosystems than the vertebrates (Gandar, 1982; Andersen and Lonsdale, 1990).

Evidence from the African savannas suggests that about one-third of total insect biomass consists of herbivorous insects. Insect populations are known to vary very greatly from year to year and predictably, the number of herbivorous insects is closely related to the rainfall which is directly related to the availability of plant biomass, again emphasizing the close relationship between herbivory and other elements of the savanna model. Insect biomass and community structure are much affected by fire. Regularly burnt savannas have lower insect populations than where burning is less frequent but of the herbivorous insects that do exist, evidence from Lamto in Côte d'Ivoire has shown that grasshoppers were more numerous where land was regularly burnt than were detritivorous, predaceous, or scavenging insects, as the grasshoppers were able to feed on the flush of fresh vegetation which followed the burn (Andersen and Lonsdale, 1990).

The relatively low nutritional value of savanna grasses is an important constraint on the size of the herds of ungulates that can be supported. Termites and grasshoppers (*Acridoidea*) are probably the most important grazing insects in West Africa, as they are in other parts of tropical savanna, though as yet there appears to be little field evidence to substantiate this. Insects may be significant grazing competitors for herbivorous mammals. However, evidence from Northern Australia also shows the existence of an inverse relationship between the impact of grazing insects and the impact of mammalian grazers. An explanatory factor might have been soil fertility with insect grazers being abundant on poorer soils. This might have been a response to reduced competition from mammals on poorer soils, or it might have been related to plant chemical defences which have been shown to be lower where nutrient supplies are poor. Hence where soils are poor plants might be less well defended against herbivores and evidence suggests that many examples of plant attack by herbivorous insects occur on nutrient-poor soils (evidence reviewed by Anderson and Lonsdale, 1990). Hobbs *et al.* (1991) emphasize the need to examine the combined effects of fire and herbivory. They argue that nitrogen lossses due to burning in the absence of grazing were almost double those observed when grazing was present and that fire and grazing act together to influence the nitrogen budget of grasslands. Grazing by ungulates is thought to conserve nitrogen which would otherwise be lost by volatilization when the savannas are burnt.

Termites and termite mounds are a feature of the West African savannas as they are of savannas throughout the world though relatively little information is available on their effect on the grazing chain in the African savannas. Some termite species are soil/humus feeders and so are not justly classified as herbivores, but many are herbivorous and as Bell (1982) demonstrated they digest a significant portion of savanna vegetation. What is more, their existence is less closely tied to the rains than are the majority of insects and their activity is not seasonal as they continue to consume dead plant material even in the dry season.

Grasshoppers, another important member of the savanna grazing chain, differ from termites in that they feed on live plant material almost exclusively. Evidence from different continents shows that grasshoppers are probably the most important grazing insects in tropical savannas. In Africa they have been found to represent between 40 and 50 per cent of total arthropod biomass. Studies by Gandar (1982) (cited by Andersen and Lonsdale, 1990) revealed that grasshoppers consume about one-third of their body weight per day and that they are major grazers even in savannas supporting significant populations of herbivorous mammals. It is thus no surprise that when swarms of grasshoppers settle on farmers' fields in the West African savannas very little plant matter remains once they have left (Baker, fieldwork 1990, 1991).

In addition to grasshoppers there are many other groups of phytophagous insects including caterpillars, beetles (*Coleoptera*), bugs, and different types of larvae in the grass layer of tropical savannas. Apart from periodic defoliation of an area by caterpillars, these members of the grazing chain are normally far less important than are grasshoppers. Phytophagous insects tend to be host specific partly because specialization is needed to overcome chemical and physical defences of plants. In similar fashion to vertebrate herbivores, invertebrates always attack new leaves first, doubtless because of their better flavour and their higher protein content.

Seed predation is a third class of herbivory which has been well researched in certain tropical forests but which, until recently, has been less studied in the savannas. Seed predators in the forest consume a large proportion of total seed production and over evolutionary time are believed to have led to widespread changes to seeds and to patterns of seed production. There is much to be learnt about the effects of seed predators in the West African savannas. Herbivory in its many forms thus has a major controlling effect on the nature of savannas and, with fire, ranks below PAM and PAN in terms of its importance as a determinant of savanna.

## Models of savanna dynamics: equilibrium and non-equilibrium

Although research has shown that a combination of plant available moisture (PAM), plant available nutrients (PAN), fire and herbivory are responsible for shaping the general form of savannas, as yet models of tree–grass dynamics are still far from being perfected. One question which has been investigated is the equilibrium behaviour of savannas. This concerns how savannas change and how much they can change before change is irreversible. How stable and/or resilient are savanna environments? This is important in savannas which are subject to large and frequent changes in production, composition, and structure, and which contain severe examples of human induced degradation (Walker and Noy-Meir, 1982). Human methods of land management in savannas are considered in subsequent chapters. With regard to equilibrium, debates have circled around whether savannas tend towards a single equilibrium position, whether they have multiple stable equilibrium states and more recently, whether they are non-equilibrium systems.

Walker and Noy-Meir (1982) conclude that the capacity of savannas to tolerate fire, drought, and floods, overgrazing by cattle and goats, the effects of fauna depletion through hunting, and the removal of trees for firewood suggests that they have a high degree of resilience. Mortimore's work (1989, 1998) in the arid savannas of West Africa also notes the resilience of the environment in the face of drought. The large underground biomass in the savannas which acts as a reserve for recovery, the dormancy mechanisms of plants and animals, and predator-switching at low population levels are all methods by which savannas retain their capacity to return, more or less, to the same equilibrium position. But, as Walker and Noy-Meir (1982) observe, the resilience of savannas is not infinite and appears in places to have been exceeded in the sahel savannas of West Africa where protracted drought has resulted in desiccation followed by land degradation as a result of which open savanna has been converted to semi-desert. The single equilibrium position model was thus challenged by changes in the sahel (Warren and Khogali, 1992).

On the other hand Walker and Noy-Meir (1982) cite the nutrient enriched zones as examples of the capacity of savannas to stabilize at more than one equilibrium position. Some of the best examples come from the savannas of southern Africa where much of the area is dominated by *Burkea africana–Eragrostis pallens* vegetation. Within this association are small patches, each maybe of no more than a few hectares, where a completely different type of savanna exists and which is dominated by *Acacia* spp. with a totally different assemblage of herbaceous species. Some of these patches have been known to exist for around 40 years during which time they have been both stable and self-sustaining, giving no indication of a change back to the *Burkea* vegetation. A possible explanation is that the *acacia* vegetation includes species which can successfully compete under higher nutrient levels than normally occur in the area. Evidence from Nylsvley in southern Africa has shown that larger herbivores in the area much prefer the *acacia* vegetation and so tend to concentrate in these areas. This maintains a higher turnover rate of vegetation and also a higher rate of nutrient release. As long as herbivore movement is not curtailed the higher nutrient status of these patches should be maintained. Without the herbivores the nutrient replacement rate would fall and as nutrient levels fell to a lower, stable concentration maintained largely by decomposition of dead plant material, the most successful form of vegetation would probably be *Burkea*.

The search for equilibria reflects the dominance of ecological thinking by the temperate zone but in the tropics and in the drylands in particular, environmental conditions have failed repeatedly to satisfy the requirements of 'northern' models (see Chapter 1). An equilibrium situation among the biotic components of an ecosystem require virtual stability in the abiotic components. While it is beyond doubt that environmental conditions the world over are far from static, nevertheless, in the savannas of West Africa abiotic conditions are characterized by far greater instability than is found in most of the temperate zone (see Chapter 2). It is the tolerance and flexibility of the biotic components of savanna ecosystems that have enabled them to cope with major fluctuations in abiotic conditions. For example, adverse environmental conditions which lead to the virtual termination of one species in an area can result in the rapid spread of another species to fill the gap. While the imbalance may not be sustainable in the long term, the system appears to be driven by a series of 'swings' or strong feedback mechanisms which continue until the extreme perturbation abates. In the case of the West African savannas it is variability of rainfall that often triggers change.

The capacity of systems to survive under these fluctuating circumstances is termed persistence and their survival by avoidance of an equilibrium makes their management extremely difficult (at least from the viewpoint of specialists geared to ecosystems of the temperate zone). While such remarks are usually focused at a large scale, similar views of non-equilibrium have also been reached by Menaut *et al.* (1990) who have attempted to model tree community dynamics in the humid savanna at Lamto, Côte d'Ivoire over a 30-year period (Menaut *et al.* 1990; Menaut and César, 1982). Fig. 5.5 shows the basic structure of Menaut's tree community dynamic model. While tree density has been related to soil water availability (Tinley, 1982) or to disturbance phenomena (Walker and Noy-Meir, 1982), little research has focused on tree distribution patterns. In the humid savannas at Lamto most trees and shrubs occur in loose clumps between which are areas of grassland where isolated trees and other woody species are to be found. Menaut *et al.* (1990) were particularly interested in clumps of trees and what limited their extent. A simulation model was built which explored three main issues: the role of dispersal and growth of individual trees within the plant community; the role of local competition on the survival of seedlings and adults; and the interaction between fire and vegetation structure.

With regard to the hypothesis that fire would limit the size of clumps of trees the results showed that fire alone did not have this ability, though it might under extreme circumstances. The simulation model showed that high winds, high air temperatures and low relative humidities might possibly convert surface fires into crown fires capable of destroying the clump of trees. However, this was rare and the spread of crown fires from one tree to another had never been recorded in the Lamto savannas.

The model provided no adequate explanation for the hypothesis that competiton between woody species would result in periodic clump reduction other than that the tallest species was always the dominant. Acknowledging that competition needed to be investigated further and that particular attention should be paid to competition for water between roots of different species, Menaut *et al.* (1990) nevertheless put forward an explanation for the distribution of trees. In the Lamto savannas a few tree species reached 12–15 m in height, well above the average height of the woody vegetation, generally below 8 m. These tall trees were always found as isolates within or outside clumps. Young individuals invariably grew on sites protected from fire, usually in the heart of a clump from which grass and hence fire was excluded. It appeared that these species established only in existing clumps, and came to dominate them. Competition between the tall species and the rest of the clump appeared to seriously reduce the growth of the understorey and increase its mortality rate. This would open up a gap which would then be invaded by grasses, be vulnerable to the effects of fire which would lead to the destruction of seedlings. The combined effects of competition and the effects of fire could then exclude the understorey and account for the occurrence of tall, isolated trees.

With regard to savanna stabilization the model provided no conclusive results. In contrast to ideas on stability in savannas such as those cited above (Walker and Noy-Meir, 1982), the simulation model for Lamto showed that there was no possibility of a stable situation being reached. The overall trend

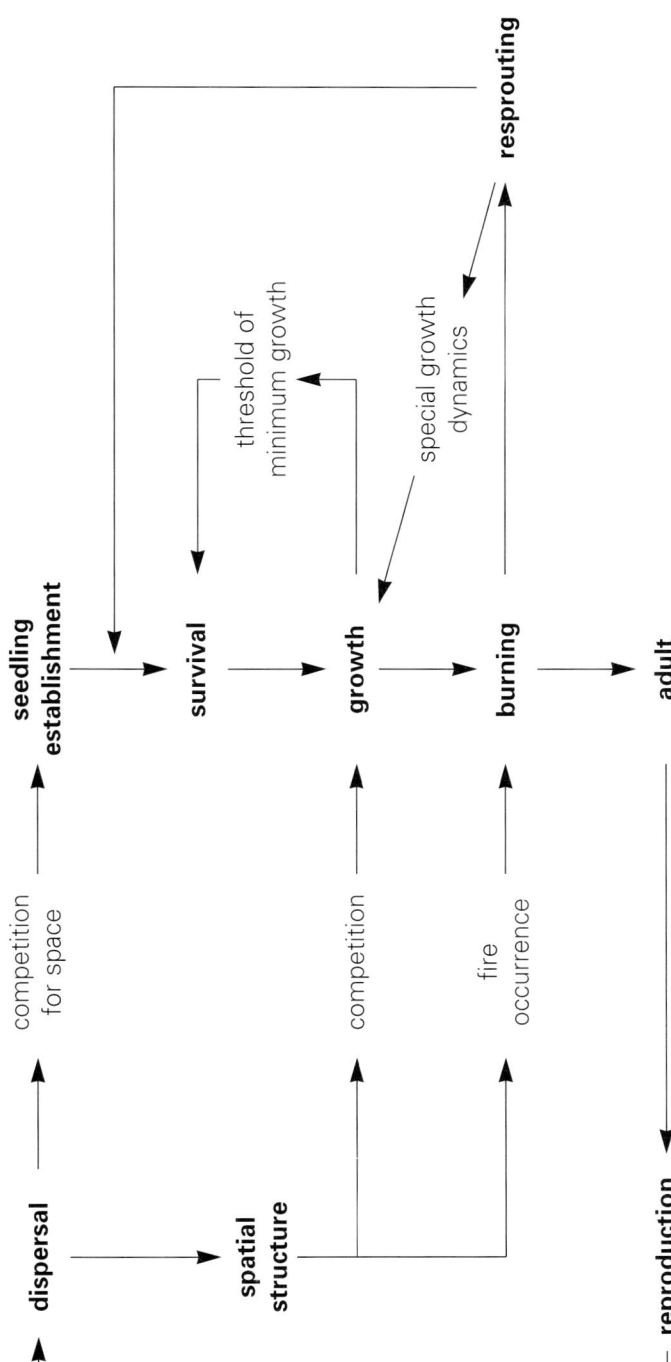

**Fig. 5.5.** Framework of a tree community dynamic model for a humid savanna of the Côte d'Ivoire

*Source:* After Menaut *et al.* (1990), 473.

was to increased invasion by woody species and indeed the predictions of the model were confirmed by field data which showed that over a 20-year period tree density in annually burnt plots in the Lamto savannas had increased by 30 per cent. However, because savanna trees can live for 50 to 200 years, and any significant disturbances might affect them for several generations, centuries of investigation may be necessary to establish patterns of change. If a better understanding of savanna dynamics is to be achieved and predictive models such as that based at Lamto are to be perfected, then it is of crucial importance, as Stott (1991) stresses, that scientific experiment is combined with ethnoecological investigation.

The variability exhibited by savannas in both space and time in response to a variety of disturbances is arguably the dominant feature of savannas and Walker and Noy-Meir (1982) argue that savannas are not only adapted to survive such perturbations but probably need them in order to maintain their resilience. Attempting to protect the savannas from disturbances can cause major problems, one example being protection from fire. This results in an accumulation of fuel and when a fire does occur as is almost inevitable, environmental destruction is so much the greater. Thus the continued search for an understanding of savanna dynamics is essential if such environments are to be managed to the advantage of West Africa's growing savanna populations.

## Savanna boundaries

Boundaries are an important feature of savannas and in West Africa they vary from fixed boundaries where, for example, the characteristic tree/grass mix of savanna meets abruptly with riverine forest, to the more gradual boundary between sahel savanna and desert. Boundaries influenced by physical factors such as changes in topography, geology, soils, and the water-table tend not to alter their positions and contrast with the northern and southern boundaries of the West African savannas which are highly dynamic and which have attracted much attention in the literature.

### The southern boundary

The southern boundary of the savanna in West Africa is a transitional zone varying from a few kilometres to over 50 km. Within this zone Adejuwon and Adesina (1992) identify three different vegetation formations: first, a mosaic of secondary forest containing species such as the oil palm, typical of forest regrowth together with species characteristic of the high forest to the south. These form islands in the savannas with very abrupt boundaries. Secondly, there is a farmland mosaic where plots with food crops such as yam, cassava, and maize are interspersed with fields in different stages of fallow regrowth and thicket. Thirdly, there is a mosaic of tree crops, mainly cocoa, kola, and citrus. Hopkins (1974) identifies a fourth formation, the presence of patches of savanna within the forest. Swaine *et al.* (1976) attribute this variation to a range of factors and inevitably human activity is high on the list. Differences in the intensity of cultivation are cited as a cause of vegetation patterns, and the significance of physical factors such as geology, soil type, and topography are also noted as important. While there appears to be an underlying belief that the derived savanna zone reflects the erosion of forest, two significant studies have revealed that this is not necessarily the case. Morgan and Moss (1965) compared the position of forest savanna boundary in Nigeria on aerial photographs taken 10 years apart and concluded that despite a marked population increase there was no evidence of widespread conversion of forest to savanna. Following this, Fairhead and Leach (1996) have shown that patches of forest in the grassy savannas of Guinea are not relics of former moist forest, but creations of the present population. Rather than causing forest destruction, their work suggests that local inhabitants are bringing about forest advance rather than the assumed forest retreat (Aubréville, 1949).

The entire boundary zone appears to be in a dynamic state with forces transforming the landscape through cultivation or allowing it to revert to its natural state through fallowing. According to Hopkins (1992), if left alone, most of this southern boundary zone would revert to forest although much of it is now classified as savanna. From an alternative viewpoint, the dynamism of this southern savanna boundary zone can be explained by the interaction of the elements of the functional model with human activity playing a major part.

## The northern boundary

On the northern edge of the savannas the boundary zone between sahel savanna and the Sahara desert is equally complex and has provoked much interest. The question here concerns whether the desert margin is moving southwards. As early as the 1930s, this was postulated by Stebbing (1937) and since the drought which began in the late 1960s, interest in the activity of the desert margin has increased. Identifying the desert margin is in itself a problem. Lamprey (1975) cited in Agnew (1995) mapped the position of the desert edge in Sudan in 1958 using rainfall and vegetation growth, and compared this with aerial surveys of the desert margin in 1975. The accuracy of this method is questionable in view of the different bases of assessment of the desert margin and the results are thus inconclusive. Lawesson (1990) mapped the structure and composition of woody vegetation in sahelian forest types in Senegal in attempt to discover whether there had been any significant movement of the vegetation boundary between sahel and desert. There was little evidence to support a floristic change from savanna to desert after two decades of drought though tree density was lower in the northern part of the sahelian zone, perhaps suggesting environmental degradation. Rather than confirming change, the study appeared to confirm the resilience of North-Sahelian vegetation in the face of fluctuating rainfall. The concept of the advancing desert seems to have lost credibility in recent years (Thomas, 1993).

While movement of the desert margin has been linked with the drought, interest has also focused on the possible causes of the drought. Are they natural, or have they been induced by human activity such as overgrazing and overcultivation of marginal land? (Grainger, 1990; Barrow, 1991; UNEP, 1992). Agnew (1995) notes that climate modelling has postulated feedback linkages between land surface moisture regimes and convectional rainfall in the sahel, and so perhaps human actions may be blamed. But human misuse of the environment may be prompted by deeper seated problems such as land tenure, debt, and inappropriate technology (Toulmin, 1993). Are such human actions causing desertification? The literature fails to distinguish between the expansion of existing deserts and the impoverishment of soils through human activities which usually occurs away from the desert margins. On this basis human contribution to the southward movement of the desert margin is likely to be marginal and much of the problem may be caused by a fluctuating climate. While much attention has been focused on land degradation, little attention has been paid to the resilience of the environment which has proved to be considerable with recovery of the rains (Agnew, 1995; Lawesson, 1990; Mortimore, 1989, 1998).

Taking a different line of explanation, the functional model outlined above may be valuable in the interpretation of this northern savanna boundary. Plant available moisture (PAM) is of critical importance here and where prolonged drought has led to desiccation and the stabilizing effects of the environment have been lost as vegetation has been destroyed, so wind-blown sand has moved southwards. In parts of northern Senegal, for example, land being cultivated in 1981 was under sand dunes in 1983 (Baker, fieldwork, 1981, 1983). In addition to the reduction in PAM in considerable areas of the sahel on account of drought, herbivory has also played a part in contributing to vegetation destruction in parts of the northern boundary area. It is conceivable, however, that with an increase in PAM the savanna boundary could move north again. The debate about the northern boundary of the sahel savannas remains inconclusive and as Agnew (1995: 143) asks, 'why focus on the desert edge anyway?' Very few people inhabit this hyper-arid region and improved management of the more productive agricultural areas to the south is of far greater consequence.

## Conclusions

This chapter has outlined the major characteristics of savanna and a functional model has been used to explain the dynamics of savannas. In an environment driven by uncertainty the savannas seem more likely to reflect non-equilibrial than equilibrial conditions, an issue highly relevant to savanna management. Savanna boundaries, as enigmatic and varied as savannas themselves, are also discussed in relation to the functional model of savanna dynamics. In the north, on the border between sahel and desert, PAM, or the lack of it, would appear to be the major influence on the boundary while in the south human activity is a greater influence on the nature of the boundary than the other elements of the model.

## References

ADAMS, W. M., GOUDIE, A., and ORME, A. (1996), *The Physical Geography of Africa* (Oxford: Oxford University Press).

ADEJUWON, J. O. and ADESINA, F. A. (1992), 'The nature and dynamics of the forest-savanna boundary in south-western Nigeria', in P. A. Furley, J. Proctor, and J. A. Ratter (eds.), *Nature and Dynamics of Forest-Savanna Boundaries* (London: Chapman and Hall).

AGNEW, C. T. (1995), 'Desertification, drought and development in the Sahel', ch. 12 in Tony Binns (ed.), *People and Environment in Africa* (Chichester: Wiley), 137–49.

AHN, PETER (1970), *West African Soils* (Oxford: Oxford University Press).

ANDERSEN, ALAN, N. and LONSDALE, W. M. (1990), 'Herbivory by insects in Australian tropical savannas: a review', *Journal of Biogeography*, 17: 433–44.

AUBRÉVILLE, A. (1949), *Climats, forêts et désertification de l'Afrique tropicale* (Paris, Société d'Editions Geographiques Maritimes et Coloniales).

BARROW, C. J. (1991), *Land Degradation: Development and Breakdown of Terrestrial Environments* (Cambridge: Cambridge University Press).

BELL, R. H. V. (1982), 'The effect of soil nutrient availability on community structure in African ecosystems', in B. J. Huntley and B. H. Walker (eds.), *Ecology of Tropical Savannas* (Berlin: Springer-Verlag), 193–216.

BELSKY, A. JOY (1990), 'Tree/grass ratios in East African savannas: A comparison of existing models', *Journal of Biogeography*, 17: 483–9.

BLACKMORE, A. C., MENTIS, M. T., and SCHOLES, R. J. (1990), 'The origin and extent of nutrient-enriched patches within a nutrient-poor savanna in South Africa', *Journal of Biogeography*, 17: 463–70.

BROOKMAN-AMISSAH, J., HALL, J. B., SWAINE, M. D., and ATTAKORAH, J. Y. (1980), 'A re-assessment of a fire protection experiment in north-eastern Ghana savanna', *Journal of Applied Ecology*, 17: 85–99.

BULLOCK, STEPHEN H., MOONEY, HAROLD A., and MEDINA, ERNESTO (eds.) (1995), *Seasonally Dry Tropical Forests* (Cambridge: Cambridge University Press).

COLE, MONICA M. (1986), *The Savannas: Biogeography and Geobotany* (London: Academic Press Inc.).

CROWDER, L. V. and CHHEDA, H. R. (1982), *Tropical Grassland Husbandry* (Harlow: Longman).

DESHMUKH, IAN (1986), *Ecology and Tropical Biology* (Oxford: Blackwell).

EDWARDS, P. J. (1984), 'The value of fire as a management tool', ch. 16 in P. de V. Booysen and N. M. Tainton (eds.), *Ecological Effects of Fire in South African Ecosystems* (Berlin: Springer-Verlag), 350–62.

FAIRHEAD, J. and LEACH, M. (1996), *Misreading the African Landscape: Society and Ecology in a Forest-Savanna Mosaic* (Cambridge: Cambridge University Press).

FURLEY, P. A., PROCTOR, J., and RATTER, J. A. (eds.) (1992), *Nature and Dynamics of Forest-Savanna Boundaries* (London: Chapman and Hall).

GANDAR, M. V. (1982), 'The dynamics and trophic ecology of grasshoppers (*Acridoidea*) in a South African savanna', *Oecologia*, 54: 370–8.
GOUDIE, A. S. (1973), *Duncrusts* (Oxford: Clarendon Press).
GRAINGER, A. (1990), *The Threatening Desert: Controlling Desertifications* (London: Earthscan).
GRITZNER, J. A. (1988), *The West African Sahel: Human Agency and Environmental Change* (University of Chicago Geography Research Paper, 226).
HEADY, H. F. and HEADY, E. B. (1982), *Range and Wildlife Management in the Tropics* (London: Longman).
HOBBS, N. THOMPSON, SCHIMEL, DAVID S., OWENSBY, CLENTON E., OJIMA, DENNIS S. (1991), 'Fire and grazing in the tallgrass prairie: contingent effects on nitrogen budgets.' *Ecology* 72 (4): 1374–82.
HOLT, J. A. and COVENTRY, R. J. (1990), 'Nutrient cycling in Australian savannas', *Journal of Biogeography*, 17: 427–32.
HOPKINS, BRIAN (1974) (2nd. edn.), *Forest and Savanna: An Introduction to Tropical Terrestrial Ecology with Special Reference to West Africa* (Ibadan and London: Heinemann).
——(1992), 'Ecological processes at the forest-savanna boundary', in P. A. Furley, J. Proctor, and J. A. Ratter (eds.), *Nature and Dynamics of Forest-Savanna Boundaries* (London: Chapman and Hall).
HUNTLEY, B. J. and WALKER B. H. (eds.) (1982), *Ecology of Tropical Savannas* (Berlin: Springer-Verlag).
JONES, M. B., LONG, S. P., and ROBERTS, M. J. (1992), 'Synthesis and conclusions', ch. 8 in Stephen P. Long *et al.* (eds.), *Primary Productivity of Grass Ecosystems of the Tropics and Sub-Tropics* (London: Chapman and Hall), 212–55.
KINYAMARIO, J. I. and IMBAMBA, S. K. (1992), 'Savanna at Nairobi National Park, Nairobi', ch. 2 in Stephen P. Long *et al.* (eds.), *Primary Productivity of Grass Ecosystems of the Tropics and Sub-Tropics* (London: Chapman and Hall), 25–69.
LAWESSON, JONAS ERIK (1990), 'Sahelian woody vegetation in Sénégal', *Vegetatio*, 86: 161–74.
LONG, STEPHEN P., JONES, MICHAEL B., and ROBERTS, MICHAEL J. (eds.) (1992), *Primary Productivity of Grass Ecosystems of the Tropics and Sub-Tropics* (London: Chapman and Hall).
MCCOWN, R. L. and WILLIAMS, JOHN (1990), 'The water environment and implications for productivity', *Journal of Biogeography*, 17: 513–20.
MEDINA, ERNESTO and SILVA, JUAN, F. (1990), 'Savannas of northern South America: A steady state regulated by water–fire interactions on a background of low nutrient availability', *Journal of Biogeography*, 17: 403–13.
MENAUT, J. C. and CÉSAR, J. (1982), 'The structure and dynamics of a West African savanna', in B. J. Huntley and B. H. Walker (eds.), *Ecology of Tropical Savanna* (Berlin: Springer-Verlag), 80–100.
——GIGNOUX, J., PRADO, C., and CLOBERT, J. (1990), 'Tree community dynamics in a humid savanna of the Côte d'Ivoire: Modelling the effects of fire and competition with grass and neighbours', *Journal of Biogeography*, 17: 471–81.
——LEPAGE, MICHEL, and ABBADIE, LUC (1995), 'Savannas, woodlands and dry forests in Africa', ch. 4 in Stephen H. Bullock, Harold A. Mooney, and Ernesto Medina (eds.), *Seasonally Dry Tropical Forests* (Cambridge: Cambridge University Press), 64–92.

MOORE, P., CHALONER, W., and STOTT, P. A. (1996), *Global Environmental Change* (Oxford: Blackwell Scientific).

MORGAN, W. B. and MOSS, R. P. (1965), 'Savanna and forest in Western Nigeria', *Africa*, 35 (3): 286–93.

—— and PUGH, J. C. (1969), *West Africa* (London: Methuen).

MORTIMORE, M. (1989), *Adapting to Drought: Farmers, Famines and Desertification in West Africa* (Cambridge: Cambridge University Press).

—— (1998), *Roots in the African Dust* (Cambridge: Cambridge University Press).

NICHOLSON, S. (1996), 'Environmental change within the historical period', ch. 4 in W. M. Adams, A. S. Goudie, and A. R. Orme (eds.), *The Physical Geography of Africa* (Oxford: Oxford University Press).

RATTRAY, J. M. (1960), *The Grass Cover of Africa* (Rome: FAO).

RAUNKIAER, C. (1937), *Plant Life Forms* (Oxford: Oxford University Press).

RICHARDS, P. W. (1952), *The Tropical Rainforest* (Cambridge: Cambridge University Press).

ROBERTS, N. (1989), *The Holocene: An Environmental History* (Oxford: Blackwell).

SANCHEZ, PEDRO A. (1976), 'Soils of the Tropics', ch. 2 in *Properties and Management of Soils in the Tropics* (New York, London, Sydney, Toronto: John Wiley & Sons) pp 52–95.

SCHOLES, R. J. (1990), 'The influence of soil fertility on the ecology of southern African dry savannas', *Journal of Biogeography*, 17: 415–19.

—— and WALKER, B. H. (1993), *An African Savanna: Synthesis of the Nylsvley Study* (Cambridge: Cambridge University Press).

STEBBING, E. P. (1937), *The Forests of West Africa and the Sahara* (London and Edinburgh: Chambers).

STOTT, P. A. (1991), 'Recent trends in the ecology and management of the world's savanna formations', *Progress in Physical Geography*, 15 (1): 18–28.

—— (1994), 'Savanna landscapes and global environmental change', ch. 13 in Neil Roberts (ed.), *The Changing Global Environment* (Oxford: Blackwell), 287–303.

SWAINE, M. D. (1992), 'Characteristics of dry forest in West Africa and the influence of fire', *Journal of Vegetation Science*, 3: 365–74.

—— HALL, J. B., and LOCK, J. M. (1976), 'The forest-savanna boundary in West-Central Ghana', *Ghana Journal of Science*, 16 (1): 35–52.

THOMAS, D. G. (1993), 'Sandstorm in a teacup? Understanding desertification', *Geographical Journal*, 159 (3): 318–31.

THOMPSON HOBBS, N., SCHIMEL, DAVID S., OWENSBY, CLENTON E., and OJIMA, DENNIS (1991), 'Fire and grazing in the tallgrass prairie: Contingent effects on nitrogen budget', *Ecology*, 72 (4): 1374–82.

TINLEY, K. L. (1982), 'The influence of soil moisture balance on ecosystem patterns in Southern Africa', in B. J. Huntley and B. H. Walker (eds.), *Ecology of Tropical Savannas* (Berlin: Springer-Verlag), 175–92.

TOULMIN, CAMILLA (1993), *Combating Desertification: Setting the Agenda for a Global Convention* (London: International Institute for Environment and Development Paper 42).

TRAPNELL, C. G. (1959), 'Ecological results of woodland burning experiments in Northern Rhodesia', *Journal of Ecology*, 47: 129–68.

—— et al. (1976), in Menaut et al. (1995), 'Savannas, woodlands and dry forests in Africa'.

United Nations Environment Programme (UNEP) (1992), *Earth Summit, Rio de Janeiro* (London: Regency Press).

WALKER, B. H. and NOY-MEIR, I. (1982), 'Aspects of stability and resilience of Savanna Ecosystems', in B. J. Huntley and B. H. Walker (eds.) *Ecology of Tropical Savannas*. Ecological Studies, 42 (New York: Springer Verlag), 556–90.

WARREN, ANDREW and KHOGALI, MUSTAFA (1992), *Assessment of Desertification and Drought in the Sudano-Sahelian Region, 1985–1991* (New York: UNSO).

WHITE, F. (1983), *The Vegetation of Africa, A descriptive Memoir to Accompany the UNESCO/AETFAT/UNSO Vegetation Map of Africa* (Paris: UNESCO).

# 6

# Farming in the Semi-Arid Domain: Adaptation to an Uncertain Environment

With the arguments for environmental non-equilibrium in place, this chapter explores indigenous responses to environmental uncertainty. Coping with the uncertainty of rainfall is one of the greatest challenges that savanna dwellers face and this occurs at several scales in time and space. First, it occurs on an annual basis, as part of the year in the savannas is, by definition, dry (see rainfall distribution at savanna sites, Fig. 1.5). Dryland or rainfed farming is thus confined to the wet season and considerable problems may arise if the rains arrive late, if they are erratic when they do arrive and if they cease before the time anticipated. A second element of uncertainty with regard to rainfall is its variability from year to year. Chapter 1 demonstrated the considerable variation in rainfall across the region in relation to the mean, and in addition to this is its unpredictability in space, both within and between seasons. Thus farmers have to be prepared to cope with whatever arrives: favourable rains, drought, or floods. A third form of uncertainty extends over a longer period and concerns the trend of declining rainfall in the sahel and northern savannas (Fig. 1.4). Since this began in the late 1960s, the spatial and temporal variability of rainfall has increased. In the early 1990s the rainy season in The Lower Gambia was about one month shorter than it was 20 to 30 years earlier (Baker, 1995), the start of the rains being late and particularly unpredictable. A reduction in the length of the rainy season has also been identified in Senegal by Khalfaoui (1991) and in Northern Nigeria by Oladipo and Kyari (1993) where the rains have tended to end sharply and earlier than in the past. A trend of lower rainfall across the savannas and sahel has led to reduced runoff and to lower water levels in many of West Africa's rivers and, as a consequence, to poorer floods which have had direct and indirect effects on farming (Amara, 1993; Park, 1993). Whether the drought is at an end remains to be seen, but although the rains have increased in recent years, unpredictability of rainfall distribution remains as much of a problem as ever (Fig. 1.8).

This chapter attempts to reveal the competence of smallholder farmers in coping with the major challenges presented by an unpredictable environment. In spite of little support and assistance from West African governments, smallholder farmers have survived many difficult years. Their autecological approach to the production of crops and animals and their land management techniques reflect a knowledge of the environment that development projects have, so far, failed to even try to emulate. Much of the material for this chapter has been drawn from the author's fieldwork which commenced in 1981, in the Western Division of The Gambia. During the past two decades the effects of the drought have become much more pronounced and farmers have learnt to cope with even greater uncertainty than hitherto. Discussion is not confined solely to The Gambia, and where relevant, is widened to include examples from other parts of the West African savannas.

From west to east, the savannas in West Africa extend over a distance equivalent to that from London to Moscow, and from north to south, the distance they cover is very little less. In an area so vast, generalization about farming systems is extremely difficult. Above all else, the seasonality of rainfall is of outstanding importance in influencing the nature of farming. Essentially, dry land farming is confined to the rainy season, which may be as much as 8 months in the south of the region, or as little as 3 months in the north (Fig. 1.5). There is a transition in the types of dryland crops grown between the humid south and the more arid north. In the humid zone root crops dominate dryland farming to the east of Côte d'Ivoire, while rice is important to the west. In the wetter guinea savannas combinations of root crops and cereals are more common, particularly the more moisture demanding cereals such as maize. The proportion of cereals increases northwards, millet and sorghum being the principal crops in the more arid sudan savannas and sahel. If maize is grown, it is in soils where the moisture content is higher. This broad pattern of crops grown on the drylands is made much more complex by cultivation of areas with moisture-rich soils. These include lands which are seasonally flooded, areas where the watertable is nearer the surface such as in *wadis* and *bas-fonds*, and artifically irrigated areas. While physical factors may determine whether more moisture demanding crops can be grown in the savannas, human factors such as availability of labour, access to land, access to market, size and function of the market and domestic needs and preferences can all influence what is produced.

Characteristic of indigenous farming in West Africa, cultivators frequently farm several plots of land, each under different ecological conditions. Broadly, these can be divided into dryland or rainfed fields, wetland fields and vegetable gardens which may or may not be on wetlands but where some form of irrigation is normally used. The organization of this chapter follows these divisions: the first section focuses on aspects of the cultivation of dryland fields to demonstrate the skill of farmers in coping with an environment not in equilibrium. The second part examines examples of wetland management throughout the region and considers the severe impact of the drought on wetland cultivation in The Lower Gambia and its consequences for local communities, in particular, for women. The third section examines the promotion of vegetable gardening by aid agencies in their attempt to compensate for the effects of the drought. This is an example of how care has been taken to ensure that vegetables, both traditional and new varieties, can be grown successfully. However, the institutional framework necessary to facilitate the disposal of the products of vegetable gardening is still not in place. Thus the initiative has not succeeded as well as it might in assisting women to overcome the virtual anihilation of much riverine wet rice cultivation by drought in the Western Division of The Gambia.

## The ecology of dryland cultivation

The variety of approaches to dryland farming in West Africa defies generalization although many systems embody similar principles, that is, the cultivation of rainfed land for a few years after which it is left to fallow. Fallowing facilitates a transfer of nutrients from subsoil to topsoil, and without an adequate fallow period or any other form of nutrient subsidy, the soil would be unable to improve its physical condition or increase its nutrient status (Kowal and Kassam, 1978). Time scales for both cultivation and fallow vary considerably but in those areas where the land is no longer left to fallow, inputs are required to maintain the productive capacity of the land. Farming on the drylands in The Lower Gambia bears many similarities to dryland farming in the more land-rich areas of south-eastern Nigeria although a major difference is that the cultivation period is much more limited in the year and is confined to the rainy season. Dryland farms are normally the responsibility of men in Mandinke villages of the Lower Gambia and are cultivated predominantly with the food crops millet and sorghum, and with groundnuts, the country's chief cash crop (Carney, 1992). Some maize, beans, cowpea (*Vigna* spp.), and possibly cotton may be grown and more recently the production of cassava and fruit, oranges and mangoes in particular, has increased (Baker, 1995). As in other parts of the country the household farm is divided into two distinct sections: a major communal area for subsistence food production using communal labour, and a smaller area of personal fields. Decisions about what to grow on the communal rainfed fields are taken by senior members of the household in consultation with the entire household. Personal fields, on the other hand, are sown mostly with cash crops though income from these may either support the household, be used for personal consumption, or marketed (Carney, 1992; Webb, 1992).

## Preparation of dryland fields for cultivation

Land preparation commences with clearance of vegetation prior to cultivation. In the Western Division of The Gambia this is not the heavy task it was even 15 years ago when the bush was left to regenerate for several years between periods of cultivation. Fallows are now down to 2 years or less and even where there is no fallow period, there is still a significant regrowth of weeds between the end of the rainy season in November and March/April, when the land is cleared for cultivation. In Gambian villages the weeds are cut and left to dry in piles, while in northern Senegal they are heaped into long narrow strips prior to being burnt. As is the custom further south, large trees and trees of economic value are left on the fields, the former in particular being valuable for the shade they cast. The trees provide shelter from the sun not only for people working in the fields, but also for animals which are allowed to roam freely in the dry season. If attracted to a plot to browse or just to stand in the shade, they can make a valuable contribution to the nutrient status of the soil through their urine and dung. Certain trees also provide products of economic value, usually from their leaves, fruit, bark, wood, and roots, and care is taken to protect these from the plough. For example, the pods of the locust bean, *Parkia biglobosa*, have a variety of culinary uses which are detailed by Schreckenberg (1996). Different species of acacia have multiple uses as does *Prosopis africana*, iron wood. Trees with medicinal properties are treated with care as are timber species such as the hardwood *Prosopis africana*, renowned for its timber and which usually appears stunted, having been severely pollarded. The varied uses of tree species has been discussed in greater detail by Madge (1994), Percival (1968), and Szolnoki (1985) in The Gambia; and Schreckenberg (1996) in the Guinea savannas of Benin.

According to Gambian farmers there are fewer naturally occurring trees on dryland fields now than there were 20 years ago. The drought and the subsequent fall in the water table is said to be the main cause of this. Similar observations have been made by the Schoonmaker Freudenbergers (1993) in northern Senegal. In response to arguments that the low density even of economically valuable trees on the drylands is due to their having been cut down by the local population, it is noteworthy that since the late 1980s when rainfall has been more plentiful, some species which had apparently disappeared have started to reappear and are being nurtured by local communities (Baker, fieldwork, 1991). This gives weight to the suggestion that removal of the trees may have been drought related and is a further indicator of the resilience of the environment and its capacity to recover after periods of stress. Working in the Maradi *Départment* of Niger, Manvell (pers. comm., 1998) has shown that the drought together with human use of trees has led to a reduction in the number of certain tree species. An additional explanation given by Gambian farmers for the reduction in the number of naturally occurring trees was the increase in the frequency of burning, which had accompanied the decrease in the fallow period. Although many savanna species are fire tolerant and the build-up of fuel is now less than it was when fallows were longer, there were limits to how many fires a tree could withstand, particularly when fields were burnt every year. Swaine *et al.* (1976), and Swaine (1992) have confirmed that the mortality of trees, particularly smaller trees, is significantly increased with repeated burning. It only requires damage to the bark by rodents or grazing animals for disease to attack or seriously increase the vulnerability of a tree to fire damage.

In contrast to examples of tree loss, the work of Fairhead and Leach (1996) in the moist savannas of the Kissidougou region of Guinea has shown that rather than reducing the number of trees on the forest margin, through the activity of the local population, patches of forest are extending into the savanna. Over 20 years earlier, Spichiger and Pamard (1973), cited in Swaine *et al.* (1976), hypothesized that farming on the savanna side of the forest–savanna boundary in Côte d'Ivoire actually favoured the spread of forest species. Such evidence confirms even earlier observations by Morgan and Moss (1965), that cultivation in patches of forest on the forest–savanna boundary in Nigeria did not necessarily lead to the destruction of forest. The examples from the more southerly savannas suggest that human effects on tree cover may not be as negative as is frequently suggested and that they may indeed be positive. This is certainly true in the Western Division of The Gambia where farmers are planting extensive orchards to mangoes and to cashew. Ten to 15 years ago oranges were being planted, but these are more susceptible to drought than mangoes and cashew and the emphasis has now changed.

One problem facing farmers clearing land for cultivation is hardening of the soil surface. In some cases, the surface, though hard, may be workable with a hoe, but in others, it can be almost impenetrable. The situation tends to be worse on soils with a high clay content in the more arid areas although throughout the savannas the unpredictable occurrence of hardpans is a problem. After the burn and shortly before the rains arrive, the plot may be hoed lightly unless it is too hard, or more rarely, ploughed with oxen in readiness for sowing. Comparatively few farmers in villages visited in the Western Division of The Gambia had access to oxen. There are apparently more ox ploughs on the north bank of the river because they are easily imported from Senegal (Baker, fieldwork, 1999). The statistics bear this out to some extent (NADC *et al.*, 1998). It has been shown that access to oxen and to weeder ploughs can significantly increase yields, though in these semi-arid parts users can run the risk of increasing soil erosion (Delgado and McIntire, 1982; Jaeger and Matlon, 1990; Adesina, 1992; Toulmin, 1992). Hardening of the surface soil can take considerable areas out of cultivation though in the Yatenga region of Burkina the introduction of *zäi* pits has done much to rehabilitate surface crusts (Ouedraogo and Kaboré, 1996). The *zäi* pits may have a diameter of 20–30 cm and a depth of 10–15 cm though they vary in size according to soil conditions. They tend to be larger on lateritic soils for example. The pits are dug during the dry season and it may take some 60 working days to produce 1 hectare of *zäi* pits. The labour input is considerable but farmers are willing to do this because the bulk of the work can be done in the dry season. Fine debris and sand accumulate in the pits and if possible, farmers add a little manure to them. This encourages the development of soil fauna which in turn aids the aeration and moisture absorption capacity of the soil. *Zäi* allegedly had their origin in Burkina but the spread of these has been encouraged by Oxfam and variants on the same theme are in evidence in Niger (Hassan, 1996).

In northern Senegal a local solution has also been found to the problem of hardened soils (Schoonmaker-Freudenberger and Schoonmaker-Freudenberger, 1993). After the harvest, millet stalks are strewn on the hard soil and through the dry season these are covered with blown sand. The land is left unused for an entire wet season and by the subsequent wet season the composted stalks have 'softened' the soil, enabling it to be cultivated once more. These are just two methods of land preparation which cope with an unpredictable physical environment. Others exist, but the point being made is that such methods reflect indigenous solutions to indigenous problems.

Before the drought began, prediction of the commencement of the rains was based on evidence from a wide range of traditional indicators including, for example, the arrival of certain species of migrant birds and baobab and other trees coming into leaf as humidity levels rose. In the knowledge that the rains were soon due to start, the seed of many dryland crops would be sown just before the rains to make the most of the moisture available. Since the drought it has become even more difficult to predict when the rains will start and in The Gambia, it is common for farmers to wait for the first rains to arrive before sowing their dryland crops. Sowing earlier is too risky, for if the rains are late, the grain becomes shrivelled, or is eaten by rodents or other predators and germination is poor as a result. In Northern Nigeria, Mortimore (1998) notes that seed is sown both before and after the rains in an attempt to reduce the risk of loss. Millet and sorghum are the main food crops planted in the drier savannas though in addition to these cowpea and beans are commonly grown as well as groundnuts. However, cultivation of the latter has declined largely as a result of disease. According to Kowal and Kassam (1978), cited in Mortimore (1998), a minimum of 348 mm is required for millet cultivation though Mortimore (1998) observes that the millet crop is vulnerable everywhere to the north of the 588 mm isohyet.

Crops sown with the rains in the Lower Gambia are harvested in September/October depending on when they mature. One response to the unpredictability of the rains is to plant seed which matures at different rates. Varieties of millet, sorghum, maize, and groundnuts which matured in 120 days used to be common prior to the drought, but in 1990/1 farmers were planting little other than varieties which mature in 90 days, in spite of their lower yielding capacity (Baker, 1995). Subsequent field research in 1999 has revealed changes. Farmers in village visited in the Western Division were, without exception, growing longer maturing varieties of cereals because consumption of the earlier maturing grains by flocks of hungry birds was extremely high. Later in the rainy season there was more food available for the birds, bird scaring was more effective and there was a greater probability of a better harvest. Had it not been for the predation by birds, farmers would not have moved back to the longer maturing varieties, for if the rains end early, they may harvest nothing at all. In 1998 the millet harvest failed for this very reason.

Manvell (pers. comm., 1998), working in the Maradi region of Niger, has witnessed the cultivation of millet varieties which mature in 45 days. Yields from these are inevitably lower than from slower maturing varieties, the heads rarely exceeding 20 cm in length, but in times when rains are very poor they do permit a harvest. With some varieties of longer maturing sorghum, farmers in The Gambia did not wait until the crop had matured but harvested early once the ears were formed but before they had reached their maximum size. This ensured that there was a harvest and reduced the risk of loss to birds. In a Northern Nigerian village, Mortimore (1998: 46) discovered that the ten crops grown for subsistence comprised forty-three varieties, seventeen of them sorghum. Each had different characteristics. Having their own private seed bank gives farmers a degree of choice, enabling them to plant seed types appropriate to prevailing rainfall patterns. Results are rarely perfect, but in most years a harvest is reaped. Toulmin (1992), working on the northern edge of the savannas in Mali, describes the use of shorter maturing varieties (60–80 days) in manured fields near the village, while further away, bush fields are sown with longer cycle millet (120 days). This both reduces the risk of crop loss and helps to ease labour bottlenecks as the times of planting, weeding, and harvesting of these varieties with different maturation rates, do not coincide. These are further examples of the reserves of the African smallholder for coping with an environment where uncertainty is the norm.

## Cropping patterns

Intercropping occurs throughout the savannas though it may be less important in the guinea savannas and on the borders of the guinea and sudan savannas further north (Yaycock, 1981). In the villages visted in The Gambia most crops were planted in single stands because higher levels of labour are required for intercropping. Very much an indigenous technique, intercropping was the focus of some 300 experiments at the International Institute of Tropical Agricultural Research in Nigeria between 1924 and 1960 but interest in it during the colonial era and in the early post-colonial era was intermittent, possibly because it was part of subsistence agriculture and of little importance to 'real' agriculture (E. F. I. Baker, 1981). However, since the 1970s interest in intercropping has increased and work by Norman (1972, 1974) in Northern Nigeria has amply demonstrated its merits. Although yields from each individual species may be lower than if the stand was grown as a monocrop, Netting (1993), Norman (1974), and Norman *et al.* (1982) have shown that the total yield from an intercropped plot may be higher. Other benefits are also evident: the variability of returns is less than from sole cropping and labour demands can be staggered, though they are high. The mix of crops helps to limit the spread of disease; it helps to limit erosion and provides variety in the diet. Plants with different physical forms, planted at different times, and with different rates of maturation do not compete excessively for resources of light, space, moisture, and soil nutrients. In consequence, the number of plants per unit area can be greater on an intercropped plot than on a pure stand. Norman *et al.* (1982) have identifed up to six different species on a single intercropped plot though a total of two or three are more common. In spite of its potential benefits, monocropping has largely replaced intercropping in the Western Division of The Gambia as the total labour demand is much higher for the latter.

Different systems of intercropping are successful for different reasons as th four following examples from the drier savannas suggest. First, in Northern Nigeria, millet and sorghum are frequently grown together on the same plot. Millet is sown very early with the first rains and sorghum is interplanted later when the rains become more reliable (E. F. I. Baker, 1981; Mortimore, 1998). The millet is planted at low density and thus capitalizes on the limited moisture in the soil from early rains. When millet is interplanted with sorghum, the plot becomes a full stand and by now moisture resources are at their most plentiful as the rainy season has reached its height. After the harvest of the millet at the end of the rainy season the sorghum alone remains on the plot to mature on the residual soil moisture. This has proved to give better returns than either sole cropping of sorghum or relay cropping of millet and cowpea (*Vigna unguiculata*), the latter having the advantage of fixing nitrogen in the soil. The major advantage of sowing a combination of early and late maturing crops is that cultivation continues over a longer period than if a monocrop is planted and greater use can be made of moisture resources (E. F. I. Baker, 1981).

Secondly, in Niger, millet and cowpea is the primary crop association and here too millet is planted as the rainy season starts, usually June to July, in rows 1–1.5 metres apart. Planting cowpeas between the rows some 4–6 weeks after the millet makes the best use of moisture resources and gets the best results (Lowenberg-De Boer *et al.*, 1991). Yields may be higher when crops are rotated rather than intercropped though two major shortcomings of monocropping are that the scope for erosion is greater, particularly in the case of cassava, and the variety of foods produced from an intercropped plot is lost.

In northern Burkina, formerly Upper Volta, Stoop (1981) has identified sorghum, millet, and cowpea as a common combination. The rationale behind this third crop combination is different from the two examples cited above. These are all photosensitive species, well adapted to early planting and to replanting in gaps where seeds have failed to germinate. When planted early, local millet and sorghum tiller profusely producing large panicles and effectively compensating for missing plants. Virtually all local fields have a small population of photosensitive cowpeas, the spreading habit of which provides soil cover, particularly in patches where millet and sorghum have not germinated. Covering areas of bare soil, cowpeas provide soil protection by reducing erosion, runoff, soil compaction due to raindrop splash and crusting of the soil surface. They also impede the growth of weeds and hence reduce the labour required for weeding. Finally, the soil benefits from the capacity of the legume to fix nitrogen (Stoop, 1981). In view of the many merits of intercropping, it is noteworthy that so much of the improved technology available has been developed for monocropping (Yaycock, 1981).

A fourth and final example of intercropping is from the Lower Middle Valley of the Senegal River, the stretch from Dagana to Boghé. Here rainfed farming is risky and farmers depend more heavily on flood recession agriculture (see below). Nevertheless, dryland fields are usually seeded in the hope of reaping a harvest. A mix of different crops is sown in the same field, crops such as sorghum and/or millet, cowpeas and watermelons in the same pocket. In addition, within the same pocket, varieties are planted with different maturation periods. Shorter maturing crops have lower yields but by mixing the crops, risk is spread. Where farmers keep small ruminants they may also include within the same pocket, tall varieties of sorghum for fodder. If disease strikes, or rainfall is insufficient to allow the development of the grain, then at least the vegetative material may be used for animal feed (Park, 1993).

## Maintaining soil quality on the drylands

Methods of maintaining the fertility of dryland fields are complex and highly varied. They are closely linked with soil erosion and the conservation of moisture. A wide range of indigenous methods exists for coping with such problems including the construction of micro-moisture catchment areas; stone bunding; contour tillage; grasses grown along field borders and in narrow strips across fields to limit erosion of soil by water and wind; the planting of crops in alternate strips along the contours to limit wind and water erosion and, where slopes are very steep, the use of terracing (Kassogué *et al.*, 1996; Hassan, 1996; Millar, *et al.*, 1996; Ouedraogo and Kaboré, 1996; Wedum *et al.*, 1996).

With particular regard to maintaining the nutrient status of soils in dryland fields the fallow system is arguably the best known means of restoring soil fertility, but in addition to this, organic inputs such as mulches and manure may be used. The use of these is closely related to the availability of the inputs, of labour and of transport. On the central plateau of Burkina, Slingerland and Masdewel (1996) have shown that mulches are one of the means by which soil fertility is increased. Mulches are also used to soften a hard crust which has formed on land formerly in cultivation but which has gone out of use. The approach bears some resemblance to that described by the Schoonmaker-Freudenbergers (1993) in northern Senegal. Mulches include a range of grasses, for example, *Loudetia togoense*, and where grass is in short supply, the dry leaves of certain tree species may be used, a particular favourite being

*Butyrospermum parkii*. Citing a study in the Sanmatenga province of Burkina, Slingerland and Masdewel (1996) note the direct relationship between the area mulched and the number of active workers per hectare. They have also shown that larger farms have larger bush fields and that these are more often mulched than fields closer to home where more manure is used. Interethnic variations also emerged with the majority of the Mossi mulching their land while only a minority of the Peul do this. The explanation offered by the authors was that the Peul, who were agropastoralists, had access to more manure and had smaller fields, while the Mossi who were predominantly cultivators had less access to animal manure so used mulches instead.

In the northern savannas in Mali, Toulmin (1992) describes a farming system where dryland fields are divided into bush fields some distance from the village, where nutrient subsidies were limited, and fields much nearer the village, where the nutrient status was much higher owing to the use of manure. In the fields at greater distance from the village where land is plentiful, a strip of bush adjacent to the field is cleared and added to the field. At the same time, a strip of land, usually on the village side, is left to fallow. Thus the bush field is made up of a series of strips each of which has been farmed for a different period and it is only when the strip of land which has been farmed the longest shows signs of nutrient deficiency that a new strip is cleared. Where new fields are being opened up, farmers tend to prefer lighter sandier soils. Millet sown in these conditions develops a good root network if the rains are favourable. While clay soils may have a greater capacity to retain moisture, they also have a greater tendency to harden in semi-arid areas and in consequence, can inhibit plant growth.

In the villages visited in The Lower Gambia little in the way of mulches was used on dryland fields, some manure was used but traditionally, fertility of the soil was maintained by a fallow period. During this phase (Fig. 3.4) the export of nutrients from the system, mainly through harvest, virtually ceases, soil structure improves, and nutrient accumulation occurs (Ahn, 1970; Sanchez, 1976; Pieri, 1992). The length of time that land should be left in fallow is debatable. According to Nye and Greenland (1960) a period of 5 years should be sufficient to see a return of the major nutrients and an improvement in soil structure. According to farmers in The Lower Gambia much depends on the quality of the soil on which the plot stands. Farmers tended not to go by the number of years that the land had lain fallow, not least because erratic rainfall could significantly affect vegetative growth. The main indicator used was the density that the vegetation cover had reached and also the return of certain indicator species. When these criteria were satisfied the land was once more ready to be cleared, burnt and planted.

Formerly, a major input into the system came from the ash which remained after the vegetation on the plot which had been cut, left to dry, and then burnt. However, in the villages visited the bush vegetation was rarely allowed to build up, the burns were very much smaller than in the past and did not represent much more than a few bonfires on the plot. Vegetation is burnt prior to cultivation of the land and the debate still rages among farmers as to whether it is better to burn late or early. Some of the more ecologically aware farmers in villages visited in The Gambia were anxious that the burn should not be too late into the dry season, as by this time, the fuel was very dry and the risk of fire spreading and causing damage to neighbouring fields and sometimes to the village itself was much increased—although the quantity of fuel was fairly low. In an earlier burn, the plant litter contained more moisture, the temperature of the fire was lower and thus, it was safer. Many did not agree with burning early as this produced less ash, much of which blew away before cultivation began. Neither did an early burn consume all the plant matter because it was not sufficiently dry and a second burn was required prior to cultivation. To some farmers this was excessively time- and labour-consuming when both were in short supply. Regardless of the arguments, the quantity of ash produced on fields where the fallows had been substantially reduced or eliminated was minimal in comparison with that on plots which had been left to fallow for several years. Having said this, it is doubtful whether long fallowed fields benefit from all the ash, as much is blown away by strong winds. In northeastern India, Toky and Ramakrishnan (1981) found that a greater proportion of ash was blown off fields fallowed for 30 years than from fields fallowed for five years. It could thus be argued that a higher proportion of nutrients was lost through wind action from plots fallowed longer. Nevertheless, experimentation has shown that on long fallowed plots the calcium content of the ash may be higher, particularly if the cut vegetation includes broadleafed species rich in calcium (Toky and Ramakrishnan, 1981). Furthermore, the soil structure would almost certainly benefit from a longer period in fallow. Observations by the author in The Lower Gambia concluded that any ash removed by the wind was also deposited, the main beneficiaries being lower lying areas either under cultivation or in fallow. Wherever it was deposited, the ash represented an additional nutrient input.

Fieldwork in the Western Division of The Gambia (1991, 1999) revealed that the fallow period had undergone major changes since the drought began. Oral testimonies of local people confirmed that land was cultivated for 4–5 years and left to fallow for approximately 8 years in the early 1980s, though inevitably there was variation. This apparently represented a decline in the fallow period over previous years, but nothing as significant as the decline that followed. By the early 1990s fallows were rarely more than 2 years and frequently there was no fallow period in either of two villages where studies were conducted: Nyofellehmedina and Kassa Kunda. A rotation system consisting of millet one year and groundnuts the next kept the land occupied in every wet season. The only relic of a fallow period was that the land remained unused during the dry season. By 1999 there were no fallows in Nyofellehmedina though in Kassa Kunda they were at least 2 years. Further investigation revealed that a significant decline in the number of animals in Kassa Kunda due to theft, had seen short fallows maintained, and in some cases increased slightly, to 2–3 years over the past decade. Nyofellehmedina had also lost animals due to theft but in Kassa Kunda the situation was worse. In Nyofellehmedina animals were tethered on the fields more regularly which suggests that the higher input of dung compensated for the loss of the fallow. As a result of the drought, crop yields in Gambian villages had declined and returns from the farms had fallen. As a consequence, many people, mostly men, had migrated to urban areas to supplement farm incomes and sent home remittances on which rural families now depended (Baker, 1995). The shortage of farm labour had become more acute as many of the farmers who remained to work the dryland fields were older men, several of them over 70 years old. Unwilling to walk long distances to their fields and lacking the labour and technology to clear well-developed bush on land over 1 to 1.5 km from the villages, they cultivated the land nearest the village far more frequently than they would have preferred. Fallows had therefore been reduced to virtually nothing and land shortage was more apparent than real, provoked by the impact of the drought and by political decisions made at household level to send migrants to Banjul and as far afield as Dakar, Ghana, and even Congo.

One of the main consequences of a reduced fallow noted by Gambian farmers was a reduction in the organic matter content of the soil. The loss of soil organic matter is of importance because it supplies most of the nitrogen and sulphur and half of the phosphorus taken up by crops which receive no additional energy subsidy. It supplies the bulk of the cation exchange capacity on acid, highly weathered soils. It improves the physical properties of the soil by aiding soil aggregation and reducing susceptibility to erosion in sandy soils (M. Adams, 1988). It improves the water holding capacity of soil and it can reduce leaching because organic matter can form complexes with micronutrients thus retaining them in the soil (Sanchez, 1976). Indigenous methods of cultivation which involve minimal use of fertilizers can suffer serious depletion of organic carbon once the period of cultivation begins but according to Sanchez (1976), soil organic carbon in such systems does not continue to fall but remains at about 75 per cent of that at the end of a long fallow period. Where pressure on the land increases through an extended period of cultivation, the carbon content can drop to 50 per cent. It is therefore conceivable that where the fallow period is not of adequate length in bush fallow systems, the organic carbon content is likely to be low. Sanchez (1976) cites evidence which shows that the annual decomposition rate of organic carbon varies according to land-use and stresses that it can be increased through good management practices. In the savannas the wet season is followed by a marked dry season when the land is rested and this in itself could be perceived as a minor fallow period. However, with limited rainfall, plant growth is constrained and the system has little chance to replenish its energy reserves in this brief period of a few months.

At face value, outputs from dryland fields of Gambian farmers appeared to exceed inputs each year thus suggesting that cultivation could not continue indefinitely without a significant nutrient subsidy to the system. Theoretically, under such circumstances, yields should decline to virtually zero. However, yield patterns are far from so simple. Farmers acknowledged that there had been a considerable decline in millet yields from around 1.5 tonnes ha$^{-1}$ equivalent, to 600–700 kg ha$^{-1}$ over the past two decades but that output was maintained at this low level as long as a millet–groundnuts rotation was maintained. A possible explanation is that groundnuts fix nitrogen in the soil and the root systems from millet which are not removed with the harvest represent an input of organic matter. According to Bationo *et al.* (1987), roots represent a significant reserve for the subsequent crop as the roots retain much of the potassium, calcium, sodium, and sulphur remaining in the plants after harvest. Some 10 per cent of the straw from the crop is either left on the plot or recycled by grazing animals and the combination of this together with nitrogen fixation and organic matter from the roots may be sufficient to permit the continued production of low yields (Pieri, 1992). Field evidence in The Gambia revealed that where farmers had cultivated a plot continuously for several years with only millet and had not involved groundnuts in the rotation, yields did fall to a level where the plot was no longer worth cultivating and it was abandoned. Furthermore, after it had been in fallow for 6 years only grasses were to be found on the plot and shrubs and saplings which should have been well established by this stage, were absent. This confirms that farmer selection of a millet–groundnut rotation is probably the most beneficial for the soil, given the very limited access that farmers have to other forms of nutrient subsidy.

Normally, it is accepted that after a plot has been fallowed for a long period, yields are relatively good. However, the relationships between length of fallow and yield patterns are rarely so straightforward. In the Casamance of southern Senegal for example, Pieri (1992) reports that crops cultivated on newly cleared land may be low for the first 2–4 years. A similar pattern was encountered by Richards (1985) in Sierra Leone where soil under long fallowed land did not produce good yields in the first 2 years of cultivation. The suggestion was that high nitrogen levels in the soil led to vegetative growth in rice at the expense of grain formation. In the Casamance, lower yields have been attributed to competition from weed growth (possibly also because of high nitrogen levels). However, persistence in combating weeds in the drier savannas can be productive and evidence from the Casamance has shown that between the fourth and the tenth or twelfth year of cultivation yields may improve. While changes in yield are comparatively slow during this phase, after 10–15 years of cultivation, yields may plummet reflecting the exhaustion of soil reserves (Pieri, 1992). While the length of the fallow period is of critical importance to crop yields, work by Kalu and Norman (1987) shows that appropriate crop combinations, crop sequences and dates of planting can also have a positive effect on the harvest.

## Balancing resources to cope with drought

The decline in returns from millet, sorghum, groundnuts, and maize on dryland fields, as a result of the drought, has seen Gambian farmers search for alternative crops. Planting of fruit trees, particularly oranges and mangoes, has increased substantially over the past 20 years and, to some extent, compensates for the loss of other tree species. By 1999 the trend had intensified and the expansion of mango orchards in particular, was significant. Both mangoes and oranges are drought tolerant, but the latter, with their fibrous rooting systems, have suffered as the water-table has fallen during the protracted drought. Deeper rooting mangoes have proved to be a better option and new trials are now being carried out by farmers with cashew—allegedly introduced by Manjago migrants from Guinea Bissau.

Market demand for oranges and for mangoes in the urban areas, and also from overseas, has helped promote the development of these crops which are convenient to grow in view of frequent labour shortages, and which can be lucrative. However, it is rare that village fruit reaches the international market as this has been cornered by the few large, capital intensive farms on the outskirts of Banjul. Fruit harvested in the villages is marketed by contractors in Dakar and Banjul. These markets are saturated rapidly and far from all the fruit produced reaches them. Much of it goes to waste in the orchards. Determined to benefit from overseas markets, some of the more progressive farmers are planting varieties suitable for export and many experiment with budding and grafting to improve the quality of the fruit. At the moment, however, there is little scope for village producers to break in to the export market, though the potential exists.

By the early 1990s dryland farmers had also discovered a growing market for cassava in the urban areas of Banjul, and Dakar in Senegal (Baker, 1995). According to local traders, this increase in demand was primarily a response to a fall in the supply of food crops to the urban market, due to the drought. Other factors explaining the increasing demand for cheap food were the squeeze on the wages of public sector workers as part of Structural Adjustment, and increasing numbers of urban dwellers, due largely to migration. By the early 1990s cassava was proving such a worthwhile venture that its status was changing from that of a compound crop to a field crop. The neatly strung barbed wire fences which surrounded and protected it were testimony to its economic value. The sturdy fences also reflected the need to protect the crop which, if left in the soil during the dry season, was vulnerable to attack by monkeys and also by cattle, sheep, and goats which are not herded at this time.

Cassava is an ideal crop owing to its tolerance of poor soils and limited labour requirement. Its tolerance to drought is reported though Mortimore (1992) notes the virtual elimination of the crop in the Kano close-settled zone between 1962/3 and 1979, due to drought. A decade later, the anticipated expansion of cassava in the Western Division of The Gambia has not been realized. The fencing problem has proved overwhelming. Farmers have neither the resources nor the labour to securely fence more than one, or at most two plots each, and fruit growing has proved a preferable option. Nevertheless, a significant market remains in cassava and as the standing crop is frequently sold to contractors who are then responsible for harvesting it, labour demands on farmers are kept to a minimum.

Yet another crop which men had begun to exploit was a type of rice which could be grown on the drylands, well away from the riverine areas. This was entirely new to the area and had been introduced by traders from Guinea Bissau. Broadcast, as was millet, and developed in similar fashion to upland rice, a major problem with this dryland rice was that it required chemical fertilizer which was too expensive for most farmers. Experimentation, innovation, and the constant search for new varieties is not limited to The Gambia. It is evident everywhere in Africa, but it is too often ignored by policy makers and by development practitioners.

One method of coping with drought and the general shortage of rainfall is outmigration from rural areas. Reduced returns from farms are increasingly being made up from a higher proportion of off-farm and non-farm income. In The Lower Gambia men have lost less than women as a result of the drought. Some have been able to exploit new opportunities as a result of which they have prospered financially. But over the past 20 years men have also increased their proportion of non-farm income, mostly from formal or informal employment in urban areas. Writing of adaptation to drought in Northern Nigeria, Mortimore (1989) notes that economic self-sufficiency, if it ever existed, has been lost and that economic survival has been possible by calling increasingly on resources outside the village system. That a very unpredictable physical environment has forced people to extend their economic horizons beyond the village is also true of The Lower Gambia.

## Wetland cultivation

Wetlands throughout Africa are highly productive environments with a multiplicity of benefits and the nature of land-use in these environments is skilfully adapted to cope with an unpredictable environment as the following examples demonstrate. The River Senegal, like the River Gambia, rises in the Guinea Highlands to the south. The annual floods are fed mainly by the heavy seasonal rains that begin in April in the Guinea Highlands but the flood does not arrive in Senegal until mid-September after which it takes about 6 weeks to reach the sea (Nelson *et al.*, 1974).

The flood-plain of the Senegal has many uses. Not only is it used for food crop production, but the floodlands are important grazing lands, the home of migratory birds, a source of fish, of fuelwood, and a variety of tree products. Game used to be important though most of the larger animals have now gone.

Without the river the more arid parts of northern Senegal would be virtually lifeless as the rains are so irregular. The ways in which the floodwaters are used to advantage by local cultivators reflect an autecological approach to crop cultivation based on a deep understanding of the environment. Along the Lower Middle Valley of the Senegal River, the stretch from Dagana to Boghé, smallholder farming has depended heavily on flood recession agriculture. Here, the wetlands are much more valuable as they are lower risk environments and usually assure a harvest although the size of the productive area varies according to the extent of the river's flood.

The flood rises up the banks of the Senegal and spills over the surrounding area. The soils are medium to fine-textured alluvial deposits with an adequate capacity for water and nutrient retention. The floods recede in December and this marks the start of the recessional cropping cycle. Sorghum and vegetables are the main flood recession crops; where irrigation facilities exist, irrigated rice and irrigated maize may be grown. Park (1993) notes that many of the villages along the river combine traditional flood recession farming with small-scale irrigated agriculture as this further reduces risk. However, in the Lower Middle Valley, remote villages which are cut off during the rainy season depend more heavily on flood recession. Sorghum which matures at different rates is sown on the moist soils, that sown furthest from the river having the longest maturation period. Where cultivators keep animals, part of the sorghum crop may be long-stemmed varieties as these provide grain for the household while the stalks provide fodder for animals. Near depressions on the floodplain where water is more abundant, or where irrigation is possible, a little maize, mainly for domestic consumption, is also grown. Planting follows the recession of the water right down to the main channel and even on the steep banks of the Senegal vegetables, including gourds and melons, are cultivated (Baker, fieldwork, 1983). Planted at the end of the flood recession period, these are watered by hand from the river as the dry season advances.

Flood recession farming finishes well in advance of the annual rains which begin in June/July in the north of Senegal and continue until October. During the annual rainy season smallholder farmers plant the high risk drylands with millet, sorghum, and cowpeas. The two cultivation periods are thus mutually exclusive with farmers cultivating different ecological environments at different times. Apart from farming, the riverine area is used by pastoralists. This is further evidence of the skilful exploitation of the local ecology.

However, the natural flood of the Senegal has been significantly reduced by the presence of two dams that stand on the river, the Diama Dam and the Manantali. The Diama anti-salt barrage, some 27 km upstream from St Louis, controls the release of water into the Atlantic and prevents the movement of saline water upstream. The Manantali dam, situated in Mali, traps the flood waters which are released at a more controlled rate in an attempt to provide adequate water for irrigation throughout the year. By controlling both the inflow and outflow, the river level has been raised by around 1 metre so that water levels never fall too low and pumped irrigation is possible all year round. Natural flooding has been reduced and the extent of the loss of this 'free' irrigation system is only just being appreciated. A system is under way to introduce an artificial flood though this continues to have its problems.

The local benefits of the Hadejia-Nguru wetlands in Northern Nigeria are detailed by Hollis *et al.* (1993). In their attempt to circumvent the problems caused by environmental unpredictability, aid donors have chosen the capital intensive approach in the form of major dams. The Tiga dam was constructed on a tributary of the Hadejia river early in the 1970s, one of the aims being to establish an irrigation project. However, the dam has reduced the extent of flooding in the wetland and this has exacerbated the effects of the drought. A barrage on the river Hadejia which was completed in 1992 has created a shallow lake upstream which is likely to have a major effect on the timing and extent of flooding in the wetlands. The Challawa Gorge Dam, completed in 1992, is also likely to adversely affect the wetlands as is the Kafin Zaki dam on the Jama'are for which the contract was let in 1993. The wetlands are of major value and calculations have revealed that the economic value of their produce far exceeds that from all the irrigation schemes for which the rivers flowing into the basin are dammed and diverted (Hollis *et al.*, 1993). Rather than being wasted, the floods which are not utilized recharge the aquifers and it is only now becoming evident that attempts to harness the floods for selected reasons are having repercussions that had not been anticipated. As Hollis *et al.* (1993) observe, prediction of the impacts of change is not easy and further study of Africa's wetlands is required as is extensive debate about the best methods of their management if they are to be developed successfully.

As yet another example of wet land cultivation which reflects innovation and skill in resource use, Kolawole (1991) describes cultivation of the floor of Lake Chad in North-east Borno, Nigeria. Lake Chad has been decreasing in size since the beginning of the twentieth century, the area it has covered fluctuating from 25,000 sq. km in 1963 to 3,000 sq. km in 1988. The 2.25 million hectares which have been exposed are much used by cultivators and pastoralists though the precise area of lake floor available to users on account of recession varies from year to year and from season to season. Cultivation of the floor of Lake Chad is not new; the lacustrine soils were being farmed in 1826 (Kolawole, 1991). Both flood advance and flood retreat agriculture are still in evidence on the floor of Lake Chad and the presence of each is determined by soil type and by the level of the water-table which in turn is controlled by the stage of the lake recession. The soils of the lake floor are hydromorphic and are of two main types: the first group are clay soils with a high capacity for water retention, high carbon:nitrogen ratio and high organic matter content. The second are the 'muddy' soils, granular in texture, low in organic matter, and well drained. The lake starts to rise in October, the flood retreat beginning in January to February. Maize is cultivated on the clay soils. This is sown in June just before the arrival of the annual rains and as it grows, it benefits from the residual moisture in the soil, the advancing lake water and from the rainfall. The maize crop is harvested in October just as the flood really begins to advance. It is thus subject to neither moisture stress nor to the effects of flooding and because of the inherent fertility of the soil, crop productivity is relatively high, even without the use of fertilizer (Kolawole, 1991).

Cowpea (*Vigna unguiculata*) is grown on the better drained, muddy soils. Planted from January to February as the waters of Lake Chad begin their retreat, cowpea is harvested from May to August. It is comparatively tolerant of dry conditions and relies essentially on residual moisture in the soil from the former flood. Cultivation of the lake floor has enabled survival of the worst years of the drought including the period from 1983 to 1988. Although crop yields were very low on the drought-prone uplands, cultivation of the floor of Lake Chad more than compensated for losses from rainfed cultivation and allowed the export of food crops from the lake floor to other parts of Nigeria (Kolawole, 1991). Increased cultivation of the lake floor has seen farmers in this area accumulate income though farmers on the uplands have fared badly over the same period. Encouraged by the success of indigenous cultivation of the lake floor, the state, through a parastatal, the Chad Basin Development Authority (CBDA), has become involved in lake floor farming. In contrast to indigenous cultivation methods, the CBDA project has involved mechanization, the use of fertilizer, herbicides, and high-yielding seed varieties. Mechanization has disturbed soil moisture balances and high yielding varieties of *Vigna* have proved susceptible to disease. If capital intensive methods are to be used then they should be more carefully adapted to local environmental conditions.

Many more examples of wetland use could be cited. Besides using *fadamas* (a Hausa term meaning 'land flooded in the wet season'), cultivators, pastoralists, fishermen, and gatherers make use of land where the water-table is nearer the surface such as in *wadis* and in *bas-fonds*, valley bottoms which are the main drainage lines for rainwater or between sand dunes such as in the Cayor region of Senegal, and in the *niayes* of Cap Vert where high water-table levels between the dunes are exploited for market gardening (Baker, 1985; Nelson *et al.*, 1974: 42–4). The skill necessary in indentifying and using a whole range of wetland environments is discussed extensively by Marchand and Toornstra (1986), Hottinga *et al.* (1991), Mortimore (1992), W. M. Adams (1992), and Hollis *et al.* (1994). A major reason for the success of indigenous land-use management of wetlands is that they make the most of a frequently changing environment rather than seeking to hold it constant, as is the pattern for so many imposed development schemes. In The Lower Gambia, however, riverine cultivation has not been sustainable because environmental change has been so extreme that most of the land which is flooded annually is no longer productive in agricultural terms. Rural livelihoods have thus altered significantly as a result.

Riverine land, cultivated with rice predominantly by women, comes in many shapes and sizes with upland rice plots normally being significantly larger than wet rice plots. Wet rice cultivation is a highly complex business with different species being matched to marginally different habitats (Webb, 1992). To make things more complicated, the area under wetland rice varies each year in response to the extent and duration of the annual flood. Under traditional tenurial systems usufruct rights to this riverine land have been passed down from mothers and mothers-in-law to daughters and daughters-in-law, rice land thus remaining within families (Dey, 1981). The wet rice or swamp lands are the most highly valued. Traditionally, rice was a minor crop cultivated by women in rain-fed swamps or depressions formed by small streams which feed into the River Gambia. Since the 1950s dryland farms have increased the proportion of groundnuts to food crops and one effect of this has been to stimulate the cultivation of rice in tidal swamps (Swindell, 1985). According to Weil (1973), the 80 per cent increase in rice production achieved by the Mandinke between 1940 and 1960 was due in the main to the extension of rice cultivation into tidal swamps.

These tidal swamps have suffered extreme damage in The Lower Gambia on account of the drought (Baker, 1995; Webb, 1992) and in consequence, the pattern of expansion of cultivation has been reversed. Lower river levels, the result of lower rainfall and runoff, have led to the movement upstream of saline water from the Atlantic and the river waters are saline for some 200 km above the mouth of the river for some 7 months in the year and over 100 km from the estuary mouth for the rest of the year (Amara, 1993; NEA, 1997; Webb, 1992). Formerly, when the floods were stronger, the saline water which moved upstream during the dry season would be washed out to sea again during the rains. Now the river water is saline in the lower reaches of the Gambia River throughout the year and when the river does flood, the land is inundated with saline water. This has rendered considerable areas of the floodplain in The Lower Gambia unusable for rice production and has increased the economic vulnerability of women farmers. (See Plates 6.1 and 6.2: Gambia flood plain in 1981 and 1991. Although the earlier photograph is in the wet season and the later one in the dry season, the substantial change in the environment is evident).

Swamp rice cultivation still exists away from the river where depressions flood during the rains. However, the extent of these areas has also diminished with the drought. Where swamps are still cultivable, women clear the plots before the river rises and shortly before the advent of the rains rice seedlings are germinated in nurseries from early June to the end of July. As soon as the flood waters begin to reach the plot the seedlings are transplanted into the rising flood. Transplanting is staggered and in consequence so is the harvest for transplanted rice which extends from December to mid-January. Swamp rice from inland depressions is particularly valued because it yields 25–33 per cent more than upland rice (see below), though this is less than from tidal swamps where yields may be two to four times greater than upland rice (Swindell, 1985).

202  *Indigenous Land Management in West Africa*

**Figure 6.1:** Paddy fields along the Lower Gambia (1981)

**Figure 6.2:** An area very close to that shown in Plate 6.1 (1991). Annual inundation with saline water has rendered the land unusable for paddy cultivation

*General note*: Although 6.1 was taken in the wet season and 6.2 in the dry season, the change in the environment and in the land use is evident over the period 1981–91.

Higher yields are probably achieved in tidal swamps because there is no moisture deficit and because soil nutrients are replenished by the annual flood waters and by decomposing plant debris. Furthermore, after the first harvest has been taken, the plants continue to benefit from high moisture levels in the soil; they grow up again and yield a second, smaller harvest.

Decreased production from the tidal wet rice land in particular, has had a serious impact on the lives of women. In ecological terms wet rice was a crop which could be planted every year, floods permitting, without causing soil degradation. Capital inputs were comparatively low and consisted of little other than seed, much of which was home produced. Furthermore, on account of the nutrient subsidy from the floods, fallowing was unnecessary. Labour inputs were quite substantial, particularly at peak periods when the plot was cleared of vegetation and the soil hoed prior to flooding. Labour was also needed to establish the nursery, transplant the seedlings, scare birds away as harvest time approached and finally, harvest the crop. Between these times labour requirements for rice were comparatively light. If a large area of wet rice was cultivated women tended to form work groups to cope with peak periods of labour demand much as they did for weeding the rice in Sierra Leone (Chapter 3). Wet rice was a valuable commodity for women, contributing food to the household and in times of plenty leaving a marketable surplus.

The upland or riverine areas which do not flood are also sown with rice but in this case, upland rice. In the the Lower Gambia, as in Sierra Leone, many of the varieties are the same as those sown in the swamps though methods of cultivation differ. Upland rice is an early rice, broadcast when the rainy season commences. It is not transplanted, and the yield from its single harvest may be as little as one-third that from swamp rice. Nevertheless, upland rice is important, first because it is the earliest rice to be harvested and second because so much wet rice land has been lost in The Lower Gambia. As the floods have decreased over the past two to three decades (NEA, 1997), so the area sown with upland rice, parallel to the river, has moved nearer the river though not so near that the crops are poisoned by saline soils. Upland rice has also been affected by the shortening of the rainy season and women have changed from growing 120-day varieties to 90-day varieties, the flavour of which is not as mellow as that of either the slower maturing varieties.

Physically and ecologically, upland rice is part-way between swamp rice and cereals grown on the drylands. Preparation of the land for cultivation is very similar to that on the drylands. The bush is cut mainly by women from the end of April until shortly before the beginning of June. Only large trees and trees of economic value such as oil palms are left. Men do help with bush clearance, particularly where there are sizeable trees to be removed. Women usually pay them for doing this. Roots are left in the soil but the cut vegetation is piled on to tree stumps. This both allows the vegetation to dry and also reduces the size of tree stumps when the dried vegetation is burnt. A similar practice is described by Richards (1985) in Sierra Leone. If the dried vegetation is abundant the entire plot burns, but if not, burning takes the form of a series of bonfires. Very shortly after the burn the seed is broadcast and hoed lightly into the soil. Not all upland plots are seeded at once; sowing is staggered over a period of about 3 weeks just at the beginning of the rains so that rice plots are in different stages of development. Germination is stimulated by the rains and the rice is weeded about 6 weeks after germination. This is extremely important because weeds start to grow almost immediately the rains arrive. Most savanna species are fire tolerant (Hopkins, 1974) and are not destroyed by the burn. The bush starts to regenerate almost immediately the plot is cleared and weed growth is always a problem. Weeding too early can be little more than a waste of time as the rice seedlings have a head start on the weeds and weeding too late can be counterproductive as unearthing larger weeds can disturb the root system of the rice and impede its growth. Timing of weeding is critical but women complained that they were almost always late with their weeding.

In the early 1990s a common pattern was for upland rice fields to be cultivated for 2 years and fallowed for 2 years. Almost a decade later, in 1999, fallows have virtually disappeared and women are anxious to have access to fertilizer to increase yields from upland rice. The problem once again is cost, owing to the removal of the fertilizer subsidy.

## Strategies for coping with drought

There is very little evidence that women have been able to do much to improve the quality of their farmlands. Some riverine villages have built mud bunds to keep the salt water at bay, but none of these has been successful. Having lost a significant part of their income, women have turned increasingly to a range of non-farm activities which include buying peanuts from their husbands, shelling and roasting them and selling them in local markets or making peanut butter for sale. Some women make soap by melting down bar soap and washing soda and re-setting the mixture. This has a ready market, particularly in the rainy season when washing is a heavy task for the women (Baker, 1995). Not everyone in the same village can make and market peanut butter or soap, for example, so most women tend to concentrate on two or three different income-generating activities.

Gathering materials from forest, field, and bush for domestic use and for sale always has been and continues to be of great importance throughout West Africa. Madge (1994), working in the southern reaches of The Gambia, identifies the vast range of bush products which are collected by village people and the uses to which they are put. Similarly, Schreckenberg's work (1996) in the Bassila area of Benin shows the continued, heavy dependence on gathered products both for domestic use and for sale. Men, women, and children may all be involved in gathering though normally, men and women collect different products. Schreckenberg (1996) notes that for children the collection of non-timber forest products is still one of their major sources of income. While this section has focused primarily on women's activities, reference is also made to the part played by men and by children as it makes little sense to separate them. Gathered products are important both for domestic consumption and as a source of income in the pre-harvest period when food is at its most scarce. But collection of bush products takes time and it was evident in The Lower Gambia that the larger families had an advantage in this respect (Baker, 1995).

The selling of fuelwood was also of economic importance to women though this involved them working for men. The selling of fuelwood requires a licence and this is usually given to men (no-one in the villages visited knew of women with such a licence). Fuelwood may be collected by men, women, and children, but when it is for sale, rather than domestic use, it is perceived as the property of the men who pay the women and children for gathering it. The actual selling is done by women who are paid by the men according to the amount of wood they sell. It is sold along the roadside or in urban areas where it commands literally twice the price it does away from the towns. For example, in April 1999, a bundle of fuelwood approximately 0.5 m in diameter and approximately 1 m in length, cost Dalassi 10 in Serrekunda and Dalassi 5 in the rural areas (at the time of field mark £1 = D16.75).

Collection of domestic foodstuffs was of great importance and this took several forms. First, there were foods that children and adults gathered for snacks. Children frequently snacked off bush products and the general belief was that these were highly nutritious and health-giving (Madge, 1994; Schreckenberg, 1996). These may have been gathered on an individual basis or, alternatively, snack foods might have been collected by women for the whole family. The fruits of the grey plum (*Parinari excelsa*) for example, were particularly popular.

Secondly, men collected wild honey, and tapped palms for the milk that made palm wine, though in The Lower Gambia these were frequently Fulani men, who also trapped or shot small game. Part of this harvest was usually consumed domestically, although some of it was sold. Schreckenberg (1996), in the guinea savannas of Benin, observes that a growing number of men supplement their farm income by collecting oil palm nuts or honey while older men make mats and baskets from local bush products. The transport of gathered products to market also provides the opportunity for younger men and those with carts to make money. Thus gathering, rather than being small-scale and 'free', results in a considerable amount of income generation.

Thirdly, foods were gathered for the stew pot. Some foods such as wood sorrel (*Hibiscus surratensis*), baobab (*Adansonia digitata*) leaves, and herbs were eaten fresh, others such as oil palm and some of the fish caught would be processed. Men, women, and children were involved in harvesting oil palm fruit and fishing though women were almost always responsible for the processing of oil palm and for the preserving and storing of fish. Seasonality greatly affected the products available through the year. Products gathered from the bush were often combined with vegetables from the garden to make a sauce to accompany the staple which was millet or sorghum based, or rice. When available, protein such as fish or meat might be added to the sauce. According to Madge (1994), over her study period of nine months in a village in the south of The Gambia, 63 per cent of meals consumed contained gathered products while in 55 per cent of cases, gathered products dominated the meal. In the food production system, however, no clear distinction was made between food from gathered products and food from cultivated crops. Moreover, distinctions were not drawn between time spent farming, time spent collecting bush products, time spent in either buying or selling at market, or time spent on domestic chores. They were all essential for survival and as Madge (1994) suggests, rural lifestyles should be considered as a mosaic of interacting activities.

A fourth category of gathered product were those collected for the production of non-food products. Medicinal products were obtained from many trees and herbs including *Adansonia digitata*, *Parinari macrophylla* (which also yields edible (but tasteless) fruit from which the nuts are extracted), *Cassia sieberiana* and *Daniella oliveri*, the West African copal. The gum from the latter, when dried, is burnt as a fumigant (Percival, 1968). The leaves of both *Piliostigma thonningii* and *Daniella oliveri* are also used for fodder, indicating the multiple use of many species. The inner bark of *Sterculia setigera* is used for making rope and the bark is stripped from *Adansonia digitata*, used as rope in fencing and for many other purposes. The leaves of *P. thonningii* may also be dried and dyes extracted from them may be used in rituals and festivals. Uses of bush products are thus numerous and varied and collectors in The Lower Gambia included both men and women. Access to trees for their products was a highly complex issue.

Eurocentricism may be blamed for preventing our appreciation of the value of bush products and of the indigenous knowledge necessary for collecting different products at different times of the year and for processing them in a variety of ways and for a variety of purposes. It is very apparent that non-cultivated products derived from forest, field, and bush are of great importance and that indigenous knowledge systems, so very important in making use of the bush, should be incorporated in development initiatives. Although the value of indigenous knowledge systems is now being recognized and acknowledged, there is still little evidence of their influencing development policies.

## Vegetable cultivation

In addition to a range of other economic and domestic activities, women in The Lower Gambia also grow vegetables. This used to be a compound activity where a small patch of vegetables would be grown alongside some cassava in the compound. Gardening was solely a rainy season activity because otherwise provision of water for the vegetables was too difficult. As women have lost a major part of their potential income with the decline of wet rice production, aid agencies have sought alternatives and have improved the opportunities for women to grow vegetables in the village both for subsistence and commercial purposes. Women have been encouraged to stop gardening in their compounds and to come together in communal gardens established by donors such as the UNDP, EU, Islamic Development Bank, Taiwan and World Bank (Carney, 1992: 78; Baker, 1995). These gardens may cover a hectare or more, depending on their location, and many are provided with permanent wells. With the provision of water, gardening is possible in both the dry and the wet seasons. Some communal gardens are not provided with permanent wells and where salt is not a significant problem, dry season gardens tend to be located nearer the River where the water-table is higher and where wells dug by hand may yield water 1–1.5 m below the surface. Rainy season gardens are slightly further away from the River because during this season riverine lands are usually used for other purposes—particularly upland rice. But here too the depth to the water-table is rarely over 1.5 m particularly if wells are hand dug. While vegetable gardening is very much the responsibility of women, men normally help with well-digging, usually for money.

Although gardening throughout the year is potentially a good idea, it has many drawbacks, not the least being that vegetables are no replacement for rice in the dietary sense. Vegetables thus have a very different role in the food production system. Vegetable production is also very time-consuming and labour intensive (Carney, 1992). The gardens, which may be as much as a kilometre from the village, have to be watered at least twice a day. Thus a significant part of labour-time is spent each day in travelling to and from the fields (Swindell, 1985). They have to be weeded, and dung is collected by the women in baskets and worked into the soil as is composted household refuse. Vegetables also have to be guarded, particularly in the dry season, from wandering animals which are not herded at this time of year as there are few other crops in the fields. Depending on the size of the garden, at least one woman and probably two have to stay on guard duty shooing away cows and goats that try to eat their way through fences made of rhun palm (*Borassus* spp.), oil palm fronds, thorn branches, live sisal plants, and any other deterrents in order to graze on the vegetables which may be the only green stuff around in the heat of the dry season. Fences round vegetable gardens can be most impressive, as much as a metre in thickness, but still they provide little challenge to hungry and determined bovines and ruminants.

Finally, marketing. There should be scope for marketing fresh vegetable produce, particularly with the significant tourist industry in The Gambia, and indeed it has been an intended aim of aid agencies that domestic vegetable production should serve the needs of the Gambian market (pers. comm., World Bank Office, Banjul, 1991). Women have been introduced to a far wider range of vegetables than they have grown in the past and so could meet at least part of the demand. But while much time has been concentrated on setting up vegetable gardens and training women in gardening techniques, less consideration has been focused on disposal of the produce. The tourist industry absorbs less than had been anticipated and where local produce is used, this tends to be from large-scale and other peri-urban producers. Gluts in local markets are a major problem for small-scale rural producers. Until a lucrative and reliable outlet can be found for their produce, gardening will remain unpopular, not providing an adequate return to labour and not yielding sufficient to purchase rice. Although the ecological needs of plants were satisfied, appropriate political decisions still needed to be made. The shortfall in rice was made up mainly from imported rice purchased with remittances sent to the village from migrants to urban areas. Having vegetables for the family diet throughout the year was an acknowledged benefit, but in villages of The Lower Gambia women gardened simply because other money-making opportunities were limited.

Development agencies have become involved in vegetable gardening mainly with the aim of increasing domestic income by increasing productivity in this area. In The Lower Gambia they had gone some way to achieving this but in other areas indigenous vegetable gardening has proved remarkably successful with little external input. Fieldwork by the author in 1981 in the Office du Niger in Mali, between Ségou and Niono, revealed that alongside rice cultivation which is the focus of Office interests, other economic activities were being carried on with a great deal more success. Returns from rice in the early 1980s were so low and communication was so bad between field-level workers on the project and those in managerial positions, that local extension agents were prepared to turn a blind eye to farmers growing millet and groundnuts on Office land, to their keeping animals and also to their growing vegetables and irrigating them with Office canal water. Vegetable production was extremely successful in comparison with rice and returns from gardening supplemented family incomes. A range of vegetables was grown for domestic consumption but of particular importance was the intensive cultivation of onions which did not perish easily and which were marketed in Niono, Ségou, and even in Bamako. The sale of onions alone earned some families as much as £800 equivalent in a season in 1981.

Vegetable gardening is not just restricted to rural areas as urban vegetable gardening is well developed throughout Africa. Small patches of land that appear not to be utilized are used as the site of gardens providing a range of crops. On the outskirts of Banjul in The Lower Gambia and also in Senegal, urban gardening has been concentrated into large communal areas at the instigation of aid agencies though small-scale initiatives are still to be found where space allows. The market problem which has limited the level of success of vegetable gardening in The Gambia is not universal as the case of market gardening in Brazzaville, Congo, shows. Although beyond the limits of West

Africa, vegetable production and marketing in the Brazzaville area by Congolese again reflects a high level of indigenous organization. The Brazzaville markets absorb produce from both urban and rural vegetable gardeners. Work by Moustier (1995) has shown that there are marked differences in the crops produced in urban and rural areas. Urban markets tend to produce leafy vegetables with short shelf lives whereas the rural areas provide more in the way of onions and other root crops, tomatoes and crops capable of withstanding the journey to market. While there is scope for improving the quality and variety of vegetables produced in Congo, the marketing problems which face gardeners are not as acute as those in The Gambia where aid has played a major part in horticultural development.

A most successful area of vegetable cultivation in West Africa is in Cap Vert, inland from Dakar. Situated in the *niayes*, depressions between sand dunes where the water-table is not far below the surface, the Taiwanese, as part of an agricultural mission in 1964–73, began working with the people of Cap Vert to improve vegetable production. Changes were made in two major directions: in the quality and number of different varieties grown, and in the techniques used for their cultivation. Fieldwork in the Cap Vert area by the author revealed that gardeners were prepared to try new varieties introduced first by the Taiwanese and later by the Chinese, but they were highly selective in what they continued to grow (Baker, 1985). The desire by the two missions to increase the size of fruit and vegetables that the Senegalese market gardeners grew in Cap Vert was not particularly successful. Market demand was for small tomatoes and small beans, for example, there being little demand for larger varieties which were progressively abandoned. One vegetable which was retained was the *navet d'hivernage*, a turnip which grew through the dry season and was harvested shortly before the rains when food was short. Although the second mission ended in 1973, the *navet d'hivernage* was still being grown in 1983. Cap Vert gardeners acknowledged that the Taiwanese and the Chinese had done much to improve local African techniques of cultivation. They had explained to vegetable growers in Cap Vert the benefits of planting crops in particular combinations, of staking tomatoes and beans rather than letting them grow along the ground, and of making and using compost from vegetative matter, household refuse, night soil, and where possible animal dung. The Taiwanese and Chinese extension workers had lived in the villages with the Senegalese, teaching by example. Innovations were debated extensively with the Senegalese before they were introduced in the field and the Senegalese were conscious that the learning process was a two-way affair. If gardeners wished to adopt the methods of the Taiwanese and later the Chinese, they had access to information and to inputs at the Mission Centre at Sangalcam (Baker, 1985).

It is worthy of note that no official record could be found in Senegal of the work of the two missions. The Chinese recall its existence but no documentation on the work of the missions could be traced and all information was obtained from oral histories. Perhaps at a time when Africans and aid donors were interested in developing the potential of large-scale, capital intensive schemes, such small-scale efforts were not perceived as 'real' development. In retrospect, the two missions seem to have achieved a great deal and they are remembered with affection and gratitude by Cap Vert gardeners though, as Bräutigam (1998) observes, little has been done to enable farmers to build on achievements at field level.

## Conclusions

The examples in this chapter show how the success of farming in semi-arid West Africa is based on an autecological approach to the production of crops which is dependent on a detailed knowledge of the physical environment. While environmental variability may be perceived as a handicap to the development of agriculture, indigenous systems have developed the capacity to cope with uncertainty and where possible, to capitalize on it. The question asked in Chapter 2 is equally relevant here. Why, if farmers are so skilful, is agriculture so technologically backward? Similar responses must be given to those in Chapter 2: the development of smallholder agriculture has been impeded by political instability and inconsistent and unhelpful policy making, by, for example, years of low producer prices for cash crops. They have been hindered by inflation, currency manipulation, and the preference until comparatively recently for developing new-style, large-scale, capital-intensive schemes rather than developing smallholder farming. Where development schemes have focused on the smallholder, achievements at field level have not been proliferated partly because the institutional framework necessary to achieve this has been absent. Furthermore, the focus on the smallholder tends to be top-down rather than bottom-up. This is certainly evident in the Western Division of The Gambia where the EU is strongly encouraging the production of groundnuts while many farmers are trying to increase the production of fruit for the export market. In view of the drought, pests and diseases that affect groundnuts, the eternal problem of labour supply, the cost of fertilizers, and the current problems with the groundnut market, which in 1998/9 saw farmers lose significantly, farmers are moving into the production of fruit and possibly of cashew nuts. However, political decision-making at a higher level is determining what should be grown. At higher levels in The Gambia there seems little awareness of changes taking place at field level, nor is there a desire to investigate these and as a consequence, the emphasis on groundnuts is perpetuated. While drought has seriously impeded the development of agriculture over the past 30 years, the resistance of African governments to promote the development of one of their assests, smallholder agriculture, has undoubtedly been the major stumbling block.

## References

ADAMS, MARTIN (1988), *Agriculture, Livestock and Foodstuffs: An Environmental Strategy for Semi-Arid Areas* (Drylands paper no. 1, London: IIED).

ADAMS, W. M. (1992), *Wasting the Rain* (London: Earthscan).

ADESINA, AKINWUMI A. (1992), 'Oxen cultivation in Semi-Arid West Africa: Profitability Analysis in Mali', *Agricultural Systems*, 32 (8): 131–48.

AGBOOLA, S. A. (1979), *An Agricultural Atlas of Nigeria* (Oxford: Oxford University Press).

AGNEW, C. T. (1995), 'Desertification, drought and development in the Sahel', ch. 12 in Tony Binns (ed.), *People and Environment in Africa* (Chichester: Wiley), 137–50.

AHN, PETER, M. (1970), *West African Soils* (London: Oxford University Press).

AMARA, S. S. (1993), 'Environmental change and flooding in The Gambia river basin' (unpubl. Ph.D. thesis, University of Reading).

BAKER, E. F. I. (1981), 'Population, Time and Crop Mixtures', *Proceedings of the International Work Shop on Intercropping, Hyderabad, India, 10–13 January, 1979* (Andhra Pradesh, India: ICRISAT), 69–77.

BAKER, KATHLEEN M. (1985), 'The Chinese agricultural model in West Africa', *Pacific Viewpoint*, 2: 401–14.

——(1992), 'Traditional farming practices and environmental decline with special reference to The Gambia', ch. 9 in K. Hoggart (ed.), *Agricultural Change, Environment and Economy, Essays in Honour of W. B. Morgan* (London: Mansell), 180–202.

——(1995), 'Drought, agriculture and environment: A case study from The Gambia, West Africa', *African Affairs*, 94 (374): 67–86.

BARRETT, HAZEL and BROWNE, ANGELA (1991), 'Environmental and economic sustainability: Women's horticultural production in The Gambia', *Geography*, 76 (3): 241–8.

BATIONO, A., CHRISTIANSON, B. C., and MOKWUNYE, A. U. (1987), 'Organic recycling of crop residue and fertilizer use for pearl millet production on the sandy soils of Niger'. Proceedings of a conference on *Farming Systems Research* held at Niamey, Niger in 1987 (Ibadan: IITA), 127–35.

BEETS, WILLEM C. (1990), *Raising and Sustaining Productivity of Smallholder Farming Systems in the Tropics. A Handbook of Sustainable Agricultural Development* (Alkmaar, Holland: AgBe Publishing).

BRÄUTIGAM, DEBORAH (1998), *Chinese Aid and African Development: Exporting Green Revolution* (Basingstoke: Macmillan Press Ltd and New York: St. Martin's Press, Inc.).

CARNEY, JUDITH A. (1992), 'Peasant women and economic transformation in The Gambia', *Development and Change*, 23 (2): 67–90.

DELGADO, C. and MCINTIRE, J. (1982), 'Constraints on oxen cultivation in the sahel', *American Journal of Agricultural Economics*, 64: 188–96.

DEY, JENNIE (1981), 'Gambian women, unequal partners in rice development', *Journal of Development Studies*, 17 (3): 109–22.

FAIRHEAD, JAMES and LEACH, MELISSA (1996), *Misreading the African Landscape: Society and Ecology in a Forest-Savanna Mosaic* (Cambridge: Cambridge University Press).

GADBOIS, MILLIE, Department of Food and Agricultural Economics, University of Reading (Pers. comm. 1997).

HARRISON CHURCH, R. J. (1980 edn.), *West Africa* (London: Longman).

HASSAN, ABDOU (1996), 'Improved traditional planting pits in the Tahoua Department (Niger): An example of rapid adoption by farmers', ch. 6 in Chris Reij, Ian Scoones, and Camilla Toulmin (eds.), *Sustaining the Soil: Indigenous Soil and Water Conservation in Africa* (London: Earthscan).

HOLLIS, G. E., ADAMS, W. M., and AMINU-KANO, M. (1993), *The Hadejia-Nguru Wetlands: Environment, Economy and Sustainable Development of a Sahelian Floodplain Wetland* (Gland, Switzerland and Cambridge: IUCN).

HOPKINS, BRIAN (1974) (2nd edn.), *Forest and Savanna: An Introduction to Tropical Terrestrial Ecology with Special Reference to West Africa* (Ibadan and London: Heinemann).

HOTTINGA, FOLERT, PETERS, HENK, and ZANEN, SJOERD (1991), *Potentials of Bas-Fonds in Agropastoral Development in Sanmatenga, Burkina Faso*, Part 3b in Ian Scoones (ed.), *Wetlands in Drylands: The Agroecology of Savanna Systems in Africa*, (IIED Drylands Programme, London: IIED).

JAEGER, W. K. and MATLON, P. J. (1990), 'Utilization, profitability and the adoption of animal draft power in West Africa', *American Journal of Agricultural Economics*, 72: 35–48.

KALU, B. A. and NORMAN, J. C. (1987), 'Crop yields under the traditional cropping patterns in a Middle Belt savanna agro-ecological zone of Nigeria', *Agricultural Systems*, vol. 24: 211–20.

KASSOGUÉ, ARMAND, MOMOTA, MAMADOU, SAGARA, JUSTIN, and SCHUTGENS, FERDINAND (1996), 'A measure for every site: Traditional SWC techniques on the Dogon Plateau, Mali', ch. 8 in Reij *et al.* (eds.), *Sustaining the Soil: Indigenous Soil and Water Conservation in Africa*, (London: Earthscan), 69–79.

KHALFAOUI, J.-L. B. (1991), 'Determination of potential lengths of the crop growing period in semi-arid regions of Senegal', *Agricultural and Forest Meteorology*, 55: 251–63.

KOLAWOLE, ARE (1991), *Economics and Management of Fadama in Northern Nigeria*, Part 3a of Ian Scoones (ed.), *Wetlands in Drylands: The Agroecology of Savanna Systems in Africa* (IIED Drylands Programme, London: IIED).

KOWAL, J. M. and KASSAM, A. H. (1978), *Agricultural Ecology of Savanna: A Study of West Africa* (Oxford: Clarendon Press).

LAL, R. (1979), 'Role of physical properties in maintaining productivity of soils in the tropics', chs. 3–5, in R. Lal and D. J. Greenland (eds.), *Soil Physical Properties and Crop Production in the Tropics* (New York: Wiley).

LAMPREY, H. F. (1975), *Report on the Desert Encroachment Reconnaissance in Northern Sudan*, UNESCO/UNEP Consultant Report, Paris (Cited in Agnew 1995).

LOWENBERG-DE BOER, J., KRAUSE, MARK, DEUSON, ROBERT, and REDDY, K. C. (1991), 'Simulation of yield distributions in millet-cowpea intercropping', *Agricultural Systems*, 36: 471–87.

MADGE, CLARE (1994), 'Collected food and domestic knowledge in The Gambia, West Africa', *Geographical Journal* 160 (3): 280–94.

MANVELL, ADAM (1998), Fieldwork in the Maradi Department, Niger as an Overseas Trainee for VSO.

MARCHAND, M. and TOORNSTRA, F. H. (1986), *Ecological Guidelines for River Basin Development* (Leiden: Centrum Voor Milieukunde Rijksuniversiteit).

MILLAR, DAVID, with AYARIGA, ROY and ANAMOH, BEN (1996), ' "Grandfather's way of doing": Gender relations and the *yaba-itgo* system in Upper East Region, Ghana', ch. 14 in Chris Reij *et al.* (eds.), *Sustaining the Soil: Indigenous Soil & Water Conservation in Africa* (London: Earthscan), 117–25.

MORGAN, W. B. and Moss, R. P. (1965), 'Savanna and forest in Western Nigeria', *Africa*, 35 (3): 286–93.

MORTIMORE, MICHAEL (1989), *Adapting to Drought: Farmers, Famines and Desertification in West Africa* (Cambridge: Cambridge University Press).

—— (1992), 'The intensification of peri-urban agriculture: The Kano close-settled zone 1964–1986', ch. 11 in B. L. Turner II, Goran Hyden, and Robert W. Kates (eds.), *Population Growth and Agricultural Change in Africa* (Gainesville: University Press of Florida), 358–400.

—— (1998), *Roots in the African Dust: Sustaining the Drylands* (Cambridge: Cambridge University Press).

MOUSTIER, PAULE (1995), 'Organization of the Brazzavillian vegetable market' (unpubl. Ph.D. thesis, University of London).

National Agricultural Date Centre (NADC), Department of Planning, Department of State for Agriculture (1998), *1997/98 National Agricultural Sample Survey Statistical Yearbook of Gambian Agriculture 1997* (Banjul: Government of The Gambia).

National Environment Agency (NEA) (1997), *State of the Environment Report—The Gambia* (Banjul: National Environment Agency).

NELSON, HAROLD D., DOBERT, MARGARITA, MCDONALD, GORDON C., MCLAUGHLIN, JAMES, MARVIN, BARBARA, and MOELLER, PHILIP W. (1974) (2nd edn.), *Area Handbook for Senegal* (Washington: American University).

NETTING, ROBERT MCC. (1993), *Smallholders, Householders: Farm Families and the Ecology of Intensive, Sustainable Agriculture* (Stanford, California: Stanford University Press).

NORMAN, D. W. (1972), *An Economic Survey of Three Villages in Zaria Province: Input/Output Study* (Samaru, Miscellaneous Papers, no. 37, Zaria, Nigeria).

—— (1974), 'Rationalising mixed cropping under indigenous conditions: The example of Northern Nigeria', *Journal of Development Studies*, 11: 3–21.

—— SONMIOUS, EMMY, B., and HAYS, HARRY M. (1982), *Farming Systems in the Nigerian Savannas: Research and Strategies for development* (Boulder, Colorado: Westview Press).

NYE, P. H. and GREENLAND, D. (1960), *The Soil Under Shifting Cultivation* (Farnham: UK Agricultural Commonwealth Bureaux).

—— and STEPHENS, D. (1962), 'Soil Fertility', ch. 7 in J. Brian Wills (ed.), *Agriculture and Land Use in Ghana* (London: Oxford University Press), 127–43.

OLADIPO, E. O. and KYARI, J. D. (1993), 'Fluctuations in the onset, termination and length of the growing season in Northern Nigeria', *Theoretical and Applied Climatology*, 47 (3): 241–50.

OUEDRAOGO, MATTHIEU and KABORÉ, VINCENT (1996), 'The *Zäi*: A traditional technique for the rehabilitation of degraded land in Yatenga, Burkina Faso', ch. 9 in Chris Reij *et al.* (eds.), *Sustaining the Soil: Indigenous Soil and Water Conservation in Africa* (London: Earthscan), 80–4.

PARK, THOMAS (ed.) (1993), *Risk and Tenure in Arid Lands: The Political Ecology of Development in the Senegal River Basin* (Tuscon and London: University of Arizona Press).

PERCIVAL, DAVID A. (1968), 'The Common Trees and Shrubs of The Gambia', unpublished, Banjul: The Gambia.

PIERI, J. M. G. (1992), *Fertility of Soils: A Future for Farming in the West African Savannah* (Berlin: Springer-Verlag).

REY, CHRIS, SCOONES, IAN and TOULMIN, CAMILLA (eds.) (1996), *Sustaining the Soil: Indigenous Soil and Water Conservation in Africa* (London: Earthscan).

RICHARDS, P. (1985), *Indigenous Agricultural Revolution* (London: Hutchinson).

—— (1986), *Coping with Hunger: Hazard and Experiment in an African Rice-Farming System* (London: Allen & Unwin).

SANCHEZ, PEDRO A. (1976), *Properties and Management of Soils in the Tropics* (New York and London: Wiley).

SCHOONMAKER FREUDENBERGER, K. and SCHOONMAKER FREUDENBERGER, S. (1993), *Fields, Fallows and Flexibility: Natural Resource Management in Ndam Mor Fademba, Senegal*, Drylands paper no. 5 (London: IIED).

SCHRECKENBERG, K. (1996), 'Forests, fields and markets: A study of indigenous tree products in the woody savannas of the Bassila region, Benin' (unpubl. Ph.D. thesis, University of London).

SLINGERLAND, JAJA and MASDEWEL, MOUGA (1996), 'Mulching on the Central Plateau of Burkina Faso: Widespread and well adapted to farmers' means', ch. 10 in Chris Reij *et al.* (eds.), *Sustaining the Soil: indigenous Soil and Water Conservation in Africa* (London: Earthscan), 85–9.

SPICHIGER, R. and PAMARD, C. (1973), 'Recherches sur le contact fôret-savane en Côte d'Ivoire; étude du recrû forestier sur les parcelles cultivés en lisière d'un îlot forestier dans le sud du pays Baoulé', *Candollea*, 28: 21–37.

STOOP, W. A. (1981), 'Cereal based inter-cropping systems for the West African Semi-arid Tropics', *Proceedings of the International Work Shop on Intercropping, Hyderabad, India, 10–13 January, 1979* (Andhra Pradesh, India: ICRISAT), 61–8.

SWAINE, M. D. (1992), 'Characteristics of dry forest in West Africa and the influence of fire', *Journal of Vegetation Science*, 3: 365–74.

—— HALL, J. B., and LOCK, J. M. (1976), 'The forest-savanna boundary in west-central Ghana', *Ghana Journal of Science*, 16 (1): 35–52.

SWINDELL, KEN (1978), 'Family farms and migrant labour: The strange farmers of Gambia', *Canadian Journal of African Studies*, 12 (1): 3–17.

—— (1985), *Farm Labour* (Cambridge: Cambridge University Press).

—— (1992), 'African imports and agricultural development: Peanut basins and rice bowls in The Gambia, 1843–1933', in Keith Hoggart (ed.), *Agricultural Change, Environment and Economy. Essays in Honour of W. B. Morgan* (Mansell: London), 159–79.

SZOLNOKI, T. W. (1985), *Food and Fruit Trees of The Gambia* (ed. Stiftung Walderhaltung in Afrika und Bundes forschungsanstalt für Forst- und Holzwirtschaft, Hamburg).

TIFFEN, M., MORTIMORE, M., and GICHUKI, F. (1994), *More People, Less Erosion: Environmental Recovery in Kenya* (Chichester: Wiley).

TOKY, O. P. and RAMAKRISHNAN, P. S. (1981), 'Run-off and infiltration losses relating to shifting agriculture (*Jhum*) in Northeastern India', *Environmental Conservation*, 8 (4): 313–20.

TOULMIN, CAMILLA (1992), *Cattle, Women and Wells: Managing Household Survival in the Sahel* (Oxford: Clarendon Press).

UDO, REUBEN, K. (1982), *The Human Geography of Tropical Africa* (London and Ibadan: Heinemann Educational Books).

VINE, H. (1968), 'Develoments in the study of soils and shifting agriculture in tropical Africa', ch. 5 in R. P. Moss (ed.), *The Soil Resources of Tropical Africa* (London: Cambridge University Press), 89–117.

VON BRAUN, JOACHIM and WEBB, PATRICK (1989), 'The impact of new crop technology on the agricultural division of labour in a West African setting', *Economic Development and Cultural Change*, 37 (3): 513–34.

WATTS, M. (1983), *Silent Violence: Food, Famine and Peasantry in Northern Nigeria* (Berkeley CA: University of California Press).

WEBB, JAMES L. A. (1992), 'Ecological and economic change along the middle reaches of The Gambia River 1945–1985', *African Affairs*, 91 (365), 543–65.

WEBSTER, C. C. and WILSON, P. N. (1966), *Agriculture in the Tropics* (London: Longmans, Green).

WEDUM, JOANNE, DOUMBIA, YAYA, SANOGO, BOUBACAR, DICKO, GOURO, and CISSÉ, OUSSOUMANA (1996), 'Rehabilitating degraded land: *Zäi* in the Djenne Circle of Mali', ch. 7 in Chris Reij *et al.* (eds.), *Sustaining the Soil: Indigenous Soil and Water Conservation in Africa* (London: Earthscan), 62–8.

WEIL, P. (1973), 'Wet rice, women and adaptation in The Gambia', *Rural Africana*, 19, 20–30.

World Bank (1989), *Sub-Saharan Africa: From Crisis to Sustainable Growth* (Washington: World Bank).

YAYCOCK, J. Y. (1981), 'Crops and Cropping Patterns of the Savanna region of Nigeria: The Kaduna Situation', *Proceedings of the International Work Shop on Intercropping, Hyderabad, India, 10–13 January, 1979* (Andhra Pradesh, India: ICRISAT), 69–77.

# 7

# Rangeland Livestock Management

## Ecology and rangeland management

Throughout West Africa the keeping of livestock is of major social and economic importance. Animals kept are of many different types and methods of their management are varied. Of major importance are herd animals such as cattle, camels, donkeys, sheep, and goats but equally important to many households are small animals such as pigs, chickens, and guinea fowl which are kept in or near the compound. The importance of wildlife, non-domesticated animals to the livelihoods of Africans is increasingly being recognized and the farming of ostrich, crocodile, deer, and a variety of rodents both supplements the protein gap in the diets of many Africans and is a tourist attraction.

The focus in this chapter is on rangeland animals kept by semi-nomadic and sedentary peoples of the savannas and sahel. Attitudes to pastoralism by African governments have for too long been negative and the political desire to change pastoralists into settled ranchers dates back to colonial times. However, research into rangeland management since the early 1970s has revealed major misconceptions about indigenous pastoralism, not least its adaptation to a non-equilibrial environment. After a discussion of the dominant traditional pastoral communities in West Africa and the nature of relations between pastoral and sedentary peoples, the chapter turns to the major problems that pastoralists have been facing over the past 20–30 years and refutes many of the arguments against indigenous pastoralism. With regard to attempts to develop pastoralism, most of which have been unsuccessful, the focus is not on schemes that have failed, but on the increasing body of evidence which explains how the misunderstanding of local ecology by development experts has been a major cause of failure. Range ecology has, for many years, been based on Clementsian theories of plant succession (Clements, 1916, 1936) (see Chapter 1). These assume that vegetation progresses in an orderly and predictable manner towards an equilibrium, a climax vegetation. Although some of the earliest work on plant communities showed that non-equilibrium rather than equilibrium was the norm, nevertheless, concepts of predictable ecological succession dependent on environmental equilibrium have been widely accepted and have masked the reality of a fluctuating physical environment (Homewood and Rodgers, 1987; Ellis and Swift, 1988; Westoby *et al.*, 1989; Behnke and

Scoones, 1992; Warren, 1995). It is increasingly being accepted that abiotic components of ecosystems, at any scale, fluctuate over both space and time though the immediate impact of such fluctuations on human activity vary. In some parts of the globe they may have negligible effect on human activity while in the semi-arid areas of West Africa environmental fluctuation is a major determinant of indigenous land management strategies. The logic behind the flexibility of indigenous systems is increasingly being appreciated at the academic level though it will take time for it to filter through to development strategies which have a history of being based on temperate zone ecology. It is concepts of a stable physical environment which have led to the development of rangeland management strategies based on animal carrying capacity (Ellis and Swift, 1988; Behnke and Scoones, 1993). Calculation of carrying capacity (see below) may be useful only when the abiotic part of the system is comparatively stable. However, in the arid and semi-arid tropics where climatic conditions fluctuate significantly, such calculations have less relevance except possibly, on a very short-term basis.

With development strategies based on concepts of equilibrium, largely inappropriate in such environments, it is not surprising that the success of livestock development programmes has been limited. It is now appreciated that the opportunistic management strategies of indigenous pastoralists, based on autecology, are more appropriate in this environment. The rationality behind selected indigenous pastoral practices is discussed below. Rational and appropriate they may be, but there is much scope for further development of the indigenous livestock industry in West Africa. The argument here is not that foreign expertise is redundant, rather that it should be adapted to existing systems, improving what is there rather than attempting to replace it with systems which have proved suitable in ecologically different environments.

## Pastoral peoples

The Moors, the Twareg, and the Tubu are major herding societies in West Africa. Among each of these there is usually a southern, sahelian group and a northerly, Saharan group. The Saharan Moors rely on oases and permanent watering points such as scarp foot springs that permit irrigated agriculture. Much use is also made of temporary water sources of which the pastoralists have extensive knowledge (Toupet, 1975). The Twareg inhabit the Sahara and sahel between Timbuctoo in Mali and Bilma in eastern Niger, InSalah in Algeria and Kano in Nigeria (Norris, 1975). The mountain massifs of Aïr and the Adrar des Iforas are important pastures for Twareg herds (Norris, 1975), while the Saharan Tubu have their 'homes' in the Tibesti mountains in northern Chad. These mountainous areas are environmentally very different from the Sahara below them. They are green oases in the desert and local people tend to move their herds over shorter distances than do their counterparts in the sahelian zone where herds are moved over considerable distances each year in search of fresh pastures.

Herds are usually mixed and consist of cattle, sheep, and goats though the proportions of each may vary from one herd to another. Camels become increasingly common towards the Sahara and at one time many of the Saharan herds consisted predominantly of camels. Camels were much in demand for trans-Saharan trade and were used by the French army until after the Second World War when most of its Saharan units were motorized (Swift, 1986). As mechanized transport in the desert and sahel grew in importance, so demand for the camel declined, and with it, a major source of income for pastoralists. The majority of herd animals are female and reproduction is managed to ensure a supply of milk throughout the year as it is on this that herding families depend. A subsistence product of fundamental importance, milk of different qualities is produced from different types of herd animals. After meeting the needs of the calves, the remaining milk is consumed fresh or is processed into cheeses and yoghurts which form major elements of the pastoralist's diet. Surplus dairy products together with other animal products such as hides and dung are important media of exchange. In addition to the female animals which outnumber males in the herd by about 50:1, a small number of breeding stock are kept together with a few castrated males as these put on weight easily and sell well in the meat market.

Of all the pastoralists in West Africa, the Fulani are the most numerous and widespread. They live mainly in the sahelian zone, migrating south to the savannas in the dry season and north again to the desert margins with the return of the rains. Those Fulani who are semi-nomadic cover vast areas in the course of a year, some $500–1,000\,km^{-2}$, often returning to 'homes' near wet season pastures in the sahel (Morgan and Pugh, 1969). Many Fulani are now settled and for these people dependence on livestock may have been reduced, but animals nevertheless remain an important part of their lives. Fulani herds consist predominantly of cattle, sheep, and goats, the cattle having particular economic and social significance. All Fulani are given a cow at birth and throughout their lives they try to build on this, for on the size of the herd depends the status of the individual in the community and the privileges received (Traoré, 1981). While Moorish herdsmen, for example, have an eye to commercial dealings, the Fulani look upon cattle as something more important than a commodity to be traded, though for the Fulani too, the importance of trade in animals should not be underestimated. Smith (1992) attributes this to Fulani belief in the divine nature of cattle which can be traced back thousands of years in West Africa. Sheep, goats, and other small stock are often overlooked because of the focus on cattle. Small stock play a vital economic role as they reproduce more quickly than cattle and are exchanged more readily in markets for grain and other needs. They are often the starting point for newly formed herds.

There are several foci of Fulani settlement in West Africa among them the Fouta Djallon, the Atacora mountains, the Jos Plateau, mountains of the Adamawa range, and the Bamenda Plateau of West Cameroon, though Fulani herders are to be found throughout West Africa. Those Fulani living in the Fouta Djallon tend to keep herds of the tsetse resistant, small N'dama cattle, whereas those to the north of the tsetse zone keep a range of types of cattle, many of them descendants of the humped zebu. The fly thrives in the humid areas where vegetation is abundant but in spite of the threat of trypanosomiasis, domestic animals have always been plentiful in the more humid areas (Morgan and Pugh, 1969). Drought in the early 1970s and 1980s together with increased vegetation removal have encouraged pastoralists to move into areas well south of traditional transhumant routes and into the tsetse fly zone. In Côte d'Ivoire for example, pastoralists have adopted new husbandry techniques which have enabled them to cope better with tsetse fly. These include crossbreeding with trypano-tolerant breeds, innoculation, and skilful movement of herds and the use of fire to avoid areas such as dense gallery forest which are prone to infestation with the fly (Smith, 1992). Although herding may be important locally in the southern part of West Africa, it is of comparatively greater importance in the tsetse free semi-arid lands, and particularly to the north of the 500/700 mm isohyets.

## Different levels of dependence on livestock

### Semi-nomadic pastoralists

The nature and degree of involvement with livestock varies within and between the main herding societies in West Africa. Few pastoralists are purely nomadic these days (Swift, 1986), semi-nomadic and sedentary pastoralists being relatively more important. The semi-nomadic pastoralists of the sahel and savannas are the most deeply involved with livestock on the rangelands. They spend most of the year moving their herds through difficult terrain, making use of limited resources. For these people herding is much more than just an economic activity. It is a way of life, part of their culture and something of which they are intensely proud. Their herds provide the greater part of their subsistence needs in the form of milk, hides, wool, transport, and young animals which can be sold to buy grain. Rarely are animals used for meat (Swift, 1975). Ensuring that milk production is maintained is a dominant and perennial concern so herds consist mainly of female animals. The proportion of subsistence needs met by the herd can vary from year to year necessitating a flexible approach to life. If, for example, animals perish on account of drought, semi-nomadic pastoralists may be forced to settle and turn to activities such as farming, fishing, trading, or craft work until sufficient resources can be mustered to re-establish their herds (Moris, 1988). It may take one year or many more until a herd is re-established and for some this may never be achieved. Semi-nomadic pastoralists classified as such today, may be reclassified as sedentary farmers next year if drought claims the greater part of their herds, and as semi-nomadic pastoralists again in 5 years' time if they have managed to re-establish their herds (Horowitz, 1975). The proportion of subsistence needs met by the livestock thus varies according to such criteria as weather, access to labour, the size and composition of the herd, health of the animals, and the number of people the herd has to support. Pastoralists have to be sufficiently flexible to take advantage of whatever opportunities are available to them.

## Sedentary keepers of livestock

Livestock have a different function in the domestic economy of sedentary cultivators. Some sedentarists may be former pastoral nomads who have settled, but many are not. Animals kept by sedentary people usually include herd animals such as cattle, goats and sheep, pigs, and chickens. Some of the cattle may be trained and used as draught animals; these are not herded, and where a family only has a few goats and chickens, usually, these would be kept in or near the compound. Although animals may be of importance in sedentary systems they usually provide a smaller part of subsistence needs than do the herds of semi-nomadic peoples. They are nevertheless a capital resource that can be liquidated in times of need. They are also a source of valuable by-products such as dung for the fields and vegetable gardens, hides, milk, and more rarely, meat.

## Urban herd owners

A third group consists of urban dwellers who have invested in cattle and some small stock purely for commercial reasons. The keeping of livestock is in no way a culture for these people, who tend to be civil servants, merchants, senior officials in the army, and others with money to invest. The shortage of home-grown meat and the restriction on imports in recent years due to the effects of Structural Adjustment policies have led enterprising people to try to increase domestic meat production. In many cases these animals, like those of sedentary farmers are entrusted to pastoralists who are acknowledged specialists in animal husbandry, but in contrast to pastoralists who manage herds which are the basis of their subsistence existence, herds entrusted to pastoralists by such urban owners are destined largely for the meat market. Furthermore, a high proportion of the animals are male which are normally heavier than the females and thus have a higher carcass value. Some animals are also kept by sedentary pastoralists and by farmers for migrants who return periodically to the village and plan to return there permanently. These herds tend to be managed much more according to traditional practices than are those with a purely commercial objective.

## Relations between pastoral and sedentary peoples

Pastoralists range from those who maintain the fewest ties possible with the land to those who have actually settled and adopted a sedentary way of life. Settled Fulani, for example, who have become cultivators are to be found throughout West Africa. Pastoral nomads have settled either out of choice or because they had little alternative. Harsh droughts in the sahel have forced Moors, Twareg, and Tubu to settle, though these people are far more limited both in numbers and in the area they cover than are the Fulani. While coming to terms with a sedentary existence has proved difficult for many former pastoralists, some have adapted quite successfully, one example being the Moors who make and sell jewellery in urban centres in Senegal, Mali, and The Gambia, though it has to be said that 'sedentary' is not an entirely appropriate term to describe these craftsmen who spend a great deal of time travelling from one West African country to another, buying silver in Sierra Leone, for example, taking it to Nouakchott in Mauritania, Dakar, or Banjul where they may have a workshop employing several jewellery makers—usually all of them Mauritanian. These jewellery makers have few direct links with pastoralism though family members frequently keep herds.

The relationship between pastoralists and non-pastoralists is symbiotic. It is commonplace for the latter to entrust their herds to the care of either semi-nomadic or sedentary pastoralists. Cultivation of the land is normally the prime interest of sedentary farmers, who rarely have adequate labour to devote to the task of caring for animals, nor do they have the skill and knowledge of animals possessed by herders (Traoré, 1981). This does not mean that they do not herd animals. In the absence of pastoral peoples, the Mandinke, for example, will herd their animals though many prefer to entrust them to the Fulani. Fieldwork in The Gambia revealed the importance of pasturing animals well away from cultivated land during the wet season where they can do much damage. They also need to be found adequate pasture in the dry season, more than may be available in the vicinity of the village. Thus pastoralists frequently play a critical role in maintaining village herds. The animals have to be watered regularly during the dry season and, particularly if water supplies are limited, this can be a lengthy and time consuming task. Writing of the Manga in Niger, Horowitz (1975) explains that it is difficult for these people to organize a household capable of handling both animals and a farm. Where farmers either have no sons to look after their herds or have a sizeable herd, their choices are few. They can either abandon cattle raising because their resources are inadequate or they can abandon farming and specialize in cattle, an unlikely change, or the cattle can be consigned to a professional herder. The terms of exchange with semi-nomadic pastoralists may vary considerably but often they include giving the Fulani, or other pastoralist groups, access to all the milk produced by the herd, particularly when the herd is nowhere near the village for considerable periods. Milk and processed milk products made by women are a major constituent of pastoralists' diets and where a surplus is available this is sold or exchanged for commodities that pastoralists cannot produce themselves. The agreement between herd owner and herder may also involve the grazing of either the owner's animals, or possibly the entire herd of the pastoralist on certain fields in the village for a certain period during which the pastoralists and their animals may also have the right to water. This is mutually beneficial as it provides some dry season pasture for the animals and also adds fertility to the farmer's soil. Manga farmers in Niger, for example, compete for the manure of the Fulani cattle. The Fulani prefer to camp on larger fields so that they do not have to move on so quickly and some farmers have found that digging wells on their fields facilitates the watering of the herds and tempts herdsmen to remain even after the surrounding pastures have become exhausted (Horowitz, 1975). Toulmin (1992) has observed similar circumstances in Mali.

While the skills of pastoralists with animals are readily acknowledged, herders are frequently less than popular with sedentary cultivators. Conflict with sedentary farmers is common over damage to crops and property by Fulani animals which have escaped. Pastoralists argue that such charges brought by sedentary farmers are invariably false and designed to extract compensation from herders. (Okaeme et al., 1988; Ayeni, 1983). Another frequent source of conflict concerns young animals born while the herd is in the pastoralists' care. On returning to the village, a Fulani herd increased by several young male animals frequently leads to accusations that Fulani herdsmen have replaced females newly born to cattle of sedentary farmers with less valuable, young males of similar age. Such accusations could be just, as in Fulani herds—and in the herds of other pastoralists—females are highly prized and outnumber males by some 50:1, a few males being retained for breeding and the rest being sold. According to Horowitz (1975) Manga farmers who 'lease' their herds to the Fulani claim that the Fulani do not give Manga animals the care they give to their own. Depriving Manga calves of milk is one such complaint, the reason for this being that the Fulani can then either use or sell the milk not used by the Manga calves. Putting a herd in the hands of one or more Fulani cattlemen is thus a risky business which ultimately depends on the good faith of the herder (David, 1993). However, for many sedentary farmers who keep animals, the risk is evidently worthwhile as the practice is very much alive.

A slightly different type of symbiosis exists between sedentary cultivators and sedentary pastoralists. In the case of the Fulani, in spite of many having settled, links with pastoralism remain strong. An important part of their lives may involve looking after the animals of the sedentary farmers. Unlike the semi-nomadic Fulani who take the herds away from the village for substantial periods, the settled Fulani tend to take the animals well away from the village during the day in the wet season in order to ensure that village crops are not destroyed and to bring the animals back at dusk. The terms of exchange usually involve allowing the Fulani a proportion of the milk from the herd and involve them pasturing their entire herd which may include the animals of both the Fulani and sedentary farmers, on the fields of certain cultivators for a specified period after the harvest. The Fulani are paid for tethering their cattle on the plots of sedentary farmers. Relations between sedentary Fulani and their neighbours are often less strained than are relations with semi-nomadic pastoralists though accusations of milk stealing by Fulani, and Fulani cattle being allowed to escape and cause damage to the property of cultivators, is commonplace. Villagers in the Lower Gambia and in southern Senegal argued that in spite of their skill with animals, the Fulani are often, though not always, less than good farmers. Their fields reflect a lack of care and poor maintenance as they much prefer working with animals. Nevertheless, their presence in a village can greatly improve the symbiotic relations between semi-nomadic Fulani who visit the village periodically, and the sedentary cultivators. 'Deals' between the two groups may be negotiated more easily, particularly when the semi-nomadic Fulani are in need of water and when farmers want substantial numbers of cattle tethered on farm land to increase the dung input (Baker, fieldwork in Senegal, 1982).

Perhaps the most important relations between pastoral and sedentary people throughout the ages has been the exchange of goods. Swift (1986) argues that because pastoralists are unable to produce all their own needs they are much more heavily involved in the market than are many sedentarists. For example, Saharan pastoralists moving northwards to the markets of southern Morocco, Algeria and Libya trade milk, milk products, small stock, and some cattle for cloth and manufactured goods, many of which come from the Mediterranean or further north. On their return to the sahelian region they pick up dates and salt, exchanging these and their milk and milk products for cereals and water with sedentary farmers, and for kola nuts, sugar and other luxury products such as radios in local periodic markets. Although trading relations have proved to be complementary over the centuries, human relations between pastoralist and non-pastoralist peoples have often been racked with conflict.

In their management of animals pastoralists make great use of local environmental resources and in contrast to the 'Tragedy of the Commons' hypothesis (Hardin, 1968), it will be argued that traditional pastoralism has been sensitive to the local ecology, sustaining the environment rather than destroying it.

## Growing problems for pastoralists

Pastoralists have suffered badly in West Africa over the past 25–30 years largely due to growing pressure on pasture land. Several factors have contributed to

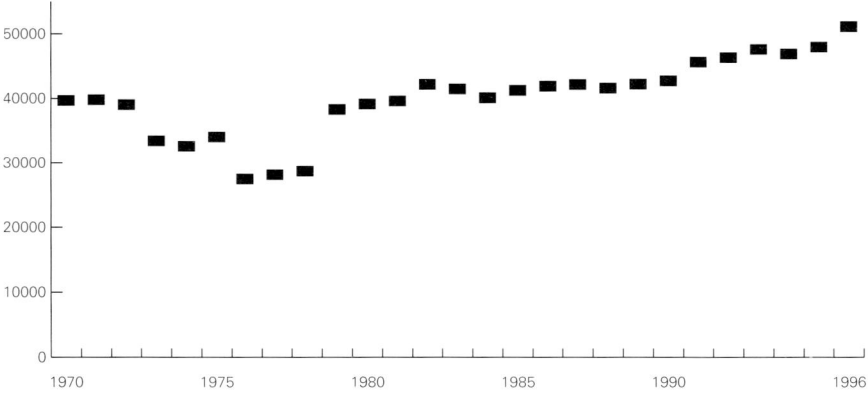

**Fig. 7.1.** Cattle stocks ('000 head) in West Africa, 1961–96

*Sources*: FAO *Production Yearbook* (1970), vol. 24; (1972), vol. 26; (1978), vol. 32; (1981), vol. 35; (1984), vol. 38; (1987), vol. 41; (1993), vol. 47; (1996), vol. 50;
FAO *Quarterly Bulletin of Statistics* (1995), vol. 8 1/2.

this. First, there has been a protracted drought in the sahel. The periods 1968–73 and 1983–5 were particularly bad years when the rains failed and thousands of animals died due to lack of water and pasture. The sharp fall in the number of cattle in the early 1970s is evident from Fig. 7.1, though the data do not adequately reflect the devastation experienced in the early 1980s in many parts of the West African sahel. In many of the intervening years rainfall has again been short, or if rainfall totals have not been low there has been a discernible change in the intra-seasonal distribution of rain increasing its unpredictability (Oladipo and Kyari, 1993; Gadbois, 1997, pers. comm.). In the more arid conditions which have prevailed since the late 1960s the water-table has fallen and in The Gambia and Senegal many permanent wells have dried up.

A second and major pressure on grazing lands has come from the migration of sedentary farmers on to range lands during the more moist conditions prior to the late 1960s. Much land in the sahel is marginal for crop cultivation and pastoralism is the most economic use of this land (Smith, 1992; Gorse and Steeds, 1987). But farmers have not moved indiscriminately on to grazing lands, they have chosen to establish gardens on prize lands such as river flood plains, *fadamas*, *bas-fonds*, *wadis*, and inter-dune depressions, areas where soil moisture was sufficient to permit plant growth even in the dry season (Touré, 1990). Moorehead (1998) describes the problems Malian pastoralists face as the extent of the floods on the Inland Delta of the Niger declines and prize land is allocated to newcomers to the area for agriculture. Pastoralists have long depended on such areas for dry season pasture and denial of access to this land is a source of growing conflict between land users (Adams, 1992;

Park, 1993). In some cases pastoralists have resorted to terror tactics to force sedentary cultivators off these lands and in some cases the more wealthy Fulani have paid people to leave (pers. comm., Nigerian agricultural extension worker, 1996).

A third encroachment on to pastoralists' dry season pastures has been large scale irrigation schemes in the major river basins such as the Senegal, the floodplain of the Niger and the large river basins of Nigeria such as the Sokoto and the Hadejia and its tributaries (Hollis *et al.*, 1993). The floodplain soils support valuable pastureland for livestock, but increasingly, pastoralists are being denied access. Irrigated agriculture has not yet lived up to expectations in economic terms but this has not stopped entrepreneurs wanting to secure land on irrigation schemes (Park, 1993).

Traditionally such lands have been held in communal tenure but with the advent of the Europeans and the establishment of colonial boundaries, much land was put into state ownership (Moorehead, 1998). As land shortage has rarely been a problem in West Africa, the state owned the land in name though communal tenure systems prevailed. As extensive areas of land were not put to any specific use by the State, it constituted 'no man's land' through which herders were free to roam. But where mechanization of agriculture and urbanization is increasing and there is growing demand for land, land tenure systems are undergoing revolutionary change (Ayeni, 1983). The state can press claims of its own on the land limiting access to traditional users, or, it can sell land into private ownership in response to the growing number of claims being made to irrigated or potentially irrigable land. Some of the effects of changing 'ownership' on land-use are evident from the example below.

The establishment of the Kainji Lake Basin Area (KLBA) in Nigeria has changed the geography of the area and created problems for the semi-nomadic Fulani for whom the area was an important source of dry season grazing on a long established transhumant route (Ayeni, 1983) (Fig. 7.2). The creation of Kainji Lake, the Kainji Lake National Park, and the earmarking of substantial areas for irrigated agriculture have reduced grazing land accessible to the Fulani. Kainji Lake National Park, a conservation area and a tourist attraction, prohibits the entry and grazing of Fulani cattle. Avoiding the Park on their transhumant route is possible but inconvenient, so many Fulani risk being caught and enter the Park illegally, grazing their cattle as they move across it. Park authorities are understandably outraged because herders readily lop branches from leguminous trees for their cattle. The Fulani also carry firearms and are not averse to shooting species of wild carnivore which are under conservation in the Park, but which pose a threat to their cattle, and which may be a source of food as well (Okaeme *et al.*, 1988). The attitude of the Fulani is understandable. For generations they have migrated into the area in the dry season and are reluctant to accept that their access is now restricted because of modern attempts at conservation.

Balancing the different demands on land is a growing problem and although

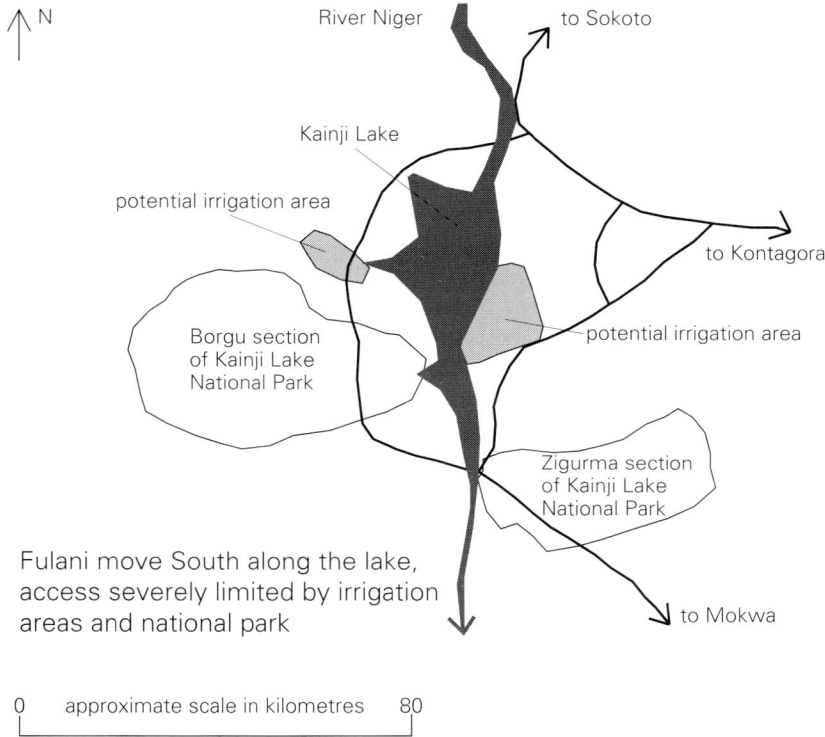

**Fig. 7.2.** Reduced access for Fulani to grazing lands due to presence of irrigated areas and National Park near Kainji Lake, Nigeria

*Source*: Based on Ayeni (1983), 240.

the Kainji developments have created problems for the Fulani, Kainji Lake provides more than two-thirds of the electricity consumed in Nigeria and has increased opportunities in commercial fishing (Ayeni, 1983). Nevertheless, if the livestock industry is to be sustained, adequate grazing land will also have to be made available. Grazing reserves and pastures are being set up in KLBA initiated by the Kainji Lake Research Institute, and supported by the State and Federal Governments but whether these will prove adequate, particularly in years of poor rainfall, remains to be seen.

Added to the problems pastoralists face through reduced access to pasture is the problem of declining pasture quality. There is evidence that the productive biological capacity of parts of the rangelands has diminished and in some areas pasture regeneration has been poor and the proportion of perennial to annual grasses has declined (Toulmin, 1990). Possible causes include the shortage of rainfall, declining soil fertility due to the effects of wind and water erosion which can be major problems in the sahel (Fones-Sundell, 1990;

Warren and Agnew, 1988), and localized overgrazing (Barrett, 1989). With improved rainfall in recent years perennial species are re-colonizing areas thought to be impoverished, a further reflection of environmental resilience and erroneous conclusions by development experts.

Overgrazing, the result of overstocking, has been widely cited as a major cause of rangeland degradation largely because of the increase in the number of animals in the region (Fig. 7.1). The overgrazing argument was seized upon because of its credibility when explained in terms of Hardin's model, cited in his paper Tragedy of the Commons' (Hardin, 1968). Here he argues from a global perspective that there is very little incentive for individuals to conserve resources to which they have communal access. Instead there is a tendency to over-exploit. Open access to water and pasture in African rangelands has thus been offered as a prime reason for their over-exploitation by increasing numbers of both people and animals in the region. The line of the argument is that pastoralists, trying to make the best living they can, have overstocked the ranges. This has resulted in overgrazing and hence degradation of pastures. In theory, the situation has been worsened by improved veterinary care and by the provision of deep wells which have changed former seasonal grazing lands into permanent pastures such as in the Ferlo of Senegal (Touré, 1990). The model is attractive and has been much used for explaining problems in the Sahel (Krebs and Coe, 1985). Proponents of the model argue that the solution to growing environmental degradation on the rangelands is to limit stocking rates so that they are within the carrying capacity of the region—whatever that may be. Secondly, they assert that tenure should be changed, placing the rangelands in private hands which would give private landowners more incentive to conserve their lands (Moris, 1988). It must also be noted that there were alternative views to the above argument which emphasized the rationality of overstocking in good years to offset the deleterious effects of lean years (Lewis, 1975) but during the 1970s these received relatively little attention in comparison with the overgrazing argument.

Overgrazing has occurred in some areas (Barrett, 1989) but recent field evidence suggests that the parallel drawn between Hardin's assertion and the sahelian rangelands is too simplistic and that many other factors are also involved. First, the term 'overgrazing' is loosely used and ill-defined. While it is applicable where vegetation declines over successive wet and dry periods, it is more often used to describe vegetation depleted temporarily during a series of dry years, but which subsequently recovers during a wet period (Deshmukh, 1986). Warren and Agnew (1988) argue that the impact of grazing may not be detrimental and that much depends on the type of grazing regime and the type of vegetation. Research evidence from the Negev/Sinai border cited by these authors revealed that the vegetation recovered rapidly on areas of stabilized dune protected from grazing. Although not an example from West Africa, it serves to show that on dune soils the water-holding capacity and seed sources had not been damaged, even by six millennia of Bedouin grazing (Warren and Agnew, 1988). This was attributed to the retention of seeds by the loose upper sandy soil, to limited run-off and erosion by water and to the less severe damage caused by trampling than would be the case on finer soils. The same authors also demonstrate that soils in the Sudan developed on former dunes have proved resilient to the effects of grazing up to a fairly distinct threshold which may be at about 35 per cent vegetation cover, the point at which movement of sand by the wind begins in large quantities and where dunes may become active again (Warren and Agnew, 1988). While dune mobilization may occur if vegetation cover falls below this threshold, sandy soils supporting vegetation were found to be quite resilient even where grazing pressure was intensified. This resilience may be explained by the existence of a feedback mechanism where animals were discouraged from grazing in an area where vegetation cover was low, thus giving the vegetation the chance to recover. It is also possible that following overgrazing, an area was recolonized with less palatable species, a situation termed 'green desertification' (Warren and Agnew, 1988). This is at odds with the work of Reichelt (1989) who argued that overgrazing, not clearly defined, leads to the reactivation of dunes and to the advance of the desert.

Overgrazing is a highly complex issue. Understanding whether or not an area has been overgrazed, being able to assess its potential for recovery, and explaining alternative or additional causes of rangeland degradation is extremely difficult and something that external experts have often failed to judge correctly. Rarely have schemes to manage the rangelands of the sahel drawn on the knowledge of indigenous pastoralists. Returning to the causes of rangeland degradation—be it temporary or permanent, it is increasingly the view that communal access and the subsequent overstocking and overgrazing are not necessarily to blame for any decrease in biological productivity (Barrett, 1989). The role of the environment has, in the past, been underestimated.

As Moris (1988) observes, one of many errors is that Hardin's model sees all pastoralists as being similarly motivated and all problems of land degradation originating from the misuse of land by nomads and agropastoralists. In some areas this may be true, but in much of the West African sahel the case can be refuted. First, pastoralists do not willingly keep all their animals in one place for if drought occurs its impact may be worse in one place than in another and in consequence, division of the herd is a preferable strategy. Keeping the herd moving to obtain access to adequate water and forage is critical for milk herds where keeping animals lactating for as long as possible is an overriding objective. Herds are thus frequently divided between members of the family both to minimize risk and to aid movement of the herd. The argument against large herds is further strengthened by the work of Moris (1988) and Swift (1986) who have shown that large herds are cumbersome, difficult to move, and can prove more of a liability than a benefit. In times of drought pastoralists tend to move into fall-back pastures, usually further south in West Africa. Because maintaining the animals in lactation is so important, pastoralists cannot remain in one place, particularly after pastures have been grazed. Thus the generalization that nomadic pastoralists aim to maximize their herds and then overgraze pastures is not strictly true. As Moris (1988) observes, such behaviour is much more typical of the 'modern', more commercialized, wealthy ranches where animals are kept for meat rather than milk. Another fallacy is that herd size is maximized and that pastoralists are reluctant to dispose of their stock in official markets. As the work of Swift (1986) has shown pastoralists are highly market oriented and quite prepared to sell their stock. The erroneous assumption is frequently made because people fail to appreciate the importance of small stock in markets (Moris, 1988). They may not have the right to sell the stock of absentee owners, but according to Sutter (1987), over 60 per cent of the subsistence needs of Fulani families in northeastern Senegal were derived from livestock sales.

## Environmental arguments vindicating indigenous pastoralism

Hardin's model which hypothesizes the abuse by the individual of commonly held resources, also raises the question of carrying capacity of the land and in terms of how many animals the rangelands can support. Growing concern for land degradation over the past 25–30 years has highlighted the impact of animals on the vegetation. One method of controlling rangeland degradation by many range officials has been to limit the number of animals on the ranges. The scientific basis of this judgement lies in assumptions about the impact of grazing on vegetation and it is this that has been used to determine the carrying capacity or the number of animals that an area can support (Behnke and Scoones, 1993). As discussed earlier, this has little relevance in these semi-arid regions. Nevertheless, the concept of plant succession has been applied to rangeland management (Behnke and Scoones, 1993). Grazing, it was thought, represented a disturbance in the successional sequence and maintained the vegetation at a sub-climax level. Range managers thus aimed to ensure that grazing pressure did not exceed the capacity of the range to regenerate. Basing static rangeland management strategies on carrying capacity has been inappropriate because carrying capacity is dependent on the characteristics of the savanna vegetation and as Chapter 5 has demonstrated, there are several determinants of savanna which make for a highly varied environment. In summary, the characteristics of the rangelands are dependent on the relationship between plant available moisture (PAM), plant available nutrients (PAN), fire, herbivory, and the actions of humans. With variation possible in all these five factors, predictability of conditions in savannas is difficult. Because changes in moisture over relatively short periods can affect the availability of nutrients and the quality of one area of grassland relative to another, so the potential carrying capacity is rarely stable or predictable and thus extremely difficult to evaluate with any accuracy.

Erratic and variable rainfall in sahelian West Africa causes particular confusion in estimating carrying capacity. As Homewood and Rodgers (1987), Ellis and Swift (1988), Westoby *et al.* (1989), and Behnke and Scoones (1993) observe, the number of animals that an area of rangeland can support is based on the concept of a stable equilibrium between animal and plant populations. Forage availability controls animal numbers which in turn control the availability of forage. Such a feedback system assumes relatively constant conditions for plant growth, but if, as in the sahel, rainfall conditions are highly variable, non-biological factors such as variability of rainfall will have a greater effect on plant growth than does the density of the animal population. As animal survival is dependent on availability of forage, it could then be argued that rainfall, via forage, is the main variable controlling the size of the grazing population, a very different argument from that which postulates that communal tenure and overgrazing due to overstocking are at the heart of rangeland degradation.

A second problem with the term 'carrying capacity' is that the definition can vary according to whether it is based on ecological or economic criteria, and also according to what a range manager wishes to extract from the environment. At ecological carrying capacity there is no surplus of either livestock or vegetation. Livestock may be numerous, but their condition may not be particularly good, nor will vegetation be as abundant as if animals were not present. If better quality of either plants or animals is required, then fewer animals should be maintained. According to Caughley (1979) cited in Behnke and Scoones (1993: 5), 'economic carrying capacity' is 'about half to two thirds of the stocking density at ecological carrying capacity . . .'. As with so much theoretical material, theories of range management were not developed specifically for Africa and for the most part have not proved applicable to African situations. There seems little value in applying figures for carrying capacity of capital intensive commercial beef ranching in North America to subsistence oriented pastoralists in West Africa whose harvest is predominantly in the form of live animal products such as milk, traction power and transport.

Thirdly, empirical evidence reveals that in non-equilibrial grazing systems, that is, where environmental conditions and in this case rainfall in particular, are prone to marked variation, mobile livestock production is more efficient than sedentary systems. The capacity of indigenous systems to move herds rapidly from one area to another to ensure that milk production is maintained enables a higher livestock population to be supported than under sedentary systems (Behnke and Scoones, 1993). In addition, the assumption that pastoralists overstock in good years can be interpreted as a security measure to ensure that their losses are not total in years of poor rainfall. The swings in animal numbers in indigenous pastoral systems are thus very much greater than in sedentary systems. Drawing on evidence from Zimbabwe, Behnke and Scoones (1993) note that over a period when rainfall has been favourable, cattle populations do approach a ceiling set by the ecological carrying capacity. However, animal numbers never reach ecological carrying capacity because of the random intervention of exceptionally stressful years when numbers fall markedly. In the long-term non-equilibrial factors tend to be the main agents of control on livestock numbers (Homewood and Rodgers, 1987; Ellis and Swift, 1988; Westoby *et al.*, 1989; Sullivan, 1998). Thus even in an area where non-equilibrial conditions predominate, there may be periods when environmental variables fluctuate within a narrower range. Nevertheless, conditions remain unpredictable. Such variability makes the calculation of carrying capacity even more difficult and its use as a model of rangeland management is thus of minimal value.

It is now accepted that rather than being inferior to temperate zone systems of rangeland management African indigenous pastoralists are highly skilled in getting the most out of their own environment, which is driven by variability. As in sedentary farming systems, pastoralists' autecological approach to livestock management is heavily dependent on environmental knowledge. Evidence by Breman and de Wit from West Africa (1983) and by researchers in East and Southern Africa (Behnke and Scoones, 1993) has shown that pastoralism through what has been termed 'opportunistic management' (Westoby *et al.*, 1989) either equals or exceeds productivity per unit land area of commercial ranching in comparable ecological environments. In the following section examples of indigenous methods of coping with a difficult and highly unpredictable environment are reviewed.

## An autecological approach to meeting the needs of cattle, camels, goats and sheep

### Transhumance

The first and overriding objective of herders is the well-being of their animals and the sustained production of milk and milk products for as much of the year as possible in order to meet domestic needs. If the herd is not to 'run dry', the animals need adequate water and pasture and in consequence, are frequently moved to ensure that these needs are satisfied. Herd movement is thus controlled primarily by environmental factors though different factors may also play a part. Pasture land may be assured by irregular movements of herds over thousands of kilometres between the sahel on the desert fringe and the sudan savannas; it may be assured through the movement of animals to higher ground as in the case of the Saharan mountain ranges, the Adamawa, the Jos Plateau, and the Fouta Djallon where movements may be far less extensive than in the sahel. For example, in the Fouta Djallon, the Fulani move their herds of N'dama cattle from the wet season pastures of the *bowé*, the high plateaux of concretionary ironstone, to the dry season pastures of the valley bottoms where former Negro slaves farm the land and exchange grains for animals products (Morgan and Pugh, 1969). In addition, pastoralists may have 'fall-back' areas such as dangerous mountain grazing or tsetse infested bush where animals could be taken in times of severe drought (Moris, 1988). An important source of dry season pasture is the wetlands, river floodplains, wadis and *bas-fonds* for example (see above).

Herders tend to congregate during the wet season in the sahelian zone or just on the desert edge where rains usually produce good pasture and where surface water accumulates frequently with a high saline content which is good for the animals. Fodder and browse are abundant and easily accessible for the short period July to September and this is normally a time of social gatherings. For the remaining 9 months the provision of water becomes much more difficult as pastoralists depend heavily on well water and the few perennial rivers. Where water and pasture are in short supply, dependence on browse increases (Gorse and Steeds, 1987).

Patterns of movement may vary according to the composition of the herd. If, for example, a herd consists predominantly of camels, its requirements will differ from a herd where cattle predominate. Mostly, however, herds are mixed and the water needs of cattle, and particularly of calves, determine when a camp should be moved. The information network among pastoralists is well developed and this aids decisions on where a herd is to be moved. Pastoralists are knowledgeable about the location of dry season pastures in floodplains, wadis and *bas-fonds* where soils tend to have a higher clay content and are usually moisture and nutrient retentive. Pastures are frequently rich in both perennial and annual species, the former being in relative abundance (Gorse and Steeds, 1987). Pastoralists are well aware of the varying characteristics of vegetation, of its declining palatability and nutritive value as the dry season advances. They are aware of the need for micronutrients, and particular care is taken to ensure that their animals have an adequate supply of salt in their diet. Lewis (1975) argues that although herd movement is dictated by environmental factors, social issues also influence decisions on when and where to move the herd. For instance, the flow of information enables groups either to meet together or to avoid relatives, friends, and patrons to whom a debt might be owed or is owing, or by whom a favour may be demanded.

## Meeting the animals' daily needs of water and pasture

### Cattle

While transhumance is the means by which pastoralists ensure that adequate water and pasture are available, very different strategies are employed on a daily basis to make sure that the best use is made of these resources by the animals. In the high temperatures of Saharan and sahelian West Africa animals need considerable quantities of water both for the retention of body moisture levels and also for cooling. Unlike humans it would seem that cattle drink not just to avoid dehydration but to cool their bodies. Breeds adapted to arid conditions can exist without water for about 2 days but depending on the temperature and on the lactation condition of the cow, lack of water for any longer can put considerable strain on the animal. Deprivation of water rapidly leads to weight loss of the animal and with this a decline in food intake. Cattle seem unable to eat once they lose weight. One means of avoiding the harshest effects of the climate is to get the cattle out to graze at first light when it is coolest. Their digestive system of multiple stomachs enables them to graze and browse when it is coolest and to ruminate, chewing the cud in the shade of a tree in the heat of midday. This is the most efficient use of the cow's energy resources (Smith, 1992).

Although lactating cows do not like leaving their offspring, they are driven well away from the camp at dawn, into the bush to graze. It is only after the mothers are out of the way that the calves are let out to graze. The cattle find their own way back to the camp in the evening because the mothers, which dominate the herd in terms of numbers, return to feed their calves. If the mothers were left with their calves all day the latter would suckle constantly, reducing the milk available for human consumption (Smith, 1992). Having said this, pastoralists are very aware of the needs of calves during the dry season and at this time adults may go without to ensure that children and calves share what is available (Swift, 1975).

The provision of water is of critical importance during the dry season and occupies much of the herdsman's day during the dry season. Water is drawn from wells often as much as 50 m deep and for this cattle are used as draught animals. Water is drawn up using forked branch pulleys and a skin bucket which contains around 20 litres (Smith, 1992). The water is put into a trough from which the animals can drink. The owners of a well both use it and maintain it and access to the water is dependent on relations with the families who own the pulleys. Use of a well usually necessitates payment either in kind or in service to whoever maintains it, but there is no guarantee that well 'owners' will be prepared to share their water, particularly in times of stress. In order to overcome this problem many aid agencies have installed boreholes which pump water continuously.

### Camels

The requirements of camels are very different from cattle and although they are far less important in numerical terms, the greater part of the pastoralists' milk in the hottest part of the year in the Saharan regions is supplied by camels (Smith, 1992; Swift, 1975). Camels can satisfy most of their moisture requirements by browsing on seemingly dry desert vegetation. After the rains they can exist for months without actually drinking, as long as the vegetation remains lush. As the dry season advances camels require more water, but never as much as cattle. They browse on a range of species among which are thorny *Acacias*. Leaves and shoots are consumed as are the seed pods which are rich in protein and highly nutritious. Where branches are beyond the camels' reach, pastoralists lop them for the animals. On account of their capacity to tolerate drought, Saharan pastoralists normally keep some camels even when market demand for beef is greatest. Camels are valuable for their milk yields. A lactating camel yields about 10 litres of milk a day and about 4 litres a day towards the end of lactation. Camels' milk is preferred by the Twareg who claim that it has special qualities for strength. The compostion of camels' milk makes it difficult to separate the milk into butter and skim. It is thus difficult to keep and so Twareg drink camel's milk fresh each day. As with the cattle, the female camels which dominate the herd in numbers are driven out to browse, away from the camp. If they are not their calves continue to suckle, consuming milk which would otherwise form part of the human diet.

In addition to browse and water, pastoralists include certain species in the animals' diet to ensure that both camels and cattle have an adequate intake of salt and that they have access to a range of other nutrients including trace elements. This both ensures good health and builds up resistance to certain ailments and disease. Chambers (1983) reveals how pastoralists associate night blindness in cattle in the dry season with the absence of certain species of plant, shown scientifically to reduce vitamin A deficiency. The depths of ethno-science are only gradually being realized.

### *Sheep and goats*

Sheep and goats constitute 62 per cent of the domestic stock of sub-Saharan Africa (Smith, 1992). They are often overlooked, but they play an important role in pastoralist economies. They reproduce relatively quickly and are readily exchanged in the market-place either for money or for other commodities. Sheep and goats are the main sources of meat for pastoralists. Herds of goats and sheep are usually looked after by Twareg boys who live on little other than their milk and on bush products. As part of the learning process young Twareg boys are given the responsibility of a herd of sheep and goats and are sent away from their families for several weeks at a time. The effect of both sheep and goats on pasture is to a certain extent complementary. Sato (1980, cited in Smith, 1992: 109) has shown that sheep used 74 per cent graze compared with 21.6 per cent browse, while goats used 52.5 per cent graze and 47.5 per cent browse. Because they are cheaper, sheep and goats tend to be owned by poorer people who cannot afford cattle. They are also a starting point for new herds or for herds that are being rebuilt.

## Indigenous veterinary care

Insect pests and diseases are a major problem for pastoralists in West Africa. One of the most virulent pests is trypanosomiasis, transmitted by the tsetse fly (*Glossina* spp.) which has a fatal effect on cattle. In tsetse-infested areas pastoralists have little option but to keep tsetse-resistant species such as the N'dama cattle which emanate from the Senegambian region, and relatives of the West African Short Horn. Other insects such as ticks, stinging flies and, in the rainy season, mosquitoes, can have a debilitating effect on animals. Even the common housefly is a constant irritant to cattle, forcing them to expend energy just keeping the flies on the move. Other diseases which are not necessarily transmitted by insects include rinderpest, bovine pleuro-pneumonia, anthrax, black quarter, and haemorrhagic septicaemia (Smith, 1992).

Over the past 20–30 years pastoralism has moved well south of the 500 mm isohyet, an approximate boundary for the tsetse fly. The reason for this change appears to have been protracted drought in the region and extensive bush clearance. The effects of tsetse have thus been to some extent reduced. It is not that tsetse fly has ceased to exist south of the 500 mm isohyet, rather that pastoralists are increasingly skilful in limiting its effects. For example, the bush in an area may be burnt before a herd is moved into it to destroy the habitat in which tsetse breeds, to rid the area of a whole host of disease-carrying insects and vermin and to promote the growth of fresh grass. The Fulani have also had some success in crossbreeding animals which are not trypano-tolerant with those that are and thus increasing immunity of their herds to trypanosomiasis (Smith, 1992). In some areas where the tsetse fly is prevalent such as the Middle Niger Valley in Nigeria, drugs such as Berenil are no longer as effective as they were formerly because of emerging drug-resistant strains of trypanosomiasis (Okaeme *et al.*, 1988).

The use of bush burning as a strategy to limit the spread of disease is a highly controversial issue largely because of the potential damage that can be done to the environment (Okaeme *et al.*, 1988). Getting to the truth of who actually sets fire to vegetation can be a problem. In the semi-arid savannas of The Gambia and Senegal there seemed little doubt among cultivators and extension workers that shortly before the rains the Fulani would burn potential pasture land to clear it of dead vegetation which was neither as palatable nor as nutritious as fresh grass. The Fulani on the other hand were reluctant to admit that they burned pastureland unless a herd had been stricken with disease when burning was necessary to 'sterilize' the area. Bush fires, according to pastoralists were more usually fires for field clearance, started by sedentary farmers, but which had got out of control. This controversy concerning the use of fire is reflected in the literature discussed by Smith (1992).

As part of a strategy of disease and pest avoidance, animals are innoculated against certain diseases such as contagious bovine pleuro-pneumonia by exposure to the disease. The Fulani and Twareg vaccinate their animals by soaking a piece of infected lung in milk for a day or two, then applying this as a poultice to an opening in the skin around the nose of the animal (Smith, 1992). Other forms of treatment involve the use of herbal plants, cutting, and bleeding. Rituals are often performed in association with these treatments to discourage pests and diseases. As part of a strategy to avoid illness in their animals, pastoralists train them not to eat certain poisonous plants. For example, the poisonous leaves of *Nerium oleander* are burnt by certain Twareg groups and placed under the noses of young camels. The animals grow to detest the smell and avoid the plant when browsing. Consumption of poisonous plants tends to increase when the rains have been poor and when pasture is limited and under such conditions pastoralists are usually more inclined to move their herds to avoid them resorting to species which they would not normally feed on and which could prove poisonous to them. This is further evidence against overgrazing.

## Rationality behind the structure and size of herds

As the above discussion indicates it is valuable to have mixed herds as these provide a range of resources and also make use of a variety of natural resources. Mace (1990) and Mace and Houston (1989) have shown that an appropriate mix of species can actually promote household survival chances in an unpredictable environment. Thus camels and cattle have different fodder and water requirements which are again different from sheep and goats. Pastoralists are accustomed to varying the composition of herds between sheep, goats, camels, and cattle so that the needs of the animals can be met in different environments (Moris, 1988). In a way, this resembles the intercropping strategies of sedentary farmers. Focusing on those Twareg who keep only camels and small stock, but no cattle, Mace (1990) shows that a balance is maintained as far as possible, between camels which are relatively resilient to drought but which reproduce slowly, and small stock which reproduce rapidly but are much less resistant to drought and perish more easily. It is noteworthy that Twareg involved in agropastoralism tend to keep more cattle than do the more pastoral members of the Twareg.

As a herd is built up by the more pastoral members of Twareg society, the number of small stock is increased until a certain number can be exchanged at a market for camels. Mace (1990) argues that the herd size at which the switch to mixed camel and small stock herding becomes optimal depends on household food and income needs. The relative food-producing abilites of camel and small stock are of importance and this is not well quantified in the literature. Dahl and Hjort (1976) show that total production of food by an adult female camel, that is milk, blood, and meat from male offspring, is equivalent to the food produced by 8.3 female goats. Thus poorer Twareg who tend to have a higher proportion of small stock are more susceptible to losses due to drought than are richer Twareg with a higher proportion of camels. Mace and Houston (1989) have shown that only a relatively small number of Twareg families live off small stock alone as this is a high risk 'boom or bust' strategy. Where cattle are also important most herders operate a similar strategy, balancing risk against household needs.

Most animals are born during the rainy season, the time when pastures are at their most abundant. In order to ensure that pastures are put to their best use, most male calves are disposed of—usually sold—within their first year of life. Because pastoralists require a supply of milk throughout the year they must ensure that animals are born throughout the rainy season to maintain lactation for as long as possible. It is also advantageous to have a few animals born after the rains though animals born in the dry season rarely prosper because they never have the benefit of good pastures. However, it does mean that the mother is lactating through the dry season and thus providing milk needed for human consumption.

Herd size varies according to a number of factors. Inevitably environmental factors will limit herd size, but a more immediate limit may be imposed by labour availability. A herd of some 300 animals requires a major input of labour. Watering the herd alone is such a major task that it can leave insufficient time for grazing. Grazing resources may be used up so quickly that the herd needs to be moved very frequently in order to maintain good supplies of pasture. This may prove a considerable strain on the herd manager. In order to overcome this herds are frequently divided into smaller groups which are more easily and more efficiently managed. Fieldwork by Dyson-Hudson (1966) in the Karimojong indicated that herds rarely consisted of over 150 animals. A herd of this size could be managed conveniently by a single family unit. Referring to East African herders, Spencer (1965, 1998) indicates that unless a family has a labour force of at least one adult male, one adult female, one fully grown boy, and two smaller boys or girls it must, to some extent, rely on other homesteads for help and this may not always be convenient or possible.

The animals within a herd may not all belong to a single person. Very often pastoralists make a point of dispersing their animals among several herds, a form of insurance against predation and disease. The ownership structure of a herd is often highly complex. Animals may be lent or borrowed in times of hardship and the beginnings of a herd may thus be based on animals loaned from other pastoralists. Among the Fulani, the herd is frequently made up of the cattle of sedentarists. In recent years with limits imposed on imports through Structural Adjustment programmes throughout West Africa, entrepreneurs who include civil servants, army officials, merchants, and sedentary farmers have spotted the need to increase meat production within the region and have thus become herd owners. Without any knowledge of herding these people have begun using indigenous pastoralists to look after their herds. In some cases where the Fulani have lost all their own cattle they may actually own none of the herd they look after.

## Pastoralists' diets

While care of the herds and decisions about where and when to move is usually the responsibility of the men, women are far from passive in pastoralist economies. Organization of food supplies is a major responsibility of pastoralist women who are known throughout West Africa for their skills in preserving milk and in making milk products. As milk is the main product of the herd, milk and milk products are a major component of the pastoralist's diet and in addition, milk processed into buttermilk, yogurts, and cheeses are sold in the market-place. In the sahelian areas where the Fulani have a fairly close relationship with sedentary cultivators such processed milk products may be brought directly to the village where they are sold or exchanged for cereals. The quantity of cereals purchased depends much on access to the market. With increasing sedentarization of pastoralists, it is often the case that members of the family will grow cereals near their 'homes'. For the Fulani and other sahelian pastoralists this may be in the vicinity of their wet season pastures, while for the Saharan pastoralists which include the Moors, Twareg, and Tubu, crop cultivation occurs near oases, or where water supplies are assured such as at the foot of escarpments (Toupet, 1975). Of particular importance are the areas of high land such as the Adrar des Iforas, the Aïr mountains, and Tibesti where there is water and where crops can be grown and pastoralists have their bases all year. Pressure on land in these parts is increasing (Swift, 1975).

Pastoralists also make use of wild grains such as *Cenchrus biflorus*, *Panicum laetum*, *Pennisetum* spp., and *Eragrostis ciliaris* (Morgan and Pugh, 1969; Moris, 1988; Smith, 1992). Seeds of *Cenchrus* spp. are particularly valued because of their high nutritive value and also species of the genus *Panicum* as this is the first to ripen after the rains and needs very little preparation. The methods of harvesting are varied and may include the pulling of baskets through the crop in order to dislodge the grains. Alternatively the grass may be cut before the seeds fall, the grass is dried and the seeds removed by threshing and winnowing. A third method involves either the cutting or the burning of the standing crop after the seed has fallen. The residues are swept into piles from which the grain is extracted by winnowing (Maiga *et al.*, 1991). These wild grains which are highly nutritious are of particular importance to the lower caste groups (Smith, 1992: 122). They are usually supplemented with seasonal fruits such as *Boscia senegalensis* which can be stored and are valuable sources of food during the dry season. Other wild foods gathered include the fruits of *Zizyphus mauritania*, the tuber *of Nymphea lotus*, and the leaves of *Maerua crassifolia* (Maiga *et al.*, 1991). Women and children play a major part in the collection of semi-cultivated and wild species and inevitably in their incorporation in the diet.

For those pastoralists who are still semi-nomadic, diet varies throughout the year. Good pasturage closely follows the rains in June/July and this is the time when the milk yield is at its highest. Grain supplies, either wild or cultivated, are at their lowest level at this time of year and represent a relatively small proportion of the diet. It is not until the end of August that the proportion of cereals in the diet begins to increase. The grain harvest is associated with the onset of the dry season. Pasture quality now declines and with it, milk yield. In consequence, cereals play an increasingly important part in the pastoralists' diet relative to milk. The end of the dry season from March to June is a critical period for herders as well as for their animals as at this time they expend much energy keeping their herds watered. Weight loss can be considerable for pastoralists, not just for animals, during the dry season when diets are poor and malnutrition levels increase among pastoralist children. This can be a time when people turn to famine foods: raiding ants nests for grains, consuming locusts and other insects, and searching through what plant resources exist for fruits, berries, and leaves.

## Conclusions

If the indigenous livestock industry is to be developed and modernized, then it is of prime importance that national governments recognize the actual and potential value of indigenous pastoralism, understand the strategies underlying their survival strategies, and base future development on improving indigenous methods of rangeland management rather than on substituting ecologically inappropriate alternatives. For example, uncertainty over tenure for pastoralists could be resolved by improving their access to grazing lands and watering points. Governments could assist with policies to improve the health and nutrition of both pastoralists and animals. They could facilitate the marketing of livestock and livestock products and improve the flow of information to pastoralists on range conditions. In times of drought, assistance with the rapid movement of herds to areas less stressed could also be of major benefit. Movement of herds should be encouraged rather than discouraged, for as Behnke and Scoones (1991, 1993) show, the ranges can support larger animal populations if they are not kept in one place.

Policies should be avoided which undercut the livelihoods of pastoralists as was the case of the milk production project in Northern Nigeria which, had it been successful, would have ruined the livelihoods of pastoralist women dependent on income from the sale of processed milk and milk products (Salih, 1991). Fortunately in this particular instance customers of the traditional Fulani vendors of milk and milk products remained loyal and livelihoods were not lost as the large-scale project failed for many reasons unrelated to the existing system of traditional milk production. Such an example shows how modern schemes have tended to ignore what exists rather than building on it. The important role that women play in agriculture has been recognized only within the last 10–15 years. Information on the role of women in pastoralist societies is only now being documented by writers such as Joekes and Pointing (1991), Toulmin (1992), Waters-Bayer (1988), and Waters-Bayer *et al.* (1995). Increased sedentarization, it would appear, has not had many benefits for women. Any reduction in livestock keeping reduces women's capacity for income generation and as social relations are restructured as a result of changes in pastoralist society, women's entitlements are diminishing, their workload is increasing and the products of their labour—the products of settled agriculture—are increasingly being controlled by men (Joekes and Pointing, 1991). Policies should attempt to ensure that pastoralist women suffer no further marginalization.

Meat production needs to be increased in West Africa. While indigenous pastoralism has not been directed towards meat production, governments might benefit from consulting pastoralists on how an increase in meat production might be achieved. It is evident that technology on ranching transferred from the temperate zone has proved inappropriate to semi-arid West Africa largely because of environmental non-equilibrium. That African governments have promoted such developments at the expense of indigenous pastoralism is a further example of political ecology: how political decisions influence environmental management. Although they have survived, the efficiency of indigenous pastoralists could probably be improved, possibly by combining the more successful elements of indigenous methods of rangeland management, which are based on autecological principles and which exploit environmental variability, with modern technology. What appears to be necessary is a reappraisal of indigenous pastoralism by national governments and a commitment to instigate development policies worked out in conjunction with pastoralist communities. Previous assessments of pastoral societies as 'traditional', in the sense of being resistant to change and isolated from the wider economy, have been shown to be inaccurate. Methods used by pastoralists are ecologically sensitive and it is these that should be incorporated into appropriate rangeland development policies.

## *References*

ADAMS, W. (1992), *Wasting the Rain* (London: Earthscan).
ADAMU, M. and KIRK-GREENE, A. H. M. (eds.) (1986), *Pastoralists of the West African Savanna* (Manchester: Manchester University Press in association with the International African Institute).
AYENI, J. S. O. (1983), 'Rangeland problems of the Kainji Lake Basin Area of Nigeria', *Environmental Conservation*, 10 (3): 239–45.
BARRETT, S. (1989), *On the Overgrazing Problem* (London: IIED/UCL London Environmental Economics Centre).
BAXTER, P. T. W. (ed.) (1991), *When the Grass is Gone: Development Intervention in African Arid Lands* (Uddevalla: Nordiska Afrikainstitutet).
BEHNKE, R. H. and SCOONES, I. (1993), 'Rethinking range ecology: Implications for rangeland management in Africa', ch. 1 in Roy H. Behnke Jr., Ian Scoones, and Carol Kerven (eds.), *Range Ecology at Disequilibrium: New Models of Natural Variability and Pastoral Adaptation in African Savannas* (London: ODI), 1–30.

BREMAN, H. and DE WIT, C. (1983), 'Rangeland productivity and exploitation in the Sahel', *Science*, 221: 1341–7.

CAUGHLEY, G. (1979), 'What is this thing called carrying capacity?', in M. Boyce and L. D. Hayden-Wing (eds.), *North American Elk: Ecology, Behaviour and Management* (Laramie: University of Wyoming Press).

CHAMBERS, R. (1983), *Rural Development: Putting the Last First* (Harlow: Longman Scientific and Technical).

CLEMENTS, F. E. (1916), *Plant Succession: An Analysis of the Development of Vegetation* (Carnegie Institute of Washington Publication, 242).

—— (1936), 'Nature and structure of the climax', *Journal of Ecology* 24: 252–84.

DAHL, G. and HJORT, A. (1976), *Having Herds: Pastoral Herd Growth and Household Economy* (Stockholm Studies in Social Anthropology, University of Stockholm).

DAVID, D. (1993), 'Droughts and the urban nomads', *The Courier*, 137, Jan.–Feb.: 19–21.

DESHMUKH, I. (1986), *Ecology and Tropical Biology* (Oxford: Blackwell Scientific Publications).

ELLIS, JAMES E. and SWIFT, DAVID M. (1988), 'Stability of African pastoral ecosystems: Alternate paradigms and implications for development', *Journal of Range Management*, 41 (6): 450–9.

FONES-SUNDELL, M. (1990), *Perspectives on Soil Erosion in Africa: Whose Problem?* Sustainable Agriculture Programme, Gatekeeper series No. SA14 (London: IIED).

FRANTZ, C. (1975), 'Contraction and expansion in Nigerian bovine pastoralism', in T. Monod (ed.), *Pastoralism in Tropical Africa* (London, Ibadan, Nairobi: Oxford University Press), 339–53.

—— (1986), 'Fulani continuity and change under five flags', in M. Amadu and A. H. M. Kirk-Greene (eds.), *Pastoralists of the Savanna* (Manchester: Manchester University Press in association with the International African Institute), 16–39.

GORSE, EUGENE JEAN and STEEDS, DAVID R. (1987), *Desertification in the Sahelian and Sudanian Zones of West Africa*, World Bank Technical Paper no. 61 (Washington DC.: World Bank).

GULLIVER, P. H. (1975), 'Nomadic movements: Causes and implications', in T. Monod (ed.), *Pastoralism in Tropical Africa* (London, Ibadan, Nairobi: Oxford University Press), 369–86.

HARDIN, G. (1968), 'The tragedy of the commons', *Science*, 162: 1243–8.

HOLLIS, G. E., ADAMS, W. M., and AMINU-KANO, M. (1993), *The Hadejia-Nguru Wetlands* (Gland, Switzerland, and Cambridge: IUCN).

HOMEWOOD, KATHERINE and RODGERS, W. A. (1987), 'Pastoralism, conservation and the overgrazing controversy', ch. 5 in D. M. Anderson and R. Grove (eds.), *Conservation in Africa: People, Policies and Practice* (Cambridge: Cambridge University Press), 111–28.

HOROWITZ, M. M. (1975), 'Herdsman and husbandman in Niger: values and strategies', in T. Monod (ed.), *Pastoralism in Tropical Africa* (London, Ibadan, Nairobi: Oxford University Press), 386–405.

HULME, M. (1996), 'Climate change within the period of meteorological records', ch. 5 in W. M. Adams, A. S. Goudie, and A. R. Orme (eds.), *The Physical Geography of Africa* (Oxford: Oxford University Press), 88–102.

JOEKES, S. and POINTING, J. (1991), *Women in Pastoral Societies in East and West Africa*, Drylands Network Programme, Paper no. 28 (London: IIED).

KREBS, J. and COE, M. (1985), 'An ecological perspective on the sahel famine', *Nature*, 317, 5 September: 13.

LAMB, R. P. (1982), 'Desertification case-study: From nomad's land to no-man's land', *Environmental Conservation*, 9 (3): 243–5.

LAMPREY, H. (1983), 'Pastoralism yesterday and today: The overgrazing problem', in F. Bourlière (ed.), *Ecosystems of the World 13, Tropical Savannas* (Amsterdam: Elsevier), 643–66.

LEWIS, I. M. (1975), 'The dynamics of nomadism: Prospects for sedentarization and social change', in T. Monod (ed.), *Pastoralism in Tropical Africa* (London, Ibadan, Nairobi: Oxford University Press), 426–42.

MACE, R. (1990), 'Pastoralist herd compositions in unpredictable environments: A comparison of model predictions and data from camel-keeping groups', *Agricultural Systems*, 33: 1–11.

—— and HOUSTON, A. I. (1989), 'Pastoralist strategies for survival in unpredictable environments: A model of herd composition which maximises household viability', *Agricultural Systems*, 30: 47–56.

MAIGA, A., DE LEEUW, P. N., DIARRA, L., and HIERNAUX, P. (1991), *The Harvesting of Wild-Growing Grain Crops in the Gourma Region of Mali*, Drylands Network Programme, Paper no. 27 (London: IIED).

MONOD, T. (1975), *Pastoralism in Tropical Africa* (London, Ibadan, Nairobi: Oxford University Press, Published for the International African Institute).

MOOREHEAD, RICHARD (1998), 'Mali', ch. 3 in Charles R. Lane (ed.), *Custodians of the Commons: Pastoral Land Tenure in East and West Africa* (London: IIED), 46–70.

MORGAN, W. B. and PUGH, J. C. (1969), *West Africa* (London: Methuen).

MORIS, J. (1988), 'Failing to cope with drought: The plight of Africa's ex-pastoralists', *Development Policy Review*, 6: 269–94.

MORON, V. (1994), 'Guinean and sahelian rainfall anomaly indices at annual and monthly scales, 1933–1990', *International Journal of Climatology*, 14 (3): 325–41.

NORRIS, H. T. (1975), *The Tuaregs: Their Islamic Legacy and its Diffusion in the Sahel* (Warminster, Wilts.: Aris and Phillips Ltd.).

OKAEME, A. N., AYENI, J. S. O., OYATOGUN, M. O., WARI, M., and OKEYOYIN, A. (1988), 'Cattle movement and its ecological implications in the Middle Niger Valley area of Nigeria', *Environmental Conservation*, 15 (4): 311–16.

OLADIPO, E. O. and KYARI, J. D. (1993), 'Fluctuations in the onset, termination and length of the growing season in Northern Nigeria', *Theoretical and Applied Climatology*, 47 (3): 241–50.

PARK, T. K. (ed.) (1993), *Risk and Tenure in Arid Lands: The Political Ecology of Development in the Senegal River Basin* (Tucson and London: University of Arizona Press).

POWELL, J. M. and WILLIAMS, T. O. (1993), *Livestock, Nutrient Cycling and Sustainable Agriculture in the West African Sahel*, Gatekeeper Series no. 37 (London: IIED).

REICHELT, R. (1989), 'Desertification in the sahel: The exposing of the 'Old Erg' of an earlier Sahara', *Natural Resources and Development*, 30: 104–13.

SALIH, M. (1991), 'Livestock development or pastoral development?' in P. T. W. Baxter (ed.), *When the Grass is Gone: Development Intervention in African Arid Lands* (Uddevalla: Nordiska Afrikainstitutet), 37–57.

——(1992), *Pastoralists and Planners: Local Knowledge and Resource Management in Gidan Magajia Grazing Reserve, Northern Nigeria* (Drylands Network Paper no. 32).

SCOONES, I. (1992), 'Why are there so many animals? Cattle population dynamics in the communal areas of Zimbabwe', ch. 4 in Roy H. Behnke Jr., Ian Scoones, and Carol Kerven (eds.), *Range Ecology at Disequilibrium: New Models of Natural Variability and Pastoral Adaptation in African Savannas* (London: ODI). 62–76.

SMITH, ANDREW B. (1992), *Pastoralism in Africa: Origins and Development Ecology* (London: Hurst; Athens: Ohio Univ. Press; Johannesburg: Witswatersrand University Press).

SPENCER, PAUL (1965), *The Samburu: A Study of Gerontocracy in a Nomadic Tribe* (London: Routledge & Kegan Paul).

——(1998), *The Pastoral Continuum: The Marginalization of Tradition in East Africa* (Oxford: Clarendon Press).

SULLIVAN, SIAN (1998), People, Plants and Practice in Drylands: Socio-Political and Ecological Dimensions of Resource Use by Damara farmers in North-West Namibia (unpubl. PhD Thesis, Department of Anthropology, University College, London).

SUTTER, J. W. (1987), 'Cattle and inequality: herd size differences and pastoral production among the Fulani of Northeastern Senegal', *Africa*, 57 (2): 196–217.

SWIFT, J. (1975), 'Pastoral nomadism as a form of land-use: The Twareg of Adrar'n Iforas', in T. Monod (ed.), *Pastoralism in Tropical Africa* (London, Ibadan, Nairobi: Oxford University Press), pp 443–454.

——(1986), 'The economics of production and exchange in West African pastoral societies', in M. Adamu and A. H. M. Kirk-Greene (eds.), *Pastoralists of the West African Savanna* (Manchester: Manchester University Press), 175–89.

TOULMIN, C. (1990), 'Pastoralists in peril: Sahelian pastoralists and their problems', Paper presented to the ASAUK Biennial Conference, Birmingham, 11–13 September.

——(1992) *Cattle, Women and Wells: Managing Household Survival in the Sahel* (Oxford: Clarendon Press).

TOUPET, C. (1975), 'Le nomade, conservateur de la nature? L'exemple de la Mauritanie centrale', in T. Monod (ed.), *Pastoralism in Tropical Africa* (London, Ibadan, Nairobi: Oxford University Press).

TOURÉ, O. (1990), *Where Herders Don't Herd Anymore: Experience from the Ferlo, Northern Senegal*, Drylands Network Programme, Paper no. 22 (London: IIED).

TRAORÉ, A. (1981), 'Herding industry or way of life', *The Courier*, no. 65, Jan.–Feb.: 60–2.

WATERS-BAYER, A. (1988), *Dairying by Settled Fulani Agropastoralists in Central Nigeria: The Role of Women and Implications for Dairy Developments* (Kiel: Wissenschaftsverlag Vaule).

—— BAYER, WOLFGANG, and VON LOSSAU, ANNETTE (1995), *Participatory Planning with Pastoralists: Some Recent Experiences* (London: IIED).

WARREN, ANDREW (1995), 'Changing understandings of African pastoralism and the nature of environmental paradigms', *Transactions of the Institute of British Geographers*, 20: 193–203.

—— and AGNEW, C. (1988), *An Assessment of Desertification and Land Degradation in Arid and Semi-Arid Areas*, IIED Drylands Programme Paper no. 2 (London: Ecology and Conservation Unit, University College London).

WESTOBY, M., WALKER, B., and NOY-MEIR, E. (1989), 'Opportunistic management for rangelands not at equilibrium', *Journal of Range Management*, 42: 266–74.

World Bank (1993), *Annual Development Report* (Washington: World Bank).

# 8

# Conclusions

Balancing resources against opportunities is a skill developed over generations by African smallholders. By exploiting variations in the local environment, both in time and in space, farmers and pastoralists minimize the risk of loss in an unpredictable environment. This is an indigenous alternative to more technologically developed systems where ecological variation is eliminated, or significantly reduced, with the aid of technology. Constantly juggling the limited resources available to them, indigenous farmers adjust to a non-equilibrial physical environment and changing social and economic circumstances by working round problems rather than by combating them directly. The case of farmers in Gambia's Western Division is a good example. In response to protracted drought and to the loss of labour from rural–urban migration, dryland farmers adjusted their cropping patterns, growing less labour-intensive and more drought resistant crops such as cassava and tree crops, particularly fruit (see Chapter 6). Irrigation and labour-saving mechanization were never options. As the drought has continued, labour remains a problem, the cost of living has risen, the cost of education, with which most families are concerned, has escalated, and farmers have begun to specialize in the production of mangoes and more recently, cashew. These tree crops are more drought-resistant than oranges and are higher in value, as long as they can be marketed. Contrary to expectations, cassava production has fallen back slightly because farmers find it difficult to successfully fence more than one, or at most two, cassava plots. Fencing is critical because cassava grows through the dry season and unless very well protected, is easy prey for animals which, in this part of the region, are not herded in the dry season. Thus adaptations to change are being made constantly but there is little evidence that such changes are being noticed either by government or by the aid agencies involved in the agricultural sector in The Gambia. This reinforces a personal view that in spite of verbal acknowledgement of the competence of smallholder producers, their skill in coping with problems receives little recognition from policy makers. Smallholders are not perceived as a resource and their potential contribution to agricultural development remains invisible to those with the power to influence change.

Throughout this book it has been argued that an understanding of ecological conditions is of fundamental importance to development and that smallholders' knowledge of the environment enables them to manage the land successfully. Using an alternative approach in support of the argument for ecology and indigenous knowledge, the following paragraphs examine what has happened on two development projects where smallholders have had little involvement in a decision-making capacity and where ecology appears not to have been a priority issue—where, in short, success has been limited. This is not intended to be a review of approaches to development nor is it intended to focus on failures. The literature is already rich with such material. The examples serve only to show that had ecology been afforded higher priority and had greater use been made of indigenous knowledge, results might have been better.

The Office du Niger in Mali promised to be one of the largest development schemes in Africa. Initiated by the French in 1934, the aim was to irrigate one million hectares of what appeared to the Colonial Administration to be little-used land on the Inland Delta of the Niger between Ségou and Mopti. An elaborate canal irrigation scheme was designed by a French engineer, M. Bélime, and this was a work of excellence as it was based entirely on gravity flow in an environment where gradients are minimal. Detracting from the technical achievement, however, was the fact that much of the canal network was built using forced labour. The French colonial administration decided to produce cotton and rice on Office land. Cotton was destined for export to France and rice was to provide food for Malians, both locally and further afield. The production potential of the Inland Delta was considerable. Experts were brought in from the Carolinas in the USA, to advise on the production of both cotton and rice. While rice cultivation appeared feasible, pilot cotton projects were beset by problems. Disregarding the advice of experts, the French colonial government introduced cotton on a vast scale. For the next 46 years the crop was produced, but at well below optimal levels, not least because it suffered from phytosanitary problems. Rice production was little better, with yields averaging around 2 tonnes per hectare or less. Achievements in terms of the area irrigated fell well below targets and even now this remains at around only 50,000 hectares. In 1980 the Office decided to abandon cotton cultivation and focus on rice production, as the potential for this seemed greater. The following discussion demonstrates not only how ecological problems were a major cause of low yields but also attempts to set this argument in the context of political ecology. The information was derived from a period of fieldwork in the Office du Niger in 1981, a time of economic difficulty in Mali when the country was rejoining the Franc Zone and the Union Monétaire Ouest-Africaine (UMOA), having withdrawn from these in 1962. At this time, Office farmers were subject to strict government restrictions prior to market liberalization.

The Office allocated farmers on the project plots of 2 hectares, a seemingly generous allocation. On this plot they could produce nothing but rice or cotton prior to 1980, and nothing but rice after 1980. There were some areas where the Office itself was experimenting with other activities such as sugar-cane production and cattle rearing, but for the most part rice was the main focus. Rice was a familiar crop along the Niger, with highly specialist flood rice being grown towards Mopti. In the early 1980s, Asian rice varieties were being advanced by the Office. On their rice plot farmers could choose whether they wished the Office to prepare the land or whether they wished to hire implements from the Office to prepare the land themselves. These ranged from tractors to a hand-held plough and bullocks. Services provided by the Office were not free and farmers had to pay in paddy at harvest time. Allowing the Office to prepare the land was fraught with difficulties: tractors arrived late, many broke down or ran out of fuel and farmers complained that those preparing the land did not devote sufficient attention to levelling the land properly for the paddy. Once water was introduced into the field, the unevenness made itself apparent and was a contributory factor to low yields. The Office supplied seed, fertilizer, insecticide and pesticide, and, if required, advice. The problem for the farmer was that there was no guarantee that any of these requirements would arrive when requested and once ordered, farmers were charged for them, whether the goods arrived or not. Payment was always in paddy. Because produce from the Office was largely destined for the urban areas and also for the army, there was great concern that every last grain of rice should be garnered. At harvest time, the army was much in evidence in the fields to ensure that smuggling was kept to a minimum. That farmers were having a difficult time was confirmed by Office officials, most of whom were powerless to alter the situation. It was widely acknowledged that returns from rice were low and one could not help but wonder why Malian farmers were prepared to continue cultivating Office land. Restrictions on them were considerable: in the villages visited they could cultivate only rice on Office land and in theory, could grow little else. Household vegetable gardens were allowed near the villages but no animals were to be kept on Office land. As local people became more confident that it was not my intention to cause them problems, evidence was provided that with the full knowledge of Office extension workers, these farmers were cultivating Office land with a range of crops other than rice, and using Office irrigation water. Millet, sorghum, and groundnuts were grown and vegetable cultivation, particularly the intensive cultivation of onions for market, was very successful.

Farmers cultivated rice because it was a route to a range of inputs: Office machinery was used for preparing private fields, fertilizers destined for rice were diverted to other crops and irrigation water was widely used, particularly for vegetables. Comparatively little attention was paid to the rice fields: these were rarely level, germination was usually uneven and weeds were a problem not only in the fields but in the canals and near the canal outlets. With such little concern for the success of the paddy harvest, yields were predictably low. Enormous threshing machines were brought to the villages at harvest time and farmers were forced to harvest and thresh their entire crop at a particular time in the presence of the army, so that none was stolen. Deductions were made on the spot equivalent in value to the inputs they had used, they were allocated a certain amount of rice per person in the household and for any surplus they were paid a fixed price, deductions being made for quality. Returns from rice were deemed so low that it was cultivated largely as a source of inputs for other, preferable crops which could be marketed privately, or in the case of millet and sorghum, consumed locally. The difference in the quality of crop on a paddy field and on a vegetable plot dominated by onions was considerable, the former being generally poor and patchy, the latter a dense green carpet—a clear reflection of the effect that incentive can have on production.

Owing to the poor quality of management, ecological problems developed and ultimately, these were the cause of low yields. The framework of suitable land, irrigation canals, extension and marketing systems were all in place for the production and the expansion of production of rice. However, the incentives for producing the crop were not considered sufficient by the farmers. As a result, they took little trouble to ensure that the ecological needs of the paddy were met.

In the Région du Fleuve of northern Senegal, the introduction of irrigated rice in the mid-1960s met with only moderate success largely because the approach, typical of many of the earlier capital intensive schemes, treated farmers as labourers and in no way used them as decision-makers. The argument used by the parastatal running the project, SAED (Société d'Aménagement et Exploitation du Delta) was that farmers were unfamiliar with rice and so were unsuitable as decision-makers. As a consequence the project was run (and badly run) by the parastatal. Farmers were allocated a plot of land which was prepared by SAED. Seed was provided by SAED and farmers were told when to sow seed in their nurseries. Fieldwork in the Région du Fleuve in 1981 and 1983 revealed that farmers had been told to have their seedlings ready to be transplanted by a certain date and yet three weeks later they were still waiting for the tractors. The original seedlings had become etiolated and could not be used. New seed had to be sown. By the time the land had been prepared the optimal date for transplanting the crop was long past and this affected yields. Supplies of fertilizer ordered by farmers through local extension workers frequently arrived late as did insecticide and pesticide. Whether or not they were able to benefit from the use of such inputs, farmers were still charged for them when the crop was harvested. Unlike farmers in the Office du Niger, they were not charged in rice but were obliged to pay in cash. The result was that many on the project accumulated considerable debts. Harvesting was also carried out by SAED and again farmers were no more than labourers. Rarely were SAED workers able to harvest the paddy at the appointed time and as a result of this, farmers incurred considerable loss. Weeding was the only element of rice production that they were encouraged to do unprompted but owing to the range of other factors that contributed to low yields, weeding was frequently neglected, farmers preferring to devote more time to their personal fields, frequently miles away. Owing to the problems caused by mismanagement, ecological conditions were rarely suitable for rice and the potential for rice production in the Delta was never realized.

When the sahelian drought struck in 1973–4, efforts to bring about agricul-

## Conclusions 249

tural development intensified. A new approach was adopted by SAED in the Middle Valley of the Fleuve. Here, access to villages along the River Senegal is much more difficult than nearer the Delta and during the wet season these villages are accessible only by river. Farmers were given a brief introduction to rice cultivation, provided with a package of seeds, fertilizers, insecticide and pesticide, and a small diesel pump which was used for pumping irrigation water from the river up on to the flood plain. Constraints on the marketing of rice were fewer in the Middle Valley than in the Delta, partly because of concern about local food shortage in the early 1970s, and partly because SAED had considerable problems collecting the crop from remote villages on the River Senegal, which were accessible only by boat during and shortly after the rainy season (Baker, 1982).

Although rice was an unfamiliar crop in this area, yields by farmers in the Middle Valley were much more encouraging than they had been on the Delta and confirmed that SAED had not been justified in using indigenous participants as little more than labour in the Delta. Yields of 4 tonnes per hectare were common in the early years in the Middle Valley. Yields have remained higher than in the Delta, though they have declined from the earliest peaks. There have been problems: pumps have caused uncertainties about irrigation, periodic shortages of diesel have been a problem and fertilizers have, from time to time, been in short supply. In addition, by 1983 farmers on these *petits périmetres* in the Matam area had identified a decline in rice yields which they could not explain other than by losses in soil nutrients which they did not know how to replenish. Once again ecological problems appear to have been the cause of lower yields and while this might seem self-evident, had development practitioners focused on eliminating ecological problems, success might have been greater.

There are, however, examples of development projects where a great deal of attention has been paid to ecological issues and yet these have still failed to bring about any significant improvement in the living standards of the people involved. One of the best examples of this concerns the work of the Taiwanese and Chinese in West Africa. The Taiwanese sent missions to the region from 1963 to 1967 and were then replaced by the Chinese who remained until 1973. In spite of political differences the approaches of the Taiwanese and Chinese bore many similarities. Their agricultural missions were very much village based and participatory. Extension workers lived in the villages, integrated themselves into village life, introduced new ideas to the farmers and spent much time in debate, not only giving out information but receiving it as well and sometimes modifying farming techniques along the lines suggested by African farmers. For the most part they received a great deal of respect and affection from the villagers and achieved considerable success in terms of increased production, for a limited period at least (Baker, 1985). In the Région du Fleuve in Senegal when SAED was extending rice cultivation to the Middle Valley, the parastatal invited the Chinese to run one of the rice perimeters. In contrast to other parts of the Middle Valley, rice farmers in the perimeter at Guédé achieved far higher yields, some 6 tonnes per hectare according to local informants. By the mid-1980s, well after the departure of the Chinese mission, yields had fallen but farmers in Guédé perimeter were still described by their neighbours as having a more 'sensible' approach to rice cultivation, the result of good teaching by the Chinese. Similarly in the market gardening area of Cap Vert the Taiwanese and Chinese missions have had a positive effect on vegetable cultivation.

However, the ecologically successful field methods introduced by the Taiwanese and Chinese do not seem to have led to any significant or lasting social and economic development. Bräutigam (1997), in her study of the Chinese in Africa, argues that although there were achievements at field level, no institutional framework was established to take these developments any further. This was particularly evident in the example of market gardeners from Cap Vert where, in 1983, the Senegalese government appeared to have no record of what the Taiwanese and Chinese missions had attempted, or achieved. Oral evidence from people who had participated in the Chinese programmes and from officials in Dakar who recalled the missions was all that was available. The extension centre at Sangalcam, complete with Taiwanese/Chinese architecture, remained, but its achievements were not integrated into Senegalese development plans.

While using ecologically appropriate methods is of fundamental importance at the production end in development initiatives, many other issues need to be considered to ensure that achievements in production can be sustained and further advanced. If any increase in the production of crops or animals is to be sustained, producers must have access to credit to enable them to obtain essential inputs: seed of good quality, fertilizer, insecticide, and pesticide. Credit is necessary for pumps for irrigation, for access to appropriate labour-saving machinery and to fuel, particularly where labour is in short supply. Producers must also be guaranteed access to the products of their efforts. Mortimore (1998) notes that land reform need not be essential as long as people benefit from the improvements in which they have invested. Governments should refrain from the arbitrary alienation of land, and women, in particular, should not lose access to resouces on account of pressure from men. In addition, systems for the disposal of products must be established, as without these, improvements in output at field level simply become fossilized, as in the case of Taiwanese/Chinese projects, and do not result in wider social and economic development.

Rather than viewing smallholders as an anachronism, whose methods should be transformed, smallholders should be perceived as a major asset by governments and by aid agencies involved in agriculture. In The Gambia community development projects which theoretically empower smallholders are under way. While this is very positive and has been well received in the rural areas visited, it is also important that the necessary institutions are established so that smallholders can dispose of their produce. Approaches which are small in scale and seemingly bottom-up do not in themselves ensure success. This is certainly the case with women's vegetable gardening in parts of The Gambia. Vegetable markets are saturated and women spend hours trying to sell their crops. Towards the end of the day they have to return home and traders agree to buy the crop at a knock-down price. They rarely agree to pay the women until they have sold the vegetables. The result is that the women have to make several trips to the market before the buyer agrees to pay and when he or she does, it is usually less than has been agreed. While problems with the actual production of vegetables have been greatly reduced by development projects, greater attention needs to be paid to crop disposal to ensure that women are not just working for nothing.

It could be argued that although vegetable gardening is not new to The Gambia, vegetable gardening projects are. Dating from the early 1980s, these have been based on externally constructed knowledges of what is good for Gambians, but clearly, implementing agencies have focused more on the production, than on the disposal of vegetables. This has created new problems for producers. However, problems with product disposal are not always project-related. Dryland farmers in the Western Division of The Gambia have moved into fruit cultivation, particularly mangoes, in response to protracted drought and to the loss of labour through rural–urban migration. Their problems of disposal are just as great as those of the women vegetable gardeners. Much of the produce is bought by traders who market the mangoes in Dakar but much of the produce is lost because local markets are saturated. Overseas markets do exist, but currently, these are supplied almost entirely by large-scale, capital intensive producers near Banjul. For the most part, smallholder mangoes are not suited to the export market as they do not have a long shelf-life after harvest. However, progressive farmers are now planting new varieties demanded by the export market and are grafting new stock on to existing mango trees in order to improve the quality of the crop. Such initiative deserves to be rewarded but there is little recognition that farmers are planting acres of mangoes, let alone that they are trying to match the varieties produced with the market. There is considerable scope in this example for the Gambian government to assist smallholder producers. For example, large-scale farms which have developed overseas marketing networks could be encouraged to use smallholder producers as contract suppliers of fruit of a specified quality. The government could assist with the processing and preservation of local produce, and, to improve levels of production there is a strong argument for the provision of subsidized inputs (Lele, 1991)—at least in the early stages of development of such linkages. However, attitudes by those in power to suggestions of support for smallholders remain uncompromisingly negative.

Hopefully, this book has drawn the reader to conclude that a new development paradigm is long overdue (Neimeijer, 1996). Ecology should be given a higher priority in development schemes than has been the case in the past and smallholder involvement should be a priority if ever agriculture is to progress. While in this final chapter it has also been argued that many other factors influence whether or not improvements in production at field level are taken further, it nevertheless remains true that without success at the production level, the development of agriculture is impossible. In view of the potential contribution that smallholders could make to development, African governments would be well advised to monitor the adaptive capacity of smallholders, fostering the more successful elements. This would base agricultural development more on local needs and expertise and less on broad brush strategies, many of which are external constructs.

## References

BAKER, KATHLEEN M. (1982), 'Structural change and managerial inefficiency in the development of rice cultivation in the Senegal River Region', *African Affairs*, 81 (325): 499–510.

——(1985), 'The Chinese agricultural model in West Africa', *Pacific Viewpoint*, 2: 401–14.

BRÄUTIGAM, DEBORAH (1997), *Chinese Aid and African Development: Exporting Green Revolution* (Basingstoke, Hants.: Macmillan Press).

LELE, UMA (1991), *Aid to African Agriculture: Lessons from Two Decades of Donors' Experience* (Baltimore, Maryland: Johns Hopkins University Press).

MCINTIRE, J. (1979), '*Coûts réels et incitations économiques dans la production de riz au Mali,*' Institut des Recherches Alimentaires, Stanford University. Rapport fait partie d'un projet subventionné par le contrat AID/AFR-C-1235 de l'Agence International de Développement (US-AID). Unpublished.

MORTIMORE, MICHAEL (1998), *Roots in the African Dust: Sustaining the Drylands* (Cambridge: Cambridge University Press).

NEIMEIJER, DAVID (1996), 'The dynamics of African agricultural history: is it time for a new paradigm?' *Development and Change* 27 (1): 87–110.

# INDEX

abiotic 21, 149
  characteristics 159
  conditions 173
  components 2, 7, 8, 173, 218
  factors 15–17, 22
  fluctuations 22
  variables 15, 22, 23
Accra-Kumasi railway 123
Adamawa range 200
  Administration Préfectorale 136
Adams, M. 194
Adams, W. M. 1, 120, 201, 225
Adejuwon, J. O. & Adesina, F. A. 176
Adesina, A. A. 187
aerial surveys 178
Africa 40, 42, 44, 46, 50–7, 61, 63–4, 68–9, 91–8, 110, 117–18, 137, 197, 209
  Central 149
  East 232, 238
  Southern 16, 50, 149, 162, 167, 231
  Sub-saharan 39, 63, 235
  West *passim*
Agboola, S. A. 80, 84, 88, 91, 92
Agnew, C. T. 177, 178
agriculture 2, 20, 32, 39, 40, 41
  colonial agricultural officers 121, 125
agricultural development 1, 2, 67, 87, 109
  exports 59
  market reforms 64
  officers 121, 124
  output 110
  practices 123
  produce 63, 68
  production 41, 58, 61
  sector 32, 59, 60–2, 64–5, 70
agroecosystems 18, 100, 104
agroforestry 105
Ahn, Peter 98, 99, 101, 120, 162, 163, 192
aid agencies 1, 184, 207–9, 245
aid donors 126, 199, 210, 234

Aïr and the Adrar des Iforas 218, 238
Akoroda N. O. *et al.* 92
Algeria 224
Allan, W. M. 67
Amanor, Kojo, Sebastian 93, 101, 122
Amara, S. S. 183, 201
Anambra state 83–5, 87–9, 92, 94–6, 102
  Awka-Nnewi 85, 87
  southern Anambra 93
Andersen Alan N. & Lonsdale W. M. 17, 170
Andrae, G. & Beckman, B. 2, 59
animals 193–4, 198, 207, 225, 229–30
  domestic 217, 220
  forage control 231
  grazing 195
  herd 217, 222
  household 217
  rangeland 217
  wild life 217
Arrow K. *et al.* 18, 19
ash 96–9, 155, 165, 192–3
Asia 51, 68
  South East Asia 51, 140–1, 144
Atacora mountains 220
Aubréville, A. 20, 157, 177
Austin, Gareth 122–6
Australia, northern 165, 171
autecology 70, 218
  of the crops concerned 70
autecological 79, 119, 183, 210
  approach 1, 2, 1–2, 90, 95, 119, 123, 169, 183, 198, 232
  emphasis 122
  methods 79, 143
  principles 241
  systems 83
  techniques 99
Ayeni, J. S. O. 223, 226, 227

Baker, E. F. I. 189, 190
Bamako 208
Bamenda plateau 220

Banjul  26, 56, 194, 196, 208–10, 222
baobab  156, 188, 206
Barbier, Edward A. *et al.*  199
Barrett, S.  228–9
Barro Colorado Island, Panama  20
Barrow, C. J.  31, 178
Barrows, Richard & Roth, Michael  87
*bas-fonds*  184, 201, 225, 233
Bates, Robert H.  1, 2, 58–9, 67, 131
Bationo, A. *et al.*  195
beans  99, 185, 188, 209
Beets, Willem C.  93
Behnke, Roy H. & Scoones, Ian  217–18, 230–2, 240
Bell, R. H. V.  162, 166, 171
Belsky, A. Joy  159
Benin  48, 55, 64, 80, 90, 133, 149
  Basilia area of Benin  86, 90, 97, 205
Benneh, George  67
Bennell, Paul  63
benniseed  90
Bennison, Hugh  55
Berg, Elliott  2, 39, 59, 60
Berkes, F.  70
Berry, Sara  61–2, 67, 117, 138, 140
Biggs, Stephen & Farrington, John  69
Bilma  21
biomass  6, 83, 101, 154, 159, 167, 172
  dead  170
  insect  170
  plant  162, 170, 172
  woody  151
biome  149, 151
  unpredictable  149
biota  8, 9, 14–15, 17, 21
biotic  8, 11, 13, 15, 18, 21–2, 149, 169, 173
  components  173
  elements  22
Blackmore, A. C. *et al.*  166–7
Blaikie, Piers  31, 70
Blaikie, Piers & Brookfield, Harold  70
Boers, T. M.  80
Bohicon  29
Borana Plateau  6
Boserup, E.  85
Bowé  233
Bramwell, Anna  4, 11

Bräutigam, Deborah  1, 2, 86, 210
bread-fruit  89
Breman, H. & de Wit, C.  232
Brokensha, D. *et al.*  68
Brookman-Amissah, J. *et al.*  169
Brown, Richard  50, 140
Burkina  42, 44, 46–8, 52, 153, 188, 191
  Burkinabé labour force  137, 138
  northern  190
  Sanmatenga province  192
  Yatenga region  187
burning  95–8, 165–6, 169, 186, 195, 204
  burn  89, 196, 236
  burnt  186
bush fallow  83–4, 104
  farming system  104
  regeneration  95
  rotational system  83, 102
bush land  100
  burning  236
  buyers  136
  clearance  204, 236
  fields  189, 192
  medicinal products  206
  products, gathered  205–6
  products, used  206

*Caisse de stabilisation des soutien des prix des produits agricoles* (*Caistab*)  135–7
Calabar  97
Camels  234–5
Cameroon, humid  61, 90, 220
  west  220
Cap Vert  201, 209
  gardeners  210
  *niayes*  201
Carney, Judith A.  185, 207
Carson, Rachel  70
cash crops  41, 58–9, 87–8, 109, 120, 185, 210
  cashew  97, 126, 187, 196, 211, 245
  cassava  40, 54–6, 84, 89–93, 98, 122, 176, 185, 190, 196, 207, 245
    meal (*garri*)  92
cattle  172, 196, 208, 219, 238–44
Caughley, G.  231
CFA franc  60–1, 65, 129

Chad, Lake, NE Borno, Nigeria  199, 200
  Chad Basin Development Authority
    (CBDA)  200
  northern Chad  218
Chambers, R.  68–9, 235
Chambers, R., Pacey, A., & Thrupp,
    L. A.  69
Chapman, G. P.  227
Chauveau, Jean Pierre  125, 135, 138
Chauveau, Jean Pierre & Léonard, Eric
    117, 135, 137–8, 142
Chinese  209
Chorley, R. J. & Kennedy, B. A.  4, 5
Citrus  89, 176
Clarence-Smith, William Gervase & Ruf,
    François  127, 140
Clements, F. E.  11, 12, 217
climate  13, 15, 155, 168
  change  22, 79
climatic:
  changes  22, 58
  climax  157
  conditions  218
  events  15, 30
Cocoa  58–60, 63, 66, 109, 117–44, 176
  Board  133
  Buying Agents (licensed) (LBAs)
    127–8
  Control Board of W. Africa  127
  Convention Peoples' Party (CPP)  128
  United Ghana Farmers' Council
    (UGFC)  128
  Ghana Cocoa Marketing Board
    (GCMB)  124–31
  International Cocoa Agreement (ICA)
    139–40
  International Cocoa Organization
    (ICCO)  139
  Cocoa Marketing Board (Ghana)
    (COCOBOD)  130, 134
  Cocoa Marketing System  136
  Cocoa Purchasing Company  128
  Cocoa Research Institute of Nigeria
    (CRIN)  126
  West African Cocoa Research Institute
    (WACRI)  124
cocoa, pests and diseases  123–5
  blackpod (fungal)  125, 132
  capsids (insects)  124
  mealy bugs  124–5
  New Juaben (virus)  124
  swollen shoot (virus)  123, 125
cocoa varieties:
  Amazonian Forastero  120, 122
  Amelonado Forastero  120, 122, 124
  Central American Criollo  120
  $F_3$ hybrids  124
  hybrid Trinitario  120
coconuts  93, 142
cocoyams  80, 93, 100, 121, 122
coffee  58, 63, 66, 93, 109, 136, 138
Coffin, Debra P. & Lavenroth, William
    K.  18
Cohen, P.  68
Cole, Monica M.  149, 158
commercial crops  87, 207
communal areas  209
Compaoré  54
compost  96, 99, 100, 102, 209
  organic  102
compound land  84–5, 88–9
  compounds  88, 102, 217, 221
  activity  297
  crop  196
  farm  98, 102
  farming system  104
  fields  93
  gardens  90–1, 93, 110, 199
Congo  194, 209
  Brazzaville  209
  market gardens in Brazzaville  209
Connell, Joseph H.  7, 16, 18, 20
Conservation techniques  93
  conservation  226
  of lands  228
  of moisture  191
  of resources  228
  of savanna species  162
Cooper, W. S.  13
Coppock, D. Layne  6
Cornia, Giovanni Andrea  63, 84, 86–7
Côte d'Ivoire  20, 30, 41, 48, 50, 55, 60,
    61, 63, 117, 119, 124–5, 127, 139–40,
    143, 153, 165, 189, 220
  Bandama river  54
  Baoulé (labour)  138

Côte d'Ivoire (*cont*):
  humid savanna at Lamto 176
  east Côte d'Ivoire 184
  south-east Côte d'Ivoire 135
  south-west Côte d'Ivoire 1, 135, 137
  west Côte d'Ivoire 184
cotton 53, 66, 90, 109
  seed 54
  seed oil 54
Cowles, H. C. 13, 22
cowpeas 91, 122, 185, 188, 191, 198, 200
Crook, Richard 136–7
cropland 1, 40
  combination 89
  density 93
  income earning 92
  plants 70
  yields 184–6
cropping patterns 87, 91, 93–4
  savanna 189–91
Cross River State (SE Nigeria) 80, 83, 85, 88, 91, 93, 97, 99, 102, 103
  eastern 84, 88, 100
Crowder, L. V. & Chheda, H. R. 156
Currencies 59–61
  devaluation 64–5
  manipulation 65, 210
  overvalued 59
  overvaluation 109, 129
Curtis, Bronwyn, N. 136, 138

Dahl, G. & Hjort, A. 237
Dakar 194, 196, 209, 222
Dale, W. T. 124–6
dams:
  Diama anti salt barrage 198
  Diama dam 198
  Hadejia barrage 199
  Kafin Zaki dam 199
  Manantali dam 198
  Tiga dam 199
Davenport, M. 54, 66
David, D. 223
Davis, W. M. 12
De Angelis, D. L. & Waterhouse, J. C. 3, 4, 14
De Frece, A. 130
De Wilde, John 1

de Zeeuw, Fons 63
decisions at farm level 109
  at field level 108
  at international level 109
Delgardo, C. & McIntyre, J. 187
density dependent 15, 20
  high crop 93
  high populaton 95, 101
  low population 95
  population 109
Deshmukh, Ian 154, 158
desiccation 17, 98, 141, 156, 172, 178
development projects 1, 2, 3, 14, 32, 184
  agencies 208
  agricultural 50, 57, 68, 87
  initiatives 2, 61, 207
  Kainji 227
  second plan 67, 128
  planning 3, 70
  policies 207
  practitioners 126, 197
  programmes 131, 661
  of reserves 143
  rural 67
  schemes 68, 136, 201
  rangeland policies 241
  vegetation 12, 13
Dey, Jennie 201
Disease 6, 58
  disease resistant 93
diseases of cattle 236
  anthrax 236
  black quarter 236
  bovine pleuro-pneumonia 236
  haemorrhagic septicaemia 236
  rinderpest 236
disequilibrium 3, 8
drought 6, 7, 17–18, 22, 26, 32, 41–2, 45–6, 53–4, 56, 58, 130, 162, 172, 177–8, 184–6, 188, 194, 196–7, 201, 203, 210–11, 220, 222, 224, 230, 233, 235–6, 240, 245
dry season 168, 186, 188, 219, 226, 233–5, 240, 245
  grazing 226
drylands 14, 15, 22, 85, 173, 197–8, 204
  crops 184, 199
  cultivation 88

degradation 156
farmers 196
farming 83, 100, 183–5
fields 88, 99, 100, 102, 184, 186, 191–2, 194, 196
dum palm 156
Dumont, R. 67
dune, stabilized 228
  former 229
  mobilization 229
  soils 229
  inter-dune depressions 229
Dunn, Justine 88, 91, 98
Dunn, Justine & Agom, Daniel 84, 87–8, 99, 102–3
Dutch disease 132
Dyson-Hudson, R. 238

Eckholm, E. 70
ecology:
  complex 158
  equilibrial 151
  non-equilibrial 16, 151
  local 217, 224
  multi-storied farming 89
  physical 83, 91, 143
  plant 11, 12
  political 91, 143
  range 217
  savanna 158
  savanna trees 170
  temperate zone 218
  tropical 10, 14, 109
ecological assumptions 14
  characteristics 87
  communities 15
  conditions 2, 79, 84–6
  criteria 23
  demands 117
  determinants 151
  environment 86, 232
  equation 95
  equilibriun 1, 10, 31
  non-eqilibrium 16
  factors 94
  function 103
  knowledge 3, 4
  models 157
  problems 120, 126, 131
  stable conditions 16
  thinking 173
  unit 70
  zone 80, 110, 149
ecologically aware farmers 193
ecologically unsound techniques 65
ecologists 32
  savanna 159
economic factors 94, 95
  decisions 109
  environment 143
  instability 139
  pressures 90
  value 186
Economist Intelligence Unit 51, 62, 127
  (Country Profile) 130, 134, 139
ecosystems 7, 8, 10, 15–19, 31, 94, 101, 104, 157, 166, 173, 218
  savanna 170, 173
  tropical 2, 31, 101
Ecuador 139
edaphic climax 157
  conditions 151
Edwards, P. J. 168–9
Edwards, P. J. et al. 19
Egbe, N. E. & Adenikinju, S. A. 126
Egypt 50
Ehrlich, Paul R. 70
Eicher, Carl K. 39
Ekanade, Olusegun 120–1
Ekanade, Olusegun & Egbe, N. E. 125–6
El Niño 139
Ellis, James E. et al. 7, 15
Ellis James E. & Mellor, A. 99
Ellis, James E. & Swift, David M. 231–2
Enugu State 89, 100
environmment:
  African 1, 2
  non-equilibrial 1, 2, 7, 19, 21, 217
  physical 1, 2, 32, 69, 80, 83, 86, 90, 118–19, 143, 210, 217
  savanna 94, 158
  unpredictable 237
  unstable 22
environmental changes 94
  conditions 2, 70, 94
  degradation 70, 157, 177

environmental changes (*cont*):
  demands 55
  destruction 176
  factors 233
  hazards 85
  improvement 107, 168
  knowledge 2
  non-equilibrium 22, 31, 65, 79, 117, 183, 241
  resilience 228
  socio-economic 83
  stablity 16, 18
  uncertainty 26, 183
  unpredictability 109
  variables 17, 117
  variability 143, 210
equilibrium 1, 3, 5, 6, 8, 11–15, 20, 30, 210
  dynamic 4
  multiple systems 7
  multiple stable states 172
  non-equilibrium 1, 3, 8, 13, 15, 22–3, 30, 80, 193
  non-equilibrium systems 172
  paradigm 3, 4, 15
  position 5–7, 18
  single position 173
  single position systems 7
  situation 173
  stable 5–6, 13, 231
  equilibrium–non-equilibrium debate 221
equilibrial 2, 6, 7, 15, 101, 149
  approach 149
  character 79
  non-characteristics 20, 79
  conditions 152
  ecology 151
  factors 232
  grazing systems 231
  nature 14, 26
  nature of environments 32, 101
  paradigm 3, 4, 15
  physical environment 1, 2, 45
  states 5–6
equilibrium, non-equilibrium; disequilibrium chapter 1 *passim*
erosion 30, 94, 99, 163, 190
  soil 31, 191
  splash 98
  water 165, 191, 228–30
  wind 165, 191, 228
  raindrop splash 190
Eshett, Ebong T. *et al.* 122, 125
Ethiopia 6
Europe 54, 117
European Union (EU) 5, 66, 104, 139, 207, 210
export quotas 139
  foodstuffs 200

*fadamas* 200, 225
Fairhead, James & Leach, Melissa 1, 21, 158, 177, 187
Falconer, Julia 103
fallow 40, 83–5, 89, 91, 94–5, 99–103, 110, 176, 185–6, 191–5, 204
  artificial fallow 103
  declining periods 101–2
  fallowing 203
  forest 88
  natural 103
famine foods 240
farming systems, *see* systems
Farrington, John & Martin, Adrienne 69
Faulkner, O. T. & Mackie, J. R. 125–6
fauna 18, 98, 149, 156
  communities 170
  soil 187
feedback 5, 7, 12, 101, 104, 167, 178, 229, 231
  mechanisms 173, 229
  systems 231
fences 93, 196–7, 208, 245
  fencing 197
  of oil palm fronds 208
  of Rhun palm 208
  of thorn branches 208
fertilizer 48, 51, 59, 62, 64–5, 68, 89, 125, 129–30, 141, 200, 204
  ash 99
fertilizer, inorganic 56–7, 80, 93–4, 96, 99, 100, 102
  chemical 197
  inorganic compounds 166

fertilizer, organic 56, 94, 96, 191
  composted 96, 100
  organic compounds 166
  matter inputs 93, 99, 103, 166–7, 171, 194, 195
  mulches 99–101
  mulches and manure 191–2
field crops 58, 88, 91, 120, 196
fire 6, 13–14, 17, 96–7, 141, 149, 151, 155, 159, 163, 165, 172, 174, 186–7, 236
  bush 130, 236
  crown 171
  firebreak 97, 168
  lightning strikes 167–8
  surface 174
  in savannas 167–70
Firmin, Kathryn & Sellers, Patrick 63
fish 206
  commercial fishing 227
Fitzpatrick, E. A. 101
floods 17, 26, 30, 51, 183–4, 200
  advance agriculture 200
  flood plains (Senegal) 197
  flood waters 203
  flood recession agriculture 191, 198
  flood recession farming 198
  flooding 199
  floods on inland delta of the Niger 225
  natural floods 199
  retreat agriculture 200
  river flood plains 225, 233
  seasonally flooded 184
  tidal swamps 203
flora 15, 98, 149
Fones-Sundell, M. 229
Food and Agriculture Organization 40, 42–3, 46, 48–9, 51–3, 55–7, 92–3, 130, 221
food crops 41, 87–8, 91–2, 102, 109, 122, 130, 134, 185, 196–7, 200–1
Forde, C. D. 92
forest 13, 18–22, 98, 100–1, 103, 121, 138, 142, 154, 177, 187
  canopy 122
  climax of forest vegetation 19
  deciduous 154
  dry 15

  environment 141
  forest/savanna boundary 21
  humid 149
  moist 18–19, 149, 151, 156, 166
  moist tropical 20, 22
  rent 141
  riverine 176
  savanna 170
  secondary 100, 176
  semi-evergreen 154
  tropical 21
  tropical rain 21
  zone 153–4
Fouta Djallon 220, 233
France 60–1, 137
francophone 119
Franc zone 60, 65, 129, 133
  non-franc zone 60
freehold 86–7, 89
fruit growing 197, 211
  fruit trees 196
fuel wood 103, 197, 205
Fulani 205, 219–20, 222–4, 226–7, 233, 236, 238–49
  cattle 223, 226
  cattle men 223
  families 230
  herds 223
  settled 222
  traditional 240
Fuls, E. R. 16, 20
Fuls, E. R. & Bosch, O. J. H. 20

Gadbois, Millie 225
Gakou, Mohamed 39
Gambia 44–5, 47, 53–6, 58–9, 62, 64, 90, 153, 188–9, 193, 197, 205, 211, 222, 225, 236, 245
  farmers 57, 186, 194, 196
  flood plain 201
  Gambian Groundnut Corporation (GGC) 62
  Lower 42, 169, 183–5, 188, 192–3, 197, 201, 203, 205–9
  River Gambia 20, 197, 301
  villages 186, 194
  Western division 56–7, 65, 110, 184, 187, 190, 193, 196, 210, 245

Gandar, M. V. 170–1
Gao 26, 29
gardens, communal 207
   compound 83, 89, 90
   vegetable 84
Ghana 19, 26, 44–8, 50, 53, 55, 59–61, 65, 67, 117, 119–25, 127, 129–31, 134–40, 142–3, 193–4
   Asante region 62
   eastern region Nankese 123
   north of Ghana 152
   north west—Brong 1
   north and west Akwapim 123
   south-east 93, 149
   southern 100
   southern Akwapim 123
ginger 91
Glaeser, Bernhard 101
Gleason, H. A. 13
goats, *see* sheep and goats
Goldman, Abe 83–5, 89, 92, 96, 100–3
Goldsmith, E. R. D. 70
Golley, Frank B. 12
Goodwin, Elizabeth 63
Gordon, Sara L. 127
Gorse, Eugene Jean & Steeds David R. 225, 233
Goudie, Andrew 30, 163
gourds 198
Grainger, A. 178
grasslands 17, 18, 157
   C3 154
   C4 154
   elephant grass 156
   grasses 154, 166, 174
   grassy savannas 163
   tropical grassland 157
grazing 15, 145–6, 168, 197, 223, 225–30, 238, 240
   lands 225
   localized overgrazing 228
   overgrazing 172, 228–9, 231, 237
   population 231
   reserves 227
   systems 1
Green Revolution 47–8
   crop 48
   desertification 229
   technology 68
Greenway, David M. W. *et al.* 61
Gritzner, J. A. 157
gross domestic product 41
groundnuts 40, 45, 51–4, 56, 58, 60, 62, 66, 91, 93, 99, 185, 188, 193, 195–6, 201, 211
   marketing 62
   millet-groundnut rotation 195
   oil 54
Grove, A. T. 22, 30, 85, 163
Guinea 21, 23, 41, 48, 50, 153
   Benin savannas 186
   Gulf of Guinea 159
   Highlands 197
   Kissidougou region 187
   southern 153
   southern savannas 153
   zones 168
Guinea Bissau 23, 26, 196–7
   Manjago migrants 196
gully formation 30
Gwynne-Jones D. R. G. *et al.* 85, 87, 90, 91, 95
Gyimah-Brempong, Kwabena & Apraku, Kofi Konadu 127–9

Haeckel, Ernst 11
Hammond, P. S. 121, 122
hard pan (impermeable layer) 30, 98, 117, 120, 143, 153, 162–3, 187
Hardin, G. 63, 70, 224, 228–30
Harrison Church, R. J. 26, 86
Hassan Abdou 143, 185, 191
Haswell, M. R. 67
He, Fangliang *et al.* 20–1
Heady, H. P. & Heady, E. B. 164
Hecht, S. B. 68
Helleiner, G. K. 1
Henderink, J. & Sterckenburg, J. J. 2, 39, 59
herbivores 157, 165–7, 170–3, 230
   grasshoppers 170–1
   insects 170–1
   mammalian 167, 170–1
   savanna 170–2
herbivory 13, 149, 151, 159, 168, 170–1
herds, mixed 220, 222–3

capital resources 221
female animals 220
large 229
movement 229
size 230
structure and size 237–8
village 222
herders 222–3
professional 222
herding 220
high yielding varieties 44, 48, 51, 56, 220
improved 47
Hill, Polly 67, 117, 123
Hodgkinson, Edith 45, 54, 137–8
Holling, C. S. 7, 18
Hollis, G. E. *et al.* 199, 201, 226
Holt, G. A. & Coventry, R. J. 164–6
Homewood, Katherine & Rodgers, W. A. 217, 231, 232
Hopkins, Brian 19, 153, 155–7, 176–7, 204
Horner, Simon 139
Horowitz, M. M. 221–3
Hottinga, Folert *et al.* 201
Houphouet-Boigny 136
Hubbell Stephen P. & Foster, Robin B. 7, 16, 18, 20, 21
humid areas 220
domain 32, 79
environment 70
humidity levels 188
savanna zone 151
south 23, 26, 30, 54, 66, 67, 85, 184
tropics 19, 22–3, 30, 32, 79, 80, 83, 98, 125, 138
zone 19, 23, 79, 82, 97, 110, 117, 142–3
Huntley, B. J. & Walker, B. H. 158
hurricanes, cyclones, downpours 30

Idachaba, Francis Suleman 68, 80
Igbokwe, E. M. 89, 92, 99, 100
Ihonbvere, Julius O. 6
IMF & World Bank 2, 39–41, 59, 60–1, 65–6, 94, 101, 129, 134, 137, 142–3, 207–8
Imo State 83–5, 88–9, 92, 94–6, 102
Uboma village 92

income non-farm 88, 197, 205–6
income earning crop 92
India, north eastern 193
indigenous people 3, 4, 64
African agriculture 69
cultivators 67, 83, 90, 110, 117, 119, 126
cultivation 200
disease control 123
farmers 68, 70, 93, 104, 118, 143, 245
farming 67, 80, 83, 87, 91, 104, 110, 120, 184
indigenous responses 183
knowledge 1, 3, 61, 67, 69, 79, 206–7
land use 201
land users 157
livestock industry 240
low input systems 101
methods of cultivation 31–2, 51, 67, 191, 194
pastoral practices 218
patoralists 169, 218, 229, 232, 238
pastoralism 217, 240–1
problems 188
skills 100
solutions 188
techniques 189
vegetable gardening 208
Indonesia 149
Insalah 218
insecticide 51, 129, 141
intercropping 46, 88–90, 92, 122, 126, 189–91, 237
combinations 190
intercropped 51, 54, 84, 88, 92
interplanting 142
strategies 237
interdisciplinary 2, 70, 118
inter-dune depressions 225
Inter Tropical Convergence Zone (ITCZ) 159, 162
International Fertilizer Industry Association (IFIA) 48
intra seasonal distribution of rain 225
irrigation 48, 51, 63, 184, 198–9, 226, 245
agriculture 1, 2, 8
areas 184
large-scale schemes 226

Islamic Development Bank 207
Iyegha, David A. 1, 2, 67–8

Jaeger, W. K. & Matlon, P. J. 187
Jarriage, F. & Ruf, F. 135, 137–8
Jewitt, S. L. 67, 69, 70
Joeckes, S. *et al.* 70
Joekes, S. & Pointing, J. 241
Johnson, C. G. 124
Jones, M. B. *et al.* 149
Jones, W. O. 1, 67
Jordan, C. F. 101
Jos plateau 220
    jewellery 225

Kalu, B. A. & Norman, J. C. 195
Kano 196, 218
Kapok (silk cotton tree) 156
Karimojong 238
Kassa Kunda 193–4
Kassogue, Armand *et al.* 1
Kenya 65
Khalfaoui, J.-L. B. 153
Killick, Tony 59, 60, 66
Kinyamario, J. T. & Imbamba, S. K. 149
knowledges, constructed 2
Koester, Ulrich *et al.* 66, 142
Kola nuts 89, 126, 142
Kolade, J. A. 121, 142
Kolawole, Are 199
Konan, G. K. 125, 135
Kowal, J. M. & Kassam, A. H. 185, 188
Krebs, J. & Coe, M. 228
Kusi, Newman K. 129

labour 137–8, 189, 222
    active workers/ha 192
    Burkinabé labour force 137–8
    farm 194
    inputs 203
    labour saving mechanization 242
    limited labour requirements (cassava) 196
    loss 251
    migrant 132
    public sector workers 196
    shortages 196
    supply 210

Lagemann, J. 85, 89, 92, 100–2
Lal, R. 98–9
Lal R, & Greenland D. 98
Lamprey, H. 177
land degradation 31, 79, 110, 172, 178, 229–30
land management 22–3, 57, 69–70, 79, 138, 159, 172, 184
    Land Use Decree of 1978 86
    land use management 201
    landowners, private 228
    landslides 17
Lane, Christopher & Page, Sheila 60
Lass, R. A. & Wood, G. A. R. 122, 124
Lastarria-Cornhiel, Susana 64
Lawesson, Jonas Erik 177–8
leached 83, 96, 98
    leaching 98–9, 154, 165–6, 169, 194
leasehold 86–7
Lensink, Robert 61, 63, 65
Leonard, Eric & Oswald, M. 135
Lepper, Alan personal communication 51, 62–4
Lewis, I. M. 233
Liberia 23, 41, 51
Libya 224
Lipton M. & Longhurst, R. 39, 40
livestock 15, 31, 51, 217, 219–21, 226–7, 230, 281–2, 240
    development programmes 218
    indusry 227
    management 232
    products 240
locust bean, pods of 186
London 133, 184
Longman, K. A. & Jenik, J. 98
Lowe, R. G. 39
Lowenberg-De Boer, *et al.* 190

Mc Cown, R. L. & Williams, John 164
Mac Gregor, Jenny 41
Mc Intosh, Robert P. 6, 11, 12
Mace, R. 237
Mace, R. & Houston, A. I. 237
Madagascar 64
Madge, Clare 185, 205–6
Magaria 29–3
Maiga, A. *et al.* 239

maize 40–2, 47–50, 60, 65, 84, 90–1, 93, 98–9, 121–2, 128–9, 167, 176, 184–5, 188, 196, 200
    fields 91
    irrigated 198
    stalks 188
Malawi 64
Malaysia 139
Mali 42, 44, 46–8, 50, 53, 59, 107, 153, 189, 222
    northern savannas 192
    Office du Niger 208
    Office du Niger development scheme 246–8
Mandinke 185, 201, 222
    villages 185
Manga in Niger 222
    animals 222
    farmers 222
mangoes 185, 187, 196, 245
Manjago migrants 196
Manshard, Walter 122
Manvell, Adam 186, 189
Maradi Department of Niger 186, 189
Marchand, M. & Toornstra, F. A. 201
market gardening 201
markets, southern Nigerian:
    domestic 65
    prices 66
    urban 65
marketing boards 62, 119, 131
marketing organizations 62, 63
Martin, Claude 19, 22, 83–5, 92
Mauritania 40, 51, 59
    Nouakchott 222
May, R. M. 6, 7, 17
Meadows, D. H. 70
medicinal products 206
Medina, Ernesto & Silva, Juan F. 166
Mediterranean 224
melons 88, 90, 198
    water melons 191
Menaut, J. C. *et al.* 151, 154, 156–7, 164, 174
Menaut, J. C. & César, J. 169–70, 174
Meteorological Office 22–5, 27–9
microbial immobilization 165
Middle Belt 23

migrants 110, 130, 194, 208
    labour 132
migration 41, 59, 88, 90, 134, 196, 225–6
    outmigration 97
    rural–urban 41, 56, 194
    routes 64
    sedentary farmers to rangeland 225
    urban–rural 62, 64
Mikell, Gwendolyn 59, 117, 127–9
Miles, J. 12, 113, 22
milk production project 240
Millar, David *et al.* 191
millet 40, 42, 44–6, 90, 184–5, 188–9, 190–8, 206
    longer cycle 189
    millet–groundnut rotation 195
    stalks 188
Ministry of agriculture 186
moisture conservation 191–3
    residual 200
    surface 178
    retentive 233
Moloney, Alfred 67
monoclimax 112
monocrop 51, 90, 92, 189, 190–1
    plots 126
monoculture 92
monopsony 127
Moody, K. 96, 101
Moore, P. *et al.* 154
Moorehead, Richard 225–6
Moors 218, 222, 229
Mopti 51
Morgan, W. B. 92
Morgan, W. B. & Pugh, G. C. 42, 45, 54, 87, 89, 98, 120, 156, 158, 168, 220, 233, 239
Morgan, W. B. & Solarz, Jerzy A. 41
Moris, J. 220, 229–30, 237, 239
Morocco 234
    southern 224
Mortimore, M. 1, 2, 8, 18, 42, 54, 172, 178, 188–90, 197, 201
Moscow 184
Moss, R. P. 2, 42, 79
Mossi 192
Moustier, Paule 209

Mouttapa, F. 99, 101
Myers, Norman 21

National Agricultural Data Centre (the Gambia) *et al.* (NADC) 87
National Agricultural Sample Survey, (the Gambia), (NASS) 64
National Environment Agency, (the Gambia) (NEA) 64, 201, 203
*naira* 132
　devaluation 134
　overvaluation 132
N'dama cattle 220, 233, 236
　humped Zebu cattle 220
Neimeijer, David 39
Nelson, Harold, D. 197, 201
Newsletter, Tropical Agriculture Association 140
Ng, F. & Yeats, A. 66
*niayes* 209
Nicholson, Sharon 158
Niger 42, 44–7, 51, 53, 190, 222, 223
　river flood plain 226
Nigeria 26, 30, 32, 40, 42, 44–50, 53–4, 57, 59–60, 62, 64–8, 80, 85–92, 117–19, 125, 127, 130–3, 134, 142–3, 155, 177, 200, 228
　Bendel State 103
　Chad, Lake, *see* Chad
　Cross River State, *see* separate entry
　Hadejia-Nguru wetlands 199
　Institute of Agricultural Research 189
　Kainji, Lake 64
　middle Niger valley 236
　northern Nigeria 52, 183, 188–90, 187, 188
　northern Nigerian farmers 65, 240
　Oyo State 132, 134
　Research Institute 227
　south-east Nigeria 100–1, 122, 123, 153, 185
　southern Nigeria 91–2, 102–3, 153
　south-west Nigeria 126, 153
　western Nigeria 122, 11, 140
Nigeria, cocoa 131–5
Niono 208
Nkrumah, Kwame 128

nomads 222
　pastoral 222
non-farm activities 93, 204
　income 197
　jewellery 222
Norman, D. W. 1, 69, 189
Norman, D. W. *et al.* 190
Norris, H. T. 218
North America 15, 18, 231
　New York 133
　prairies 165
　U S A 54, 139–40
Noy-Meir, I. 6, 7, 15, 16
Nugent, Paul 128–9
nutrient retention 198, 233
　enriched zones 173
　release 173
Nyanteng, V. K. 125
Nye, P. H. & Greenland, D. 83, 85, 100–2, 192
Nye, P. H. & Stephens, D. 103
Nyofellehmedina 193–4

oases 218
　permanent 218
　temporary 218
Odemerho, F. O. & Avwunudiogba, A 92
Odum, Eugene P. 11
Oduwole, O. O. 121
Ofomata, G. E. K. 80
oil boom 131
　dollar revenues 182
　wealth 134
　world prices 132
oil palm 84, 88–9, 91–2, 126, 142, 176, 204, 206
　carob 138
　coconut palm 89
　coconuts 93, 142
　kola nuts 89, 126, 142
　palm oil 117
　palm wine 205
oils and fats 66, 109
Okaeme, A. N. *et al.* 223, 226, 236
Okafor, Francis C. 83–5, 87–90, 92–3, 96, 100, 102
okra 88, 90, 99, 122
Oladipo, E. O. & Kyari, J. D. 183, 225

Olagbaju, J. O. & Adeseun, G. O.  66
Olaloku, F. A.  87
Oluwasanmi, H. A. *et al.*  92
Onafowora, Olugbenga & Owoye, Oluwole  66
O'Neill, R. V.  16, 17
Onimode, Bade  2, 67
onions  208–9
Organization for Economic Co-operation and Development, (OECD), countries  66
organizations:
   indigenous  209
   International Cocoa (ICCO)  135, 139
   marketing  43
   problems  137
oranges  186, 187, 196, 245
Osemeobo, G. J.  87, 103
Ouedraogo, Matthieu & Kaboré, Vincent  187, 191
oxen  187
Oyejide, T. Ademola  61

paddy  40, 41, 50–1
palynology  22
PAM (plant available moisture)  149, 151, 159, 162–4, 168, 170–2, 178–9, 230
PAMSCAD (Programme of action to mitigate the social cost of adjustment)  61
PAN (plant available nutrients)  149, 151, 159, 163–4, 168, 170–2, 230
Park, Thomas K.  63, 183, 191, 198, 226
Pastoralists  6, 63–4, 168, 198, 206, 217, 219–38, 240–1, 245
   diets  239–40
   peoples  218–20
   problems  224
   sedentary  222
   society  240
pastoralism  32, 155, 217, 225, 236
paw paw  121
Pélissier, Paul  1, 68
Pellow, Deborah & Chazan, Naomi  60
pepper  122
Percival, David A.  186, 206
Perham, M.  67

pests  236
   birds  45
   insect  41, 58, 122
pesticide  51, 141
Peter, Gregor & Runge-Metzger, Artur  88
Peulh herders  97, 192
Pieri, J. M. G.  96, 101, 192, 195
pigeon peas  90
Pitty, A. F.  101
plant classification  155
   litter  193
plant succession  11–13, 157, 217, 230
   climax state  11, 12, 157
   primary  11
   secondary  11
plantain  12
policies:
   agricultural  53, 61
   development  206
   discriminatory  12
   government  54
   makers  197
politics, local  86
   attempts to control  143
   conditions  95
   decisions  109, 119–20, 143–4, 194, 208, 211
   desire  217
   ecology  91, 241
   environment  143
   factors  87, 90, 94–6
   instability  41, 139, 210
   involvement  118
polyclimax  12
population  40–1, 56, 64–5, 70, 80, 83–6, 88, 90, 92–3, 95, 99, 101–2, 109, 149, 186–7
   growth  41, 56
   invertebrate  170
   local  186–7
   mammals  156, 170
   plants  190
   rural growth  83
potatoes, sweet  91, 93
Potts, Deborah,  61
Poulson, Rudolph A. & Spencer, Dunstan S. C.  90

Pound, R. & Clements, F. E. 6, 13
prices:
  depressing effects 142
  instability 139
  international 127
  marketing board 119
  producer 119, 129–30, 132, 134, 135–7
  rising producer 143
  reference system 136
  world market 118–19, 130, 132
primary:
  production 2, 149, 167
  products 65–6
private sector 62–3, 65
  public sector 54
programmes:
  development 131
  economic recovery: (ERP 1) 129–30; (ERP 2) 130
  livestock development 218
  Structural Adjustment *see* under 'S'
pumpkins 88, 99

Raikes, Philip 39
raindrop splash 191
rainfall 2, 8, 14, 16, 18, 22–3, 26, 30, 32, 40–2, 44, 47–8, 51, 55, 58, 79, 82, 85–6, 99, 120, 130, 141–2, 153, 150, 167, 170, 173, 177, 185, 189, 191, 200, 225, 227–8
  convectional 178
  declining 183
  frontal 162
  irregular 198
  range 30
  seasonality 184
  unpredictability 79, 225
  variability 23, 26, 173, 183, 231
rainfed 42, 51, 83–6, 91, 185, 188, 191, 200, 225
  cereals 42
  crops 188
  cultivation 200
  farming 183
  fields 99, 184
  lakes 85
rainy (or wet) season 86–7, 97, 183–5, 188, 238–9

activity 207
ranchers, settled 217
rangelands 16, 18, 31, 32, 217, 220, 225, 227, 128, 230–1
  African 228
  animals 217
  degradation 228–31
  management 14, 32, 217, 230, 232, 240–1
Rattray, J. M. 156
Raunkaier, C. 155
Redclift, Michael 70
Reichelt, R. 229
Reij, Chris *et al.* 1
resilience 18–19, 172, 176–8, 186, 228–9
  environmental 228
resistance 17, 19
rice 41, 50–1, 54, 60, 65–6, 84, 86, 88, 91, 93, 96–7, 184, 187, 206, 208
  dryland 197
  floating 51
  flood 51
  irrigated 199
  rainfed 51
  riverine wet 185, 201, 203
  swamp 203, 204
  upland 51, 85, 197, 201, 203–4, 207
Richards, Paul 1, 67–8, 80, 83, 85, 87–91, 95–6, 103, 195, 204
Richards, P. W. 157
Richardson, J. 57, 65
Rietbergen, Simon 21
river basins 226
  Hadejia 226
  Senegal 226
  Sokoto 226
riverine lands 84, 197–8, 201, 207
  areas 197–8
  cultivation 201
  villages 204
Roberts, H. 158
root crops 54–6, 99, 184, 209
rotations 88
rubber 117, 126
Ruf, François 120, 123, 135, 140–2
run-off 30, 85–6, 93, 99, 162, 183, 191, 229

rural areas 209
　development 67
rural rapid appraisal (RRA) 60
Ruthenberg, Hans 1

Sahara desert 40, 159, 176–8, 218–19
　margin 177–8
　Moors 218
　mountain ranges 232
　northerly groups 218
　pastoralists 239
sahel 16, 18, 22–3, 32, 42, 53, 152, 162,
　　172, 176, 183–4, 217–20, 222, 224–5,
　　228–9, 231–3
　savannas 172, 177–8
sahelian 41, 224
　pastoralists 239
　rangelands 228
　southern group 218
　vegetation 177
　West Africa 231, 234
　zone 233
Salih, M. 240
saline water 198, 201
　anti-salt 198
　salt 207, 233, 235
　soils 203
Sanchez, Pedro A. 96, 101, 103, 192, 194
Sankara, Thomas 53
Saris, A. & Shams, H. 67
savannas 16, 21–2, 32, 42, 53, 60, 90, 94,
　　97, 120, 149–79, 183, 186, 189, 217,
　　219
　African
　cropping patterns in 189–91
　derived 155–6, 177
　drier 190
　drier northern Guinea 156, 167
　drier Sudan 151, 153, 156, 167–8, 184
　dynamics 176
　ecology of 153, 158
　ecosystems 164
　fire in 167–70
　floristic zones in 156
　herbivory in 170–2
　humid 151, 194
　moist Guinea 151, 153, 167, 184, 186
　Nylsvley 164, 166, 169, 173

　plant cycle 155
　pure grass 153
　sahel 153, 156, 162, 167, 172, 176–7,
　　183
　semi-arid 162, 168
　soils in 166–7
　Southern Africa 162, 173
　southern Guinea 155–6
　stabilization 174
　Sudan 151, 153, 167, 184, 186, 232
　tree 20, 21, 196
　tropical 159, 170
　vegetation 230
　West African 153, 163, 184
　woodland 151
　woody 163, 166
　zone 58
savanna boundaries 176–8
　between Guinea and Sudan 189
　between sahel and desert 176, 179
　fixed 176
　forest—savanna 21, 151, 158, 187
　northern 177–8
　southern 176
scale 183–218
　landcape 16
　large 40, 210, 226
　larger 16, 21, 59, 68
　local 80
　multiple 16, 17
　quadrat 16
　regional 39, 58, 70
　smaller 16, 18, 68
　spatial 19
　time 16, 22, 185
Schaff, Thomas 122
schemes, agricultural 68
　capital intensive 210
　large scale irrigation 226
Schimper A. F. W. 11, 157
Scholes, R. J. 166–7
Scholes, R. J. & Walker, B. H. 149, 151,
　　153, 162, 164–6, 168–9
Schoonmaker-Freudenberger, K. & S.
　　186, 188
Schreckenberg, K. 90, 99, 168, 186, 205
Schultz, T. 68
Schumacher, R. 70

Scoones, Ian 7, 14–16
sedentary peoples 217
    cultivators 221, 226, 239
    farmers 232–1, 236–7
    pastoralists 220–2
seed predation 172
Ségou 208
semi-arid areas 15, 18, 22–3, 26, 30, 44, 58, 192, 220
    Africa 120, 210
    domain 183
    ecosystems 19
    environment 6, 15
    grassland 18
    savannas 236
    tropics 14, 19, 22, 79
    West Africa 18, 58, 210, 218, 241
    zone 79
semi-nomadic peoples 217, 219, 221
    Fulani 226
    pastoralists 200, 221–2
    pastoralists v. sedentarists 222–4
Senegal 42, 44–5, 49, 51, 53–4, 56, 59–60, 62–4, 153 177, 183, 187, 196, 209–10, 222, 225, 236
    Casamance 195
    Cayor region 201
    Dakar 152, 194
    Dakar and Kaolack 56
    Ferlo 230
    lower middle valley, River Senegal 191, 198
    northern 152, 178, 186, 188, 191, 198
    Région du Fleuve, irrigated rice project 248–9
    River Senegal 51, 63, 197–8
    river basin 226
    river delta 59
    southern 23
shade 121, 186
    nurse shade 121–2
sheep and goats 196, 235–7
    goats 100, 172, 207
Shimada, Shuhei 56, 92–3
Sierra Leone 23, 44–5, 50, 59, 80, 88, 85–6, 90–1, 94, 96, 99–100, 102, 153, 203–4, 222
Sivakumar, M. V. K. 22, 26

Sjaastad, Espen & Bromley, Daniel 63
Slingerland, Jaja & Masdewel, Mouga 191–2
smallholders 1, 2, 21–33, 30–2, 39–40, 51, 56, 58–9, 61, 65–70, 79, 84–5, 87, 89–93, 95, 97, 101, 103, 109, 110, 121, 123, 125, 127, 134, 141–2, 189, 195, 210, 245
    agriculture 68–9, 109, 134, 210–11
    community development projects, Gambia 250–1
    farmers 30, 183
    indigenous farmers 70, 79, 198
    knowledge 245
    production 68
Smith, Andrew B. 1, 219–20, 225, 234–6, 239
social changes 94
socio-economic factors 69, 87, 90, 94, 95, 118, 149, 158
socio-political factors 94, 98
soils 86, 88–91, 96–101, 153, 163, 165, 188
    alluvial 198
    compaction 191
    declining fertility 228
    degradation 91, 98, 104, 205
    deterioration 88
    disease 92
    erosion 30–1, 88, 92–3, 187, 190
    factors 158
    fertile 55, 80, 92, 100–2, 163, 166, 171, 228
    hardened 188
    high gravel content 85–6
    higher silt fraction 85–6
    humidity 151
    lacustrine 200
    loamy 120
    loss 30
    moisture 162, 184, 198
    nitrogen content 166
    nutrient deficiency 62
    organic matter 94, 101, 162, 169
    poor 42, 54, 83, 92, 171, 196
    productivity 103
    reserves 195
    sandy lithosols 163

structure 100
sub-soil 185
surface 155, 187
top 330–1, 96, 154, 185
type 90, 162, 200
water erosion 191
wind erosion 191
sorghum 40, 42, 46–77, 90, 184–5,
 188–91, 196, 198, 206
 stalks 198
Sousa, W. P. 7, 17, 19
South America 117, 140
 Bahia 141
 Brazil
 Colombia
soy bean 54
 oil 54
Sparks Donald, L. 66
Species 16, 19, 21, 93, 101, 121, 154–6,
 173–4, 229
 forest 187
 herbaceous 173
 indicator 192
Spedding, C. R. 69
Spencer, Dunstan S. C. 238
Spichiger, R. & Pamard, C. 187
Stebbing, E. P. 67, 157, 177
Stevenson, G. G. 70
stochastic 8, 10, 15
Stocking, M. 31
Stott, P. A. 11, 151, 154, 157–9, 167, 176
strategies 11, 2, 11, 14, 39, 110
 corporate 96
 development 1, 2, 14, 39, 79, 110, 218
 intercropping 237
 land management 70, 79, 218
 rangeland management 218, 240
 rural 79
Structural Adjustment 40, 44, 50, 57, 58,
 60–2, 64–5, 109, 119, 130, 134,
 136–7, 140, 142, 196, 218, 221, 238
St Louis 198
subsistence crops 84, 120, 207
 existence 221
 needs 220, 230
 oriented pastoralists 231
substrate 94
Sucres et Denrées 137

Sudan 177, 229
sugar cane 84
Sullivan, Sian 1, 15, 16, 20, 252
sunflower 54
 oil 54
Sutter, J. W. 230
Swaine, M. D. 151, 158, 167, 170, 177,
 186
Swaine, M. D, et al. 187
Swift, J. 3, 220, 224, 229–30, 234
Swindell, K, 69, 96, 201, 203, 207
Synge, Richard 134
systems 2, 4, 6, 8, 16, 18, 101, 117, 119
 agricultural 2, 117–19
 autecological 83
 bush fallow farming 83, 104, 194
 chit 128
 compound farming 104
 cropping 126
 dynamic 168
 ecological 10
 environmental 8, 18
 equilibrium 8
 fallow 99, 191
 farming 36, 67, 69, 80, 83, 91, 110,
  117, 120, 170, 184, 192
 farming research (FSR) 677
 food production 206
 high input capital intensive 80
 indigenous farming 99, 104, 117, 210,
  218, 231
 indigenous knowledge 207
 indigenous pastoral 232
 intensive 96
 isolated 4
 land tenure 68
 land use 63
 low input 80, 101
 marketing 63, 128
 multiple stable equilibrium 172
 non-equilibrium 172
 non-equilibrial grazing 223
 nutrient cycling 104
 physical 4
 root 154
 rotation 193
 savanna landscape 158
 sedentary 231–2

systems (*cont*):
  tenure  64, 86, 90, 154, 201
  tropical forest  104
  village  197
Szereszewski, R.  122
Szolnoki, T. W.  186, 210

Taiwan  207
Taiwanese  209
  development project in West Africa by Chinese and Taiwanese  249–50
Tansley, Sir A. G.  11, 12
temperate zone  3, 12, 138, 173, 232, 241
  ecology  218
tenure, land  64, 86–7, 109, 178, 226, 228, 240
  communal  63, 64, 66, 87, 226, 231
  formal  86
  private  63
  state owned land  226
  tenurial situation  89, 90
  tenurial system  90, 201
termites  154, 165–6, 171
  mounds  154, 167, 171
  terrace cultivation  93
  terracing  191
Thomas, D. G.  177
Thresh, J. M. *et al.*  121, 124
Tibesti mountains  218, 238
tidal swamps  201, 203
Tiffen, M. *et al.*  70
timber  186
  acacia  185
  hardwood  186
  iron wood  186
  medicinal purposes  186
Timbuctoo  219
Tinley, K. L.  162–3, 174
tobacco  84, 91
Togo  48, 129, 140
Toulmin, Camilla  1, 178, 187, 189, 192, 223, 227, 241
Toupet, C.  218, 239
tourist indusry  208
  attractions  217
  wildlife  217
Touré, O.  225, 228
tractors  57

transhumance  232–3
  routes  220
  village cattle, daily  223
trans-Saharan trade  217
  camel caravans  147
  camels  219
Traoré, Amadou  219, 222
Trapnell, C. G. *et al.*  154
tree crops  40, 176, 245
tropics  3, 10, 14, 15, 18–20, 22, 69, 173
  forests  19, 20–2
  humid  22, 80
  semi-arid  14, 19, 22
  rangelands  15
trypanosomiasis  220, 236
tsetse fly:
  mosquitoes  236
  stinging flies  236
  ticks  236
tsetse zone  100, 220
  fly zone  220
  infested bush  200
  resistant species  236
tubers  100
Tubu  218, 222, 239
turnip (*navet d'hivernage*)  209
Twareg  218, 222, 235

Udo, Reuben  68, 83
United Nations Environmental Programme, (UNEP)  178
United Nations Food and Agriculture Organization Fertilizer Yearbook (1995)  64
upland rice  85, 90, 94, 201, 203–4, 207
  rice farms  99
Upton, Martin  1, 56, 67, 88
urban areas  91, 95, 132, 194, 196–7, 208
  centres  222
  diets  41
  dwellers  196
  herd owners  221
  market  196
  vegetable gardening  209
usufruct rights  86, 201

vegetable gardens  84, 91, 184–5, 208, 221
  commercial  207, 208

herbs 89
oils and oilseeds 109
subsistence 207, 220–1
vegetables 91, 122, 185, 198, 206–8
vegetable oils and oilseeds 109
vegetation 11–13, 15, 18, 22, 30, 95–9, 101, 103, 149, 151, 153–6, 165–8, 177–8, 185, 192–3, 203, 228, 230, 235–6
   climax 217
   cut 99
   desiccated 154
   herbs 89
   patterns 151–2, 177
   savanna 149
   vegetables 189
Vine, H. 96
Volta 129–30
von Bertalanffy, L. 4
von Buren, Linda 13

*wadis*, 184, 201, 225, 233
Wallace, Tina 1, 68
Walker, B. H. & Noy-Meir, I. 6, 7, 15, 172, 174, 176
Warming, Eugenius 11, 13
Warren, Andrew 218
Warren, Andrew & Khogali, Mustafa 173
water table 95, 184, 186, 196, 200–1, 207, 209, 225
   fallen 225
   rising 85
Waters-Bayer, A. 241
Waters-Bayer, A. *et al.* 241
waves 17
Weaver, J. E. & Clements, F. E. 12
Webb, James L. A. 185, 201
Webster, C. C. & Wilson, P. N. 96
Wedum, Joanne *et al.* 191
weeder ploughs 187

Weil, P. 201
wells 207, 223, 225, 233–4
   deep 228
   dry 225
Westoby, M. *et al.* 217, 231–2
wetland 84, 86, 88, 197–8, 200, 201–33
   fields 184
   tidal swamps 201, 203
   wet rice (swamp) 90, 201, 203–4, 207
   wet season 183, 188, 233
Wharton, A. L. 125
wheat 41, 60
White, F. 156–7
Whitmore, T. C. 98
Whittaker, R. H. 12
Wiens, John A. 4, 7, 8, 10, 13, 15
Williams, Donald C. 86
Wilson P. N. 170
winds:
   dust laden 14
   high 17
women 63–4, 89, 91, 94, 123, 184–5, 197, 201, 203–8, 223, 239–41
   work groups 95, 203
World Resources Institute, 21, 40

yams 55–6, 84, 88, 90–93, 97–8, 103, 176
   mounds 98, 122
   water 122
   white 122
Yaycock, J. Y. 191
yield patterns 195
Young, Anthony 99

*zai* pits 187, 188
Zambia 154
   Brachystegia (*miombo*) woodland 154
Zimbabwe 15, 232
zone of semi-deciduous forest 120
Zurich, Montpellier 12